TESTING RESEARCH HYPOTHESES USING MULTIPLE LINEAR REGRESSION

By
Keith A. McNeil
Francis J. Kelly
Judy T. McNeil

SOUTHERN ILLINOIS UNIVERSITY PRESS
Carbondale and Edwardsville

Feffer & Simons, Inc.
London and Amsterdam

Library of Congress Cataloging in Publication Data

McNeil, Keith.
    Testing research hypotheses using multiple linear
regression.

    Bibliography: p.
    Includes indexes.
    1. Statistical hypothesis testing.   2. Regression
analysis. I. Kelly, Francis J., 1925–   joint author.
II. McNeil, Judy, joint author.   III. Title.
QA277.M33   519.5'36   75-6639
ISBN 0-8093-0732-4
ISBN 0-8093-0755-3 pbk.

"Meeting the Goals of Research" is from K. A. McNeil, "Meeting the Goals of Research
with Multiple Regression Analysis," *Multivariate Behavioral Research,* Vol. 5 (1970),
pp. 375-386, and is used by permission of the journal and the author.

"Selecting a Harem" is from R. E. Christal, "Selecting a Harem—and Other Applications of the Policy Capturing Model," *Journal of Experimental Education,* Vol. 36 (1968),
pp. 24-27, and is used by permission of the journal.

"Suggested Inferential Statistical Models" is from F. J. Kelly, K. A. McNeil, and I. Newman, "Suggested Inferential Statistical Models for Research in Behavior Modification,"
Journal of Experimental Education, Vol. 41 (1968), pp. 54-63, and is used by permission
of the journal and the authors.

"Miscellaneous Data Transformation Notions" and "N=60 Data Set Format" are from
Judy McNeil and Keith McNeil, *Manual of DPLINEAR for Use with McNeil, Kelly, and
McNeil Testing Research Hypotheses Using Multiple Linear Regression,* Copyright ©
1975 by Southern Illinois University Press, and are used by permission.

Contents

Preface

During the past few decades, behavioral scientists have been expanding theory beyond the simple bivariate problem of one variable as it relates to another variable. Theories of behavior have been constructed which reflect complex functional relationships among numerous variables as they relate to particular behaviors.

While behavioral scientists have been developing multidimensional theories, statisticians have been producing mathematical models based upon multivariable and multivariate analyses. On the surface, such a match should produce the best of all worlds--a match between theory and model. Yet when one reviews the recent literature relating to behavior, the statistical analyses used do not reflect the complex functional relationships which are inherent in the problem under investigation, and many of which may even be described and expected by the investigator (but not reflected in the analyses). Bivariate correlation analyses and simple factorial analyses of variance are the most typical statistical techniques used. It appears that, although the researcher may propose complex relationships, he is unable to use a design that would reflect this complexity. To appreciate the problem one needs only to look at the newer design texts and methodological

articles. The statistician is confusing his trade with the notation of his trade. Matrix operations abound throughout the texts,
and Greek letters are introduced with abandon. These writings
communicate well to fellow statisticians, but the typical behavioral scientist is not a genius when it comes to manipulating
mathematical symbols. It would be nice if the typical behavioral
scientist had overlearned matrix algebra in order to read the
statisticians who have so much to offer, but seldom is this the
case.

The intent of this book is to present one powerful technique
of analyzing research hypotheses which can reflect hypotheses
(particularly hypotheses that reflect theoretically complex functional relationships) in a language that is understandable to the
non-mathematically trained behavioral scientist. It is hoped that
this endeavor can help to reduce the gap between behavioral science theory and statistical analysis.

Most traditional statistics texts present a myriad of techniques without attempting to show the underlying similarity of
those techniques. The emphasis in this text is on that underlying
similarity. Indeed, statistical tests are viewed as tools for
answering research hypotheses. With that view in mind, Generalized Research Hypotheses are presented throughout the text. In
some cases computer solutions (on data found in the Appendix)
referred to as Applied Research Hypotheses, accompany those Generalized Research Hypotheses. The reader is repeatedly encouraged
to state the Research Hypothesis of interest, rather than rely on

one of the ones in the text. The flexibility of the multiple regression approach is discussed throughout, particularly in Chapter Eleven. Chapter Eleven presents "the most general model"--a
notion which ties together all the material in the first ten
chapters. Chapter Eleven could well be read before any of the
other chapters. Indeed, we have not been able to fully convince
ourselves that eleven comes after ten.

A number of regression texts have recently appeared. We are
encouraged, because we see regression as a technique more generalizable than the "analysis of variance" and "correlational"
techniques. Indeed, those are restricted subsets of multiple linear regression. The present text differs from most other regression texts in two general areas.

First, we believe that the Research Hypothesis must be
stated first. A researcher cannot afford to let statistics do the
thinking. Once the Research Hypothesis is stated, the remaining
steps in hypothesis testing flow with surprising ease. This focus
on the Research Hypothesis has led to the firm belief that most
Research Hypotheses will be directional in nature. Two-tailed or
non-directional research hypotheses have little value for the
researcher. Encouraging researchers to ask questions they are interested in leads to the inclusion of continuous variables, and
curvilinear and interacting functions of those continuous variables. Consequently, relatively more space in this regression
text is devoted to such notions.

The second general area of difference concerns the mathe-

matical aspects of statistical analysis. Few researchers have
ever calculated a statistic by hand, especially since computers
have been available. The presentation in this text assumes a computerized regression program is available, preferably one which
allows models to be defined. The matrix solution is not presented.
A simpler approach, that of vectors, is utilized.

Given that Research Hypotheses are designed to answer real-
world questions, this text discusses many real-world problems.
Ramifications of small sample sizes and missing data are discussed. Approaches to building predictability are discussed, along
with the desire to do such. Several of the real-world problems,
such as not being able to meet statistical assumptions and lacking reliable and valid measures, are put in their perspective. It
should be noted here that the authors of this text take a different view towards the above notions than do most statistical authors.

Various persons have contributed to the development of this
text. The work of Bob Bottenberg, Joe Ward, and Earl Jennings
planted the seed. The text written in 1969 by the three of us in
collaboration with Don Beggs and Tony Eichelberger stands as our
first iteration. We simply have more to say now, and feel less
tied to a particular regression program.

Some of our former students and colleagues have made valu-
able contributions to this text, especially Mike McShane and
Irene Hawley. Jack Byrne, Jim Carlson, Jack Hiller, Conrad Krauft,
Isadore Newman, John Pohlmann, and Tom Pyle read over the manu-

script and made valuable suggestions. We also want to thank the graduate students who tried to learn multiple regression from a draft of this text. We taught several classes and received many valuable suggestions. We hope that we have been faithful to those students and have incorporated their notions.

Chris Schmidt labored over the original draft which we used in class. Doris Dixon has done an excellent job in typing the final manuscript.

We, of course, take responsibility for the final version. We do appreciate, though, the assistance we have received.

January 1975   K. A. M.

                                                       F. J. K.

                                                       J. T. M.

Testing Research Hypotheses

1

Introduction to Multiple Linear Regression

Investigators into human behavior take many forms and are conducted for various reasons. Some want to describe, others to predict, and still others to improve (control) behavior. Underlying these activities is some expectation held by the investigator regarding a causal network behind observed behavior. Research designs and statistical procedures apparently have been developed to help the investigator make decisions regarding the adequacy of the description, prediction, or improvement in relation to expected outcomes. These three functions are illustrated in the following simplified example.

Description

"Description" refers to reporting data which has been obtained and applies only to the particular people (or other entities) from which that data comes. In description, no inference is made to other data or to other people.

A social scientist may intuitively suspect that children from "lower-class" homes seem to have less success in school than children from "middle-class" homes. This research investigator, based upon his observations as filtered through his theoretic frame of reference, would expect children from lower-class homes to have lower grade point averages (GPAs) when compared with the GPAs of their classmates from middle-class homes.

In order to describe the expected relationship, the investigator will usually (1) select a representative number of children from middle-class and lower-class homes (based upon some operational definition), and (2) then collect data on the criterion (dependent variable) selected to represent school success (GPA in this case). Suppose that observed data reveals that 50 children from lower-class homes have a mean GPA of 1.52 (on a five-point scale), and 50 children from middle-class homes have an observed mean GPA of 3.68. The expectation is confirmed: the children from lower-class homes do have lower GPAs on the average than children from middle-class homes. It seems that lack of preparation in the home causes less school success. Further inspection of the individual GPAs may reveal that there is some overlap between the two groups. Some children from lower-class homes have higher GPAs than some children from middle-class homes. In fact, some children from lower-class homes have higher GPAs than the average child from the middle-class home, and conversely, some children from middle-class homes score lower than the average child from the lower-class home. Lack of preparation attributed to home activity does not always cause poor school success; there is some overlapping of observed scores among groups. Perhaps other predictor or independent variables also influence school success;

this matter will be discussed shortly.

Another way to describe this data is to report the strength of the relationship between home preparation and school success using a squared bivariate correlation coefficient (r^2). This value (r^2) represents the proportion of the total observed GPA variance which is explained by group membership. Suppose in this case that $r^2 = .45$. Then the percent of observed variance (variation from subject to subject) in GPA which is explained by home background is 45%. The large mean differences and percent of explained variance between groups may be sufficient for the investigator to tentatively accept some causal relationship between home preparation and GPA for this specific group of children.

Prediction

"Prediction" is the process of using data from a smaller group (sample) of people (or measurable entities of any kind) to make estimates about a larger group (population) from which such data has not been gathered. The position taken by the authors of this text is that a researcher uses data to establish a prediction only after there is descriptive data to support that prediction.

For example, a second researcher may read the descriptive data on the relative school success of lower- and middle-class children that was reported by the researcher described above. This second researcher may further have noted that schoolroom activities approximate the activities in the middle-class homes, and that lower-class homes provide few of the schoolroom-type of activities. The lack of preparation in the lower-class home could cause these children to be at a competitive disadvantage with the prepared middle-class child.

In an example such as the one just given, the investigator usually desires to do more than describe a particular set of children; typically, he wants to generalize from the sample of children he measured to a theoretical population of similar children. The researcher then would choose children for the study that adequately represent the two theoretical sub-populations (children from lower-class and middle-class homes), and then the information from that sample of children can be used to generalize to all children from lower- and middle-class homes.

In order to know what generalization to make, one must determine how likely it is that the observed difference between means (or the r^2 value) would occur due to sampling variation. Suppose that this researcher finds with his sample that the home preparation and school success has an r^2 value of .47, and the mean GPA values are 1.59 for the lower-class children and 3.73 for the middle-class children. The r^2 value of .47 can be subjected to a statistical test to determine how likely it is that the observed mean differences can be attributed to sampling variation (discussed in Chapter Two). Given this sample of 100 children or more, the observed r^2 and mean differences are highly unlikely to be due to sampling variation. Indeed, based on a statistical test on these two means, the investigator would conclude that home preparation apparently causes differential school performance. If a new set of children from the two types of homes were selected, based upon this researcher's predictive study, he would predict (or estimate), with some confidence, lower GPAs for children from lower-class homes when compared with the GPAs of children from middle-class homes. The researcher may conclude that the type of home "causes" the level of GPA.

The causal interpretation just given is based upon the theoretic network developed by the investigator. The theory, bolstered by observation of differential past home experiences in lower-class and middle-class homes, led to specific expectations. As with most research, other causal relationships may be operating. The causal interpretation is always subject to modification based upon subsequent information.

Perhaps the difference between children from lower-class and middle-class homes is due to nutrition, or to expectations of teachers. These possible "competing explainers" have not been controlled in the present study. Some possible competing explainers can be eliminated by the design of the study; some can be statistically controlled; most must be logically eliminated by the researcher.

Please note that this is a process of logic and research design rather than of statistics. Statistical verification is a necessary but never sufficient condition for a causal interpretation. More on causality is presented in Chapters Five and Eight. Knowing the causers allows a meaningful change in the situation.

Improvement

"Improvement" involves making a change in a given situation
that has been expected and verified to be a "cause." Many applied behavioral scientists want to improve the lot of humans and
will attempt to manipulate the environment in order to bring
about controlled outcome states. For example, the researcher
discussed above is able to predict that lower-class children will
have lower GPAs than middle-class children. But he may wish to
"upset the prediction" and do something to change the situation
so that the children for whom he would have predicted low GPAs
will in fact have higher GPAs.

He cannot, however, be very successful in improving the
situation until he is certain of the cause of the low GPAs. The
researcher found that home preparation was related to GPA, but
changing a child's home preparation may not raise his GPA (if
social class itself is not the cause).

Improvement on a large scale cannot be attempted until the
variables which are found to be predictors and are suspected of
being causes are manipulated in a sample and found to be related
to improvement.

Manipulation to search for causal factors

Suppose the researcher investigating the performance of the
children from lower- and middle-class homes wanted to upset the
predicted "poor" school success for the children from lower-class
homes (LCH). He therefore must manipulate (change) one of the

factors suspected of causing poor school success. He may first select 100 children who represent the same theoretical population of children from LCH and on some random basis (Cochran and Cox, 1957) select from these 100 LCH children 50 children for an intervention program and 50 children to act as a natural control group. If home experience is the relevant causal variable, then a change in the home environment so as to introduce classroom-like experiences theoretically should result in higher GPA performance than expected.

If these classroom-like experiences in the home were provided for the 50 experimental LCH children and not for the 50 control LCH children, a GPA difference between these two groups favoring the experimental group should theoretically be found. When the data on the criterion is examined, it is found that the mean GPA for the experimental group is 2.98, and for the control group the mean GPA is 1.54 (see Figure 1.1). The bivariate r^2 of .30 and its associated $\underline{F}$ value are very unlikely to be due to sampling variation. Apparently, the improved home experiences cause better school performance; the improved GPA is 2.98, compared to what might have been expected for these children on the basis of the descriptive study above (mean GPA was 1.52 for LCH children). However, note that the improved GPA does not equal 3.68, the GPA observed for children from middle-class homes in the earlier study. Apparently, the investigator has isolated part of the causal factors underlying lack of success, but some additional influence or set of influences are also operating.

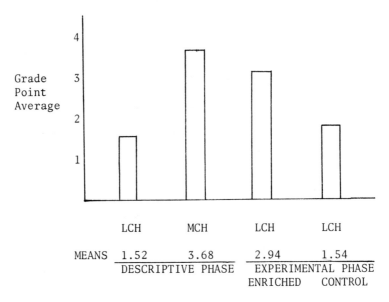

Figure 1.1. Grade Point Average means for children from Lower
            Class Homes (LCH) and Middle Class Homes (MCH)
            obtained in the descriptive phase and the improve-
            ment phase.

It could be that the experimental treatment was not as strong
as the natural middle-class environment, or that some other variable or variables were operating to influence GPA. Are children
from middle-class homes a bit brighter? Are they more motivated
for success? The data to answer these questions is lacking in the
present study.

    Another thing the investigator may note here is that the
control group's GPA of 1.54 was close to the GPA of 1.52 observed for children from lower-class homes in his initial study.

There really is not much difference, and he would likely conclude
that the two samples represent the same population. With these
sample values (1.52 and 1.54) he still does not know the population GPA mean, but he might assume that it is close to the average (1.53) of the two sample means.

Complex Determinants of Behavior

The simple example just presented was provided in order to
illustrate the descriptive, predictive, and improvement aspects
of research. At the present level of theoretical sophistication
in behavioral research, such bivariate (single predictor and
single criterion) research surely would be labeled naive. From
past research it is "known" that there are many other variables,
such as test anxiety, measured intelligence, peer expectations,
content of the material to be learned, and past special learning
(e.g., as measured by standardized achievement tests) which enter
into performance relating to such broad constructs as grade point
average. Furthermore, variables such as those just listed may not
be additively related to performance. For example, Castaneda,
Palermo, and McCandless (1956) demonstrated that the way anxiety
level affected a student's performance on a task depended upon
task difficulty. Given a difficult task, low-anxious students
tended to perform better than other students; but given an easy
task, high-anxious students tended to perform best. This effect
is referred to as a "difficulty-by-anxiety interaction." Statisticians use the term "interaction" for situations wherein the

effect of one variable depends upon another variable.

In addition, to the interaction discussed above, one might expect task difficulty to depend upon the relative brightness of the performing student. What may be operationally defined as a difficult task for the average student may not be difficult for the exceptionally bright student. If one were to extend the Castaneda et al. study to include a task difficulty-by-anxiety-by-"intelligence" interaction, one would expect to explain more of the criterion variance (and have less unexplained criterion variance).

Given this complex state of affairs, it seems that research-ers who want to describe, predict, and improve need complex theo-retical models and flexible statistical procedures which accu-rately reflect the complex models. Multiple Linear Regression Analysis as presented in this book is one flexible statistical procedure which meets the needs of a broad spectrum of behavior-al researchers.

A conceptual model

The details of this flexible statistical procedure occupy most of the rest of this text. First, however, a conceptual model is given which provides a way of considering variables which re-late to the behavior one is investigating. One of the major rea-sons for the grouping of variables discussed below is to stimu-late the search for multiple variables, since consideration of multiple variables is usually necessary to account for variance in any criterion variable. The variable groupings presented here are applied to behavioral research; other groupings could well be developed for this or other areas of research.

A critical review of the behavioral literature reveals that the variation on the criterion is caused by a network of inter-related predictor variables. These predictor variables can be arbitrarily grouped into three categories: (1) person variables; (2) focal stimulus variables; and (3) context variables.

Person variables (characteristics of the behaving individual)

Concurrent with studies specifying the effects of focal stimuli upon response, a body of data has been gathered regarding individual characteristics. Studies of the intellect over the last 50 years show that knowledge of an individual's "intellective functions" can aid prediction of performance across a broad class of tasks. In addition, Cattell (1966) reported a number of personality characteristics (for example, ego strength or excitability) which aid prediction of school achievement over and above the prediction obtained from intelligence measures. Manifest anxiety (Taylor, 1953) has been reported to be related to complex behavior such as school success. Furthermore, refinements upon the anxiety studies show that anxiety level interacts with ability (Katahn, 1966; Denny, 1966) and task difficulty (Sarason and Palola, 1960; Castaneda et al., 1956). These and many other person variables may be considered when constructing a model for explaining behavior. In attempting to account for differing performances, one would want to specify characteristics of the individual that may influence her degree of success.

Focal stimulus variables (characteristics of the task)

Focal stimulus variables are characteristics of the instrument or other task that is used to measure the criterion behavior. There are many ways to measure any construct that a researcher wishes to consider as a criterion variable. The ways in which those measures differ from one another are the "focal stimulus variables."

The notion that human behavior is a function of focal stimulus variables has existed for a long time. Early animal learning studies focused upon discrimination and generalization of stimulus properties. More recent psychological theorizing which centers upon focal stimulus properties deals with more complex aspects of stimulus-response (S-R) relationships. Gagné (1965) has proposed a learning structure for number operations which hypothesizes a learning hierarchy that is composed of complex S-R chains. Guilford (1950, 1959, 1966) reported a number of content characteristics which might be relevant to problem solutions. Suppes (1966) has investigated process characteristics of arithmetic problems (e.g., the number of transformation, operation, and memory steps) which account for variation in the proportion of correct answers.

In attempting to account for differing performances, one would want to specify dimensions of the task to be performed that may influence an individual's degree of success. See Figure 1.2 for examples of focal stimulus variables.

Context stimulus variables (characteristics of the situation, other than the task)

A number of non-task-related stimuli can influence an individual's behavior. For example, Whiting and Child (1953) and McGuire (1956) have reported that peer and adult expectancies are related to school success. Such expectancies are neither part of the task nor within the person; they are part of the context in which the criterion behavior occurs. In a sense, the context provides information regarding "payoff" or reinforcement associated with performance.

In a comparison of an innovative school with control schools, Foster (1968) hypothesized that a number of elements in each program might interact with student characteristics to produce systematic variation in achievement. For example, an innovative school procedure which includes a large amount of independent study might result in exceptional academic growth for the bright high-autonomous student and only mediocre growth for the bright low-autonomous student. Conversely, the bright low-autonomous student might perform exceptionally well when the structure of the school is highly organized and supervised.

In the educational setting, other context variables which might interact with student characteristics to influence response acquisition might be the physical condition of the class-

room, the reward system for good grades, and peer expectations
(if the student's friends value school success, he might work
even though he himself does not value school achievement).

Symbolic representation of the conceptual model

In view of the preceding discussion, one can say that complex behavior of a person is a function of (1) person characteristics, (2) focal stimulus characteristics, and (3) context characteristics. The preceding statement can be reduced to a quasi-
mathematical model as a shorthand notation:

$$Y = f(P,S,C)$$

where:

Y = the criterion;
P = person characteristics;
S = focal stimulus characteristics;
C = context characteristics;
f = the functional relationship of the
variables in the three classes (P,S,C)
as they relate to Y.

The equation can be read: The criterion behavior of a number
of individuals is a function of the characteristics of these
individuals, the focal stimulus characteristics, and the context
characteristics.

> Note: Don't panic at the symbols. Symbol notation
> is used to simplify the expressions so they can be pre-
> sented in the form of quasi-mathematical models, a
> practice which becomes extremely useful as this text
> goes on. But there is no need to remember the particu-
> lar symbols used here, or for that matter to understand
> fully the variables and examples used in this chapter.
> They are from only one content area and are presented
> to show how one might go about grouping and exploring
> variables which may contribute to prediction.

Figure 1.2 presents the whole conceptual model for the study

of complex behavior with examples of person, focal stimulus, and

context variables.

$$Y = f(P,S,C)$$

where: Y = scores on a math achievement test for a set of individuals;

P = person variables;

Examples of person variables for this criterion are:
1. convergent thinking ability (as measured by most
   standardized intelligence tests)
2. divergent thinking ability (as measured by tests
   of creativity)
3. symbol aptitude
4. motivation
5. sex role identification (masculinity-femininity)
6. past learning relevant to success on the task at
   hand

S = focal stimulus variables;

Examples of focal stimulus variables (characteristics
of the criterion measure) for this criterion are:
1. length of the test
2. ordering of items by difficulty
3. numerical items vs "word problems"
4. type of mathematics problems contained in the test
5. number of operations required by each item

C = context variables;

Examples of context variables for this criterion are:
1. peer expectancies
2. adult expectancies
3. physical plant (light, noise, heat)
4. reward conditions

Figure 1.2. An illustration of variables in the three predictor categories of the comprehensive model,
for the criterion of math achievement.

The functional relationships among the three classes of variables and the criterion (Y) can take many forms depending upon the research expectations of the investigator. Figure 1.2 presents the quasi-mathematical model with examples for each of the three classes which might explain observed behavior. The examples would, of course, vary for any given criterion behavior. One of the tasks of research is to quantify the variables and then attempt to account for the observed criterion variation in terms of some weighted combination of the predictor variables. It is premature to delineate what functional relationships might exist between (P,S,C) and Y. However, the following example is provided to give some ideas as to how one might operationalize the three categories.

Accounting for complex behavior

Consider the task of predicting complex behavior such as success on a job-training program. If the training program is costly, it would be worthwhile to be able to predict the successful and unsuccessful trainees in terms of cost effectiveness (to avoid spending the cost of training on many individuals who are unlikely to be successfully trained).

In view of the proposed conceptual model just presented, a number of aspects of the situation should be investigated. One might first examine the set of behaviors that are related to the criterion (Y), the observed terminal behavior. What are the task characteristics (represented by S in Figure 1.2) that might be relevant to the criterion? Such an examination might give the

investigator a few notions regarding relevant human character-

istics (represented by P in Figure 1.2) needed as prerequisite

skills (e.g., are specific abilities and/or special previous

learning necessary for training success?). Furthermore, what are

the conditions (represented by C in Figure 1.2) surrounding the

learning and testing setting? Is there a great deal of peer pres-

sure for success? Are the trainers placing pressure on the train-

ees? If so, will trainee anxiety be relevant? What are the inter-

mediate payoff schedules? Must the trainee work through the pro-

gram before some reinforcement is given? Will the need to achieve

be relevant to the reinforcement schedule? If he has failed often

in the past, does he have an expectancy to fail in this new task?

If so, can the training context be manipulated to minimize these

effects? Familiarity with the univariate research literature, as

well as with the multivariate taxonomies provided by Cattell

(1966) and Guilford (1966), should provide additional suggestions

regarding the relevant variables which account for complex be-

havior. Indeed, the whole research endeavor is centered around

selection of variables based upon theoretical expectations and

past empirical findings regarding the relationships among sets of

variables.

One might hypothesize that the desired terminal behavior

(criterion) of the training program is related to: spatial abili-

ties (SA); anxiety level (Anx); ability to manipulate symbols

(Sy); the amount of written work the job to be trained for will

require (Wr); the length of the training program (Ln); and the
number of trainees in the group being trained for any particular
job (Nt). Note that variables 1, 2, and 3 are person variables;
4 is a focal stimulus variable; and 5 and 6 are context variables. This set of variables can be cast as:

$$Y = f(SA, Anx, Sy, Wr, Ln, Nt)$$

where:
Y = observed terminal behavior of the trained
    individuals, and

SA, Anx, etc., are defined as above.

The function sign (f) implies that there are some mathematical functions which express the relationship of the predictors to
the observed criterion. Types of commonly used functions are:
additive, multiplicative (allowing for interaction), squared
(second degree polynomial), square root, cubic, and trigonometric functions (e.g., sine and cosine).

One may assume that the scores on the training task (Y) are
a sum of the six weighted predictor variables (a weighted additive function). Equation 1.1 expresses this function:

Equation 1.1 $Y = a_1 SA + a_2 Anx + a_3 Sy + a_4 Wr + a_5 Ln + a_6 Nt$

The weights a_1, a_2, ... a_6 might rationally be chosen or
empirically derived using some mathematical solution. In the domain of any science, researchers have seldom found a set of
weights which satisfies the equality (=) expressed above. The
reason for lack of equality is that theory, measurement, and expressed functions typically are not perfect. Errors of prediction
are made due to incomplete or erroneous theory, inadequate mea-

surement tools, and lack of ability to adjust relationships per-

fectly. What this means, then, is that the task of behavioral

research is to develop more comprehensive theories, better mea-

surement, and more appropriate quantitative procedures--so as to

minimize errors of prediction.

The equality expressed in Equation 1.1 is an ideal which is

not observed in most research endeavors. There will usually be

some errors of prediction. That is, no matter how "good" the

weights are, the sum of the weighted predictors will hardly ever

be equal to the criterion score. Equation 1.1 can be adjusted to

account for errors of prediction and therefore satisfy the ex-

pressed equality expression. Equation 1.2 reflects the inclusion

of such an error component which will thus satisfy the expressed

equality.

Equation 1.2: $Y = a_1 SA + a_2 Anx + a_3 Sy + a_4 Wr + a_5 Ln + a_6 Nt + E$
$Y = \hat{Y} + E$
where:
E = the difference between the predicted
score using the sum of the weighted
scores ($\hat{Y}$) and the observed terminal
performances (Y); $E = Y - \hat{Y}$.

The Multiple Linear Regression procedure derives the set of

weights (a_1, a_2, ..., a_6) to minimize the sum of the squared dif-

ferences between the observed criterion scores (Y) and the pre-

dicted criterion scores ($\hat{Y}$). The squared discrepancy between an

observed score and a predicted score can be expressed symboli-

cally as $(Y - \hat{Y})^2$, or as E^2. The quantity $\{\Sigma(Y - \hat{Y})^2\}$, or ΣE^2, has

interesting properties which satisfy a number of statistical dis-

tributions (e.g., the F statistic). Thus, with some manipulation

the least squares solution can be used inferentially for decision-making purposes. Chapters Two and Four will treat statistical inference in greater detail.

Accounting for complex behavior with alternate functions

Suppose the weighted additive function expressed in Equation
1.2 yielded a rather large ΣE^2 for the group of n individuals.
(The sum of the squared errors of prediction for the individuals
is large). Some of this error might be due to selection of inappropriate functions. Knowledge of past research might suggest
to the investigator that the additive function is not quite adequate. He might conclude that Wr is geometrically related to Y.
(The criterion scores are approximately the same for very low to
middle Wr levels; but the criterion scores increase rapidly as Wr
scores increase from medium to high). Furthermore, the investigator might expect anxiety (Anx) and length of training (Ln) to
interact such that anxiety might adversely influence terminal behavior especially when the training period is long. These two circumstances can be reflected by adding two additional variables to
Equation 1.2: (1) a squared function of Wr (Wr^2, reflecting the
geometric expectation) and (2) a multiplicative function for
anxiety and length of training (Anx*Ln, reflecting the interaction expectation). The expanded equation would be:

Equation 1.3: $Y = a_1 SA + a_2 Anx + a_3 Sy + a_4 Wr + a_5 Ln + a_6 Nt$

$$+ a_7 Wr^2 + a_8 (Anx*Ln) + E_2$$

where: E_2 = a new error value $(Y - \hat{Y})$.

(The error (E) is subscripted differently
in Equation 1.3 because two additional
variables are used for prediction. The
asterisk (*) will be used to indicate multiplication in this text.)

If the eight variables in Equation 1.3 represent the func-

tional relationship more adequately than the variables in Equa-

tion 1.2, then the predicted behavior ($\hat{Y}$) should be closer to the

observed behavior (Y) in Equation 1.3 than that found using Equa-

tion 1.2, thus Equation 1.3 should yield a smaller $\Sigma(Y - \hat{Y})^2$.

Pictorial Representation of Accounting for Variance

Let us now take another look at the prediction of the cri-

terion variable presented in Equation 1.2. A number of predictor

variables were included because it was suspected that the criter-

ion was complexly caused. If only spatial abilities (SA) had been

used as a predictor, a certain degree of predictability might

have been observed, say $r^2 = .20$ (which is an overlap of 20% be-

tween predictor and criterion variance) as in Figure 1.3. Each of

the Venn circles in Figure 1.3 represents one unit of variance.

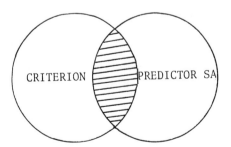

Figure 1.3. A pictorial representation of a situation
            wherein the predictor accounts for 20%
            (the shaded area) of the criterion variance.

The unshaded area of the criterion represents 80% of the criterion variance, which is unaccounted for. Often this variance is labeled as "error variance" and considered to be due to measurement error. But it may very well be due to not including all of the relevant (and necessary) predictor variables. Adding another predictor variable to the prediction scheme (say Sy, ability to manipulate symbols) might produce the situation depicted in Figure 1.4.

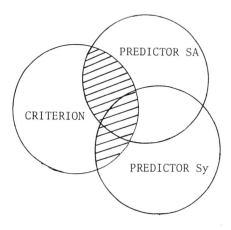

Figure 1.4. A pictorial representation of a situation where the predictors are correlated and account for 30% of the criterion variance.

Note in Figure 1.4 that the second predictor (Sy) accounts also for approximately 20% of the criterion variance; but since the two predictor variables are correlated with each other, the new predictor adds only 10% new overlap with the criterion.

Predictor variables will most likely be correlated in the real world. The concern is not how highly they are correlated; the concern is how much of the criterion variance they account for together. Has the set of predictor variables come close to the goal of accounting for 100% of the criterion variance? If not, then what other predictor variables could be considered? There are many other within-person variables, but including other within-person variables may yield diminishing returns. That is, there is criterion variance that is accounted for by variables other than within-person variables. A criterion will be a function of

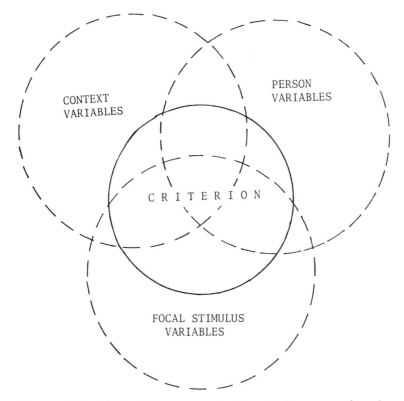

Figure 1.5. Pictorial representation of the comprehensive model presented in Figure 1.2.

relevant focal stimulus variables, context variables, and person
variables. Figure 1.5 represents the ideal situation (accounting
for 100% of the criterion variance), with the three sets of predictor variables being represented by dashes because each set
would not always be expected to account for the same amount of
criterion variance. Which variables to use from each set is derived from theory, and the proportion of criterion variance accounted for by each set is always an empirical question to be answered by the data.

2

Hypothesis Testing

In Chapter One, it was suggested that research design and statistics are tools to help an investigator make decisions regarding the adequacy of his description, prediction, and/or improvement of behavior. When testing expectations, one seeks to generalize from a relatively small group of individuals to other individuals who are similar. Most readers of this book will be familiar with sampling theory and inferential statistics. However, a brief discussion of sampling theory is presented for review purposes. Figure 2.1, which is on the next page, contains the steps in hypothesis testing which most statisticians accept.

The population of interest

In a typical research situation, a population is defined as a collection of individuals about which one is seeking information. The population information to be estimated is assumed to have a true value. For example, theoretically there is a true average value ($\Sigma X/N$) for the height of all men in the United States aged 21 and over. This value is called a parameter (a population fact). Furthermore, there is a true value (another parameter) which represents the dispersion of the height of men

25

1. Statement of the <u>research</u> <u>hypothesis</u> - a statement about the population which the researcher is hoping to support

2. Statement of the <u>statistical</u> <u>hypothesis</u> - a statement about the population which is antithetical to the research hypothesis

3. Statement of <u>alpha</u> - the risk (probability) the researcher is willing to make in rejecting a true statistical hypothesis

4. Collection of the data from a representative sample of the population

5. Calculation of the test statistic (t, <u>F</u>, χ^2)

6. Comparison of test statistic with tabled value

    a. If test statistic greater than or equal to tabled value, reject the statistical hypothesis and accept the research hypothesis

    b. If test statistic less than tabled value, fail to reject the statistical hypothesis and fail to accept the research hypothesis

Figure 2.1. Steps in hypothesis testing.

from the true mean (population variance, $\frac{\Sigma(X-\bar{X})^2}{N}$).

With most populations of interest, it is usually not possible to find and measure every individual, so it is not possible to arrive at the parameter value. A researcher therefore selects a subset of the population of interest and measures the individuals in the subset in order to make some statement about the population.

A <u>sample</u> is defined as a collection of individuals who are a subset of individuals representing the research population. The sample information to be collected is assumed to be an <u>estimate</u> of the population parameter.

Whenever a random sample is drawn, the mean of that sample is the best estimate of the population mean. The sample variance is a biased estimate of the population variance. An explanation of that statement can be found in other statistical texts. It is sufficient here to indicate that $\frac{\Sigma(X-\bar{X})^2}{N-1}$ is a better estimate of the population variance than $\frac{\Sigma(X-\bar{X})^2}{N}$. Both the sample mean and the sample variance are subject to sampling error. For example, if a population is defined as one hundred men in a particular location, one can randomly select twenty men from this population, measure their heights, and calculate a mean value for these twenty men. This mean value is the best estimate of the average height of the population of 100 men that can be obtained from 20 men. Another sample of twenty men from this population can be randomly selected and measured, and a mean value of the measures can be calculated. The likelihood that these two means are exactly the same is very small, yet one would expect them to be relatively close in value. Furthermore, if one measured the total population and calculated the population mean, one would expect this parameter value and the two sample mean values to be very similar but not exactly the same. The difference between a population mean and a sample mean is assumed to be due to sampling error--of course this may not be exactly true because there may be systematic errors of measurement in the sample. But if one assumes there are no systematic measurement biases, then the random errors of measurement should cancel out, and the sample mean would approximate the population mean. Errors of measurement are

treated extensively elsewhere and are beyond the scope of this
chapter (See Chapter Seven).

In real life situations, the researcher is typically unable
to obtain population parameters for his population of interest.
For example, in a particular school system a researcher may be
interested in the relative merits of Method "e" when compared
with Method "f" for seventh grade boys. He could give Method e
to half of the seventh grade boys and give Method f to the other
half. He could then use the mean scores of the two groups (on
some test designed to measure success in the subject matter being
taught) to describe the relative success of the two treatments
for these seventh grade boys. Usually, however, he would wish to
use the data from these two groups to generalize to (or predict
the success of these two methods for) a much broader population
of students. In this example he may wish to use the results of
the methods for the two sample groups to generalize to seventh
grade boys in this school for the next, say, two years. He would
be assuming that this year's seventh grade class represents a
random sample of the population of seventh grade boys in the
school district for the three years of interest (this year plus
the next two years). Of course, something may happen in the community or school to change the composition of the next two years
of seventh grade boys so that the boys this year do not represent
that changed population. His confidence in his ability to predict,
or generalize, would then be reduced by an unknown amount. Possibly his generalization might be totally invalid; although without

further data and analysis of that data, his original sample means
are still the best information from which to predict the effects
of the treatments.

The researcher wants to make a decision regarding which
method to use for the next two years for seventh grade boys.
Since those two (groups) of boys are not yet seventh graders, he
cannot expose them to the two methods and then decide. He therefore wants to use this year's seventh grade boys to predict for
the next two years of seventh grade boys. Suppose there are one
hundred boys in the present year's seventh grade class. Each boy
can be assigned a number from 1 to 100; and using a table of random numbers the researcher can select 50 boys for Method e and 50
boys for Method f. (For a description of this procedure see
Edwards, 1964, p. 250). This sampling procedure results in two
treatment samples which are likely to be representative sub-sets
of the total population being investigated. Random assignment
does not guarantee that the samples are representative of the
population, but it does minimize the chances that the samples are
biased sub-sets of the population.

These boys can be subjected to the treatment to which they
were assigned, and then the criterion performance can be measured
on some task the methods were designed to teach. Suppose the following information is obtained: a mean value of 55 and variance
of 25 for the sample receiving Method e, and a mean value of 35
and variance of 25 for the sample receiving Method f. Originally
the two groups represented the same population; as a consequence

of the treatment, do they still represent the same population?
Are the means of 55 for Method e and 35 for Method f likely to be
due to sampling variation?

Analysis of variance

The analysis of variance technique analyzes estimates of the
population variance and yields a probability statement indicating
how likely it is that observed differences between means are due
to sampling error. If the differences are highly unlikely (say 1
time in 1,000), one may be willing to conclude that, as a result
of the two treatments, the two groups of boys now represent different populations. The matter of risk and cost must enter into
the decision of what level of probability to select. These matters will be discussed later.

The $\underline{F}$ statistic is based upon the ratio between two estimates of the population variance. In research designs there are a
number of ways to estimate population variance, some sensitive to
group mean differences and others which are insensitive to group
differences. In our present example, two estimates are of interest:

(1) within group estimate of the population variance, and

(2) among group estimate of the population variance.

Within group estimate of the population variance $(\hat{V}_w)$

The estimate of population variance using within group information in analysis of variance usually is called mean square

within (symbolized in many texts as MS_W). The preference in this text is to refer to this variance estimate as the "within variance estimate" and to use the symbol $\hat{V}_W$, where (1) the capital V is used to represent the population variance, (2) the hat ($\hat{}$) indicates an estimate, and (3) the w indicates how the estimate is calculated (within group).

$\hat{V}_W$ is an estimate of population variance based on how much the criterion scores of persons in each group differ from the mean score of their group. $\hat{V}_W$ is calculated in several steps. First, the mean of one group is calculated, and that group mean is subtracted from each of the observed criterion scores (X) of the members of that group. Second, each of these discrepancy scores is then squared, and the squared scores are summed. This procedure is followed for each of the k groups. (In the above example k is 2, Method e and Method f.) The sums of squared discrepancy scores for all groups are then added together (sum of squares within, or SS_W) and divided by the degrees of freedom within all of the groups. The result is $\hat{V}_W$.

As typically presented, "degrees of freedom" is difficult to understand, even though it is a simple concept. A more elaborate explanation of this matter is given later, but a brief discussion is presented here because degrees of freedom relate to each estimate of the population variance.

A variable is by definition an attribute which may take on any of a range of numerical values. A variable is then free to vary. Within group e (Method e) there are 50 boys and there are

50 observations on the criterion variable. Each of the 50 observations or scores is free to vary. The scores are considered free
to vary although each represents one of the 50 boys and is therefore dependent on the performance of that boy.

After obtaining the 50 scores for group e on the criterion
variable, one can calculate the mean of those 50 scores ($\Sigma X/N$).
When the differences between the observed scores and the mean are
calculated, are all 50 difference scores free to vary? (Note that
it is a property of the mean that the sum of the difference
scores between the mean and the observed scores is equal to zero.) The first score is free to vary--that is, it can be any
value. Likewise, each subsequent score through observation 49 is
free to vary. But what about the 50th calculation? That score is
not free to vary because each of the preceding 49 scores is specified and the group mean is known. Given the 49 scores and the
mean, there is only one value the 50th difference score can be.
It is not free to vary.

The following three observed scores: 4, 3, and 2, will help
to demonstrate degrees of freedom. The mean of these three scores
is $\Sigma X/N = 9/3 = 3$. Now calculate the differences between each observed score and the mean. Given that the mean is 3, if one did
not already know the actual scores, one might think that those
three scores could take on an infinite number of values to get a
mean of 3. If the first observed score is 4, then the difference
score for that observed score is 4-3 = 1. (That observed score
was free to vary; it happened to be 4.) If the second observed

score is 3, the difference score for it is 3-3 = 0. That score
was also free to vary. But now that it is known that two of the
scores are 4 and 3, there is only one value the third score can
be and still yield a mean of 3 for the three scores (and still
have the difference scores sum to zero). That third observation
score must be 2, and the difference score would of course be
2-3 = -1. Of the three scores, only two are free to vary once the
mean is known. To generalize: when the mean is known, one less
than the total number of scores (N-1) is free to vary.

When the problem of calculating $\hat{V}_W$ was presented, it was
given that for Method e, the mean score = 55 and the sample variance = 25; for Method f, the mean score = 35 and the sample variance = 25. One could go back to the 100 observed scores (50 for
each group) and calculate $\hat{V}_W$ from the procedure described above.
It will be easier, however, to use the knowledge of the group
means and variances. The variance for Group e (25) was calculated
by finding the difference between each score and the group mean,
squaring and summing the difference scores, and dividing the sum
by the number of scores (50), or $\Sigma(X-\overline{X}_e)^2/N$. To obtain the un-
biased estimate of the population variance $(\hat{V}_W)$ one does not divide by N, but by degrees of freedom; and the separate sums of
squares for each group are summed together before dividing by degrees of freedom, so the sum of squares for each group is needed.
If the Group e variance was 25 and N = 50, then from the above
formula one can see that the sum of squares for Group e must
equal 1250 (25 * 50). The variance and N are the same for Group f,

so the sum of squares for Group f must also equal 1250.

$\hat{V}_W$ is calculated by adding the sum of squares of the k

groups and then dividing by degrees of freedom. This problem con-

tains two groups, and the sum of squares for each is 1250. The

total degrees of freedom is equal to the sum of the degrees of

freedom for each group. In each group a mean was calculated, so

the degrees of freedom for each group is 50-1 = 49, and the total

degrees of freedom would be 98. One could also arrive at the to-

tal degrees of freedom by looking at the total number of observa-

tions in both groups (N) and subtracting the number of means cal-

culated (k), or in this case 100-2 = 98.

$\hat{V}_W$ is therefore calculated for the example problem as:

Equation 2.1 $$\hat{V}_W = \frac{\Sigma(X-\overline{X}_e)^2 + \Sigma(X-\overline{X}_f)^2}{N-k}$$

$$= \frac{1250 + 1250}{100-2} = \frac{2500}{98} = 25.5$$

$$\hat{V}_W = 25.5$$

This value (25.5) is the best estimate of the population

variance because it is not influenced by group mean differences.

It is best in the sense that repeated calculations on random

samples from the same population would result in variance esti-

mates which deviate the least from the actual population vari-

ance. If the group mean differences are small, then the estimate

of the population variance using among group information ($\hat{V}_a$ -

discussed next) should be close to the within group estimate ($\hat{V}_W$).

Among group estimate of the population variance $(\hat{V}_a)$

The among group estimate of the population variance is symbolized here as: $\hat{V}_a$. The subscript denotes that it is an estimate
using among (or between) group information.

$\hat{V}_a$ is derived by subtracting the mean of all the boys regardless of group membership (this is often called grand mean,
$\bar{X}_g$) from each of the k group means. For each group, this value is
then squared and multiplied by the number of scores in the group.
The product for each group is summed over groups (SS_a) and divided by degrees of freedom, which for this variance estimate is
k-1.

Degrees of freedom for $\hat{V}_a$ equals the number of groups (k)
minus one (1). Following the logic given for $\hat{V}_w$, each group mean
is an observation free to take on any value, but once the grand
mean is calculated using these means, only k-1 means are free to
vary. In this example, $\bar{X}_e$ = 55, $\bar{X}_f$ = 35, $\bar{X}_g$ = (55 + 35)/2 = 45.
For this case, $\bar{X}_g$ can be derived by getting the average of the
two means because the groups are of equal number. When groups
differ in number, the above procedure is not applicable, but $\bar{X}_g$
can be obtained by summing all scores and dividing by N.

$\hat{V}_a$ is therefore calculated for the example problem as:

Equation 2.2 $$\hat{V}_a = \frac{\{(\bar{X}_e - \bar{X}_g)^2 \times N_e\} + \{(\bar{X}_f - \bar{X}_g)^2 \times N_f\}}{k-1}$$

$$= \frac{\{(55 - 45)^2 \times 50\} + \{(35 - 45)^2 \times 50\}}{2-1}$$

$$= \frac{(100 \times 50) + (100 \times 50)}{1}$$

$$= \frac{5,000 + 5,000}{1}$$

$$\hat{V}_a = 10,000$$

The F ratio

Two estimates of the population variance, $\hat{V}_w$ and $\hat{V}_a$, have just been discussed. Within group estimates of the population variance are unaffected by group mean differences. Among group estimates of the population variance are sensitive to group mean differences because they are based upon group mean variations from the grand mean. If there is little difference between group means, then $\hat{V}_a$ will be a good estimate of the population variance, and would be expected to be about equal to $\hat{V}_w$ (the best estimate of the population variance).

Consider the discussion of random sampling. When two samples are randomly drawn from a population, the sample means are expected to vary somewhat from each other. The difference between sample means is referred to as sampling error. The larger the number of subjects in a sample, the closer the sample means tend to be. By drawing two more samples, calculating the two means, then replacing the members and drawing two more samples, calculating means, and then continuing this process many times, it would be noted in most cases that the two sample means are relatively close to each other. Occasionally, however, fairly large mean differences will be observed.

The F ratio represents the distribution of the ratios of two

independent estimates of the population variance. If the means of

the two (or more) groups are close together, the ratio of $\dfrac{\hat{V}_a}{\hat{V}_w}$ will,

on the average, be 1.00. It can be mathematically shown that

the expected F ratio is 1.00 if the two samples are representative

of a single population. Nevertheless, occasionally the randomly

selected groups will have large mean differences due to sampling

variation. Therefore, occasionally an F ratio much greater than

one should be expected. If the ratio of the two estimates is dis-

tributed as F, then for any set of degrees of freedom, the likeli-

hood of an observed F ratio being due to sampling variation can be

determined.

Figure 2.2 is an approximate distribution of F ratios with

1 (k-1) and 98 (N-k) degrees of freedom. The area shaded to the

right of 3.94 represents the 5% of the observed F ratios which

will occur solely from random sampling variation. An F ratio of

3.94 or greater would be observed 5% of the time due to sampling

variation. An F ratio of 6.90 or greater would be observed 1% of

the time due to sampling variation. These values of 3.94 and 6.90

can be found in F tables in many statistics books. In the example,

Method e and Method f, the estimated population variance using

within group data was: $\hat{V}_w$ = 25.5; and the among group estimate was:

$\hat{V}_a$ = 10,000. The F ratio is:

Equation 2.3 $F = \dfrac{\hat{V}_a}{\hat{V}_w} = \dfrac{10,000}{25.5} = 392.15 \quad df = 1,98$

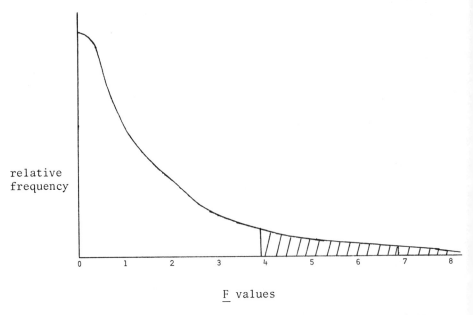

F values

Figure 2.2. An approximate distribution of F with 1 and 98
degrees of freedom

Mathematical statisticians tell us that an F of 392 with 1
and 98 degrees of freedom would be observed less than 1 time in a
hundred due to sampling variation. Since $\hat{V}_a$ is so very large in
relation to $\hat{V}_w$, one can conclude that the difference between the
two observed means (Method e mean was 55, and Method f mean was
35) is very unlikely to be due to sampling variation. Most likely,
the two treatments have resulted in producing two populations--one
of seventh grade boys exposed to Method e and another population
of seventh grade boys exposed to Method f.

On the other hand, one must be aware of the fact that one
time in a hundred, such a large mean difference is expected due to
sampling variation. Suppose you are the decision-maker regarding
the selection of a method of instruction. What would you recommend?

Traditionally, one would state a rival (statistical or null) hypothesis which is, for this example:

"The mean of the criterion measures is the same for each treatment population."

When should one reject the statistical hypothesis? This depends upon the amount of risk one wishes to take. In the case just presented, the researcher asks himself, "If my observed sample mean differences are that one-in-a-hundred resulting from random sampling variation, how much will that mistake cost? On the other hand, if the sample results are indicative of the population, how much will it cost to not act as if they are?" The probability that the present results came from a population wherein the two means were the same is very small. Whether this probability is tolerable must be determined by the researcher.

Figure 2.3a represents the two samples drawn from a theoretic larger population. When statistical significance is obtained, as a result of differential treatment effects, it can be said with some degree of confidence that the two samples no longer represent the same population with respect to the criterion measure of concern. Those two samples now represent two populations, Treatment e and Treatment f populations.

Figure 2.3b represents the other state of affairs. Statistical significance has not been obtained, and consequently the differential treatments have not produced a large enough difference to conclude that they have different effects on the criterion measure of concern.

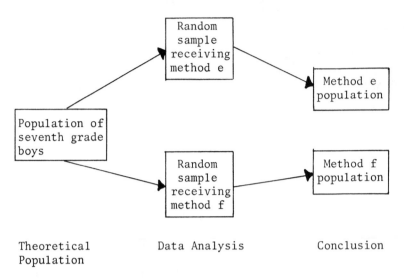

Theoretical   Data Analysis   Conclusion
Population

Figure 2.3a Conclusion, when statistical analysis <u>indicates</u>
           <u>significance</u>.

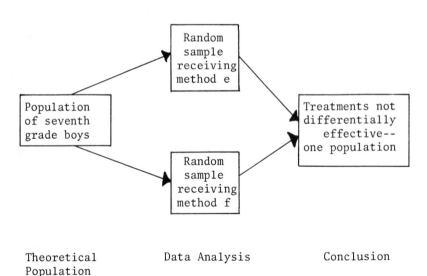

Theoretical   Data Analysis   Conclusion
Population

Figure 2.3b Conclusion, when statistical analysis <u>does</u> <u>not</u>
           <u>indicate</u> <u>significance</u>.

Type I errors

The decision to reject or fail to reject the statistical hypothesis is essentially a decision of risk-taking. Usually, the probability level of either 5 times in 100 or 1 time in 100 (alpha = .05 or alpha = .01) is the decision point used by researchers in the behavioral sciences. That is, they are willing to risk making 5 errors (or 1) out of 100 rejections of a true statistical hypothesis (a Type I error). The choice of alpha is up to the decision-maker and the degree of risk he wishes to take. Also, the decision-maker must recognize that taking a conservative risk (e.g., one time in a thousand rather than one time in a hundred) regarding Type I error increases the probability of making a Type II error.

Type II errors

The decision-maker has two possible conclusions he can make from the statistical hypothesis testing procedure. He can reject the statistical (rival) hypothesis, or he can fail to reject the statistical hypothesis. One hopes that a correct decision is made, but since one never knows the population value(s), one can never be sure that the correct action has been taken. The Type I error (rejecting a statistical hypothesis when it should not be rejected) is discussed above. The researcher has complete control over the probability of such error by choosing his alpha level. If alpha is .05, then 5% of the time the statistical hypothesis will be incorrectly rejected (if the statistical hypothesis is indeed true).

On the other hand, the other decision of failing to reject
the statistical hypothesis may also be in error (Type II error).
If the statistical hypothesis is not true of the populations, but
the sampled data does not provide enough evidence to reject it,
then an error is made. The probability of the Type II error decreases as sample size increases, and it also decreases as the
alpha (probability of a Type I error) increases. The probability
of Type II error also decreases as the discrepancy between the
statistically hypothesized population value and the true population value increases. But since the true population values are
never known, the probability of a Type II error being committed
cannot be determined. Because the consequences of making a Type II
error are usually less devastating (e.g., failing to accept some
new fact that is correct) than those of making a Type I error
(e.g., accepting a new fact that is incorrect), the alpha levels
chosen by most researchers are quite low (e.g., .05 or .01, or
.001). Actually, the content of the research hypothesis determines
the cost of making either a Type I or Type II error. Even though
the probability of making a Type II error cannot be calculated,
the researcher needs to be aware of the concept in order to realize that when he fails to reject the statistical hypothesis he may
be making an error.

Assumptions related to the F test

A few conditions must be met before the ratio between any
two estimates of the population variance is distributed as F.

There are four basic conditions of concern:

(1) All subjects in the treatment groups were originally drawn at random from the same parent population.

(2) The variance (V) of the criterion measures is the same for each of the treatment populations.

(3) The criterion measures for each treatment population are normally distributed.

(4) The means of the criterion measures are the same for each treatment population.

Figure 2.4 illustrates assumptions 2 and 3, for the case where two populations are under consideration.

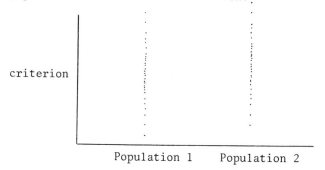

Figure 2.4. Illustration of two populations which meet assumptions 2 and 3, but not 4.

Note that these assumptions refer to the populations. Since the populations are never available, the validity of the assumptions is never known for sure.

> Note: Tests of significance are available to test assumptions 2 and 3; but when using such tests, the researcher hopes to accept the statistical hypothesis rather than the research hypothesis. This is opposite to the usual hypothesis testing procedure and should require different values of alpha. The desire in assumption 2 is to show that the variances are equal and in assumption 3 that the population distribution is equivalent to a normal distribution. In both cases, the

traditional hypothesis testing process is reversed in
that an equivalent population state of affairs is desired. The statistical hypothesis, rather than the research hypothesis is hopefully true. If one wants to
protect against falsely accepting this hypothesis, then
alpha should be set at a much higher level (say .60).

Investigations into these assumptions have generally con-

cluded that assumptions 2 and 3 can be violated in most cases with-

out seriously distorting the stated alpha level. The position tak-

en in this text (and defended more fully in Chapters Seven and

Eight) is that attempting to obtain an R^2 of 1.00, along with rep-

lication of findings, is a necessary and sufficient guard against

any violation of assumptions. The reader who is more concerned

about this issue may wish to read a review regarding these assump-

tions (Glass, Peckham, and Sanders, 1972).

The fourth assumption is really the statistical hypothesis

(often referred to as the "null hypothesis") being tested:

(4) The means of the criterion measures are the same for
each treatment population.

The assumptions can be viewed in this fashion: when a signi-

ficant F is obtained, it could be due to any one of the four (or a

combination) of the assumptions not being true. If the researcher

is reasonably sure of assumptions 1, 2, and 3 being true, then

assumption 4 can be rejected. For example, with 1 and 100 df, an

F of 6.90 or greater is observed one time in 100 due to sampling

variation. If one specifies .01 as the amount of risk one will

tolerate to reject a true statistical hypothesis, given an F larg-

er than 6.90 and faith in the first three assumptions, one would

reject the condition that the criterion mean is equal for the two
treatment populations.

Elaboration upon assumptions

The assumptions given above are applicable when one is dealing with treatment groups; however, they can be extended to cover
linear fits of various sorts when the independent (predictor) variable is continuous. Assumptions 2 and 3 regarding equal variance
and normal distribution can be extended to the situations pictured
in Figure 2.5.

Very often one wishes to establish the fact that there is a
linear relationship between Ability and Performance (as depicted
in Figure 2.5b). This is a research hypothesis, and one may construct a rival hypothesis (statistical hypothesis) which takes the
form: There is no linear relationship between Ability and Performance.

It is assumed that for each data point on the Ability (X)
continuum the criterion values have equal variance and are normally distributed in the population. Given a sufficiently large sample (say, greater than 100), violation of the assumptions of equal
variance and normal distribution in the population criterion
scores across all data points of interest of the X variable is not
too serious.

Figure 2.5c and 2.5d are two other examples where the assumptions of equal variance and normal distribution are applicable.
These cases will be discussed as they are developed in Chapters

Four, Five, and Seven.

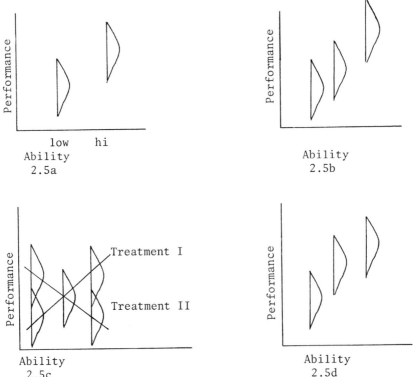

Figure 2.5a. A relationship between two groups and the
criterion.

   2.5b. A linear relationship between two continuous
variables.

   2.5c. A linear relationship between two continuous
variables for each of two treatment groups.

   2.5d. A curvilinear relationship between two continuous variables.

R^2--the proportional estimate of sums of squares

The $\underline{F}$ ratio can also be derived using R^2, which is a propor-

tional estimate of the total observed sums of squares (SS_t) for

each source of variance. In the two-group case of Method e and
Method f, an among group sum of squares (which was 10,000) and
a within group sum of squares (2,500) was calculated. A total observed sum of squares could also have been calculated by subtracting the grand mean from each of the 100 students' scores,
squaring the difference scores, and then summing these 100
squared scores. This total sum of squares would be 12,500 (equal-
ing among sum of squares plus within sum of squares).

"R^2" is an expression of the proportion of total sample
criterion variance accounted for by a particular set of predictor
information. The range of R^2 is from .00 to 1.00. The total sample variance can be calculated by dividing the total sum of
squares by the number of subjects: total variance = SS/N. The
variance accounted for by knowing group means can be calculated
by dividing the sum of squares among by the number of subjects:
variance accounted for by knowing group means = SS_a/N. R_a^2 (which
uses among group data) is simply SS_a divided by SS_t. Therefore,
R_a^2 is:

$$R_a^2 = \frac{SS_a}{SS_t} = \frac{10,000}{12,500} = .80$$

In this example SS_a is 80% of SS_t. The SS_a relies upon knowledge of which group (e or f) the subject is in; thus knowledge
of group membership accounts for 80% of the total observed sum of
squares (SS_t). The reader should note that such a high R^2 is usually not associated with treatment group.

R_w^2 (which uses within group data) is SS_w divided by SS_t.

Therefore, R_w^2 is:

$$R_w^2 = \frac{SS_w}{SS_t} = \frac{2500}{12,500} = .20$$

The sum of squares within reflects the variation within the groups that is not directly due to the two treatments. SS_w is 20% of the total sum of squares; thus 20% of the observed SS_t is due to unknown sources.

What does introducing the notion of R^2 provide that was not known when using estimates of the population variance ($\hat{V}_a$ and $\hat{V}_w$)? The R^2 notion provides a different perspective regarding variation and research designed to explain variation. Indeed, this different perspective is one of the fundamental reasons for this book.

The R^2 perspective

Given any one sample from a population, the best source from which to estimate the population variance is the observed variability of the sample. In the sample of 100 subjects given Methods e and f there was a total sum of squares of 12,500. The estimated population variance ($\hat{V}_t$), based on the sample, was found by dividing SS_t by degrees of freedom. The grand mean had to be calculated in order to get SS_t; therefore, degrees of freedom must be N-1. Once the grand mean is known, only 99 scores are free to vary; the last score is fixed. $\hat{V}_t$ then equals:

$$\hat{V}_t = \frac{SS_t}{N-1} = \frac{12,500}{99} = 126.2$$

Using the R^2 perspective, given only a sample and a sample mean, 100% of the variance in the criterion scores is due to unknown sources. A researcher's goal is to find out what accounts for variance in the criterion--why some subjects score high and some score low. Some of the variance might be due to different treatments, ability, motivation, etc. The task of research is to seek out information that will account for all of the criterion variance, or for as much of it as is temporarily satisfactory.

In the case of the 100 seventh-grade boys, an estimated population variance $(\hat{V}_t)$ of 126.2 was observed. Given this state of affairs, one might say, "These boys certainly vary on their criterion scores." The researcher may then bring into the picture the information that these boys were randomly assigned to two different methods of instruction. One might now wonder, "Hey, does this treatment information explain some of the criterion variance?" As was just shown above, R_a^2 was based upon knowledge of which boys received which treatment. It was found that this treatment information explained 80% of the total variance (R_a^2 = .80).

Now that it is known how much criterion variance can be accounted for by using knowledge of treatment groups, can it be determined how likely it is that this large an R^2 value would be due to sampling error--as was done with the F ratio previously? As a matter of fact, it can; the F ratio can be calculated from R_a^2 and R_w^2 as well as from $\hat{V}_a$ and $\hat{V}_w$.

Using R_a^2 and R_w^2 (two proportional parts of the total sample

variance), it can be determined if the knowledge of group membership (Methods e and f) will explain a non-chance amount of the
observed sample variance. The following formula for the $\underline{F}$ test
can be used:

Equation 2.4 $\underline{F}(df_n, df_d) = \dfrac{R_a^2/df_n}{R_w^2/df_d}$

> where: R_a^2 = the proportion of unique observed variance
>   due to knowledge of group membership;
>
> R_w^2 = the proportion of unique observed variance
>   due to unknown sources;
>
> df_n = the number of group means free to vary once
>   the grand mean was calculated; and
>
> df_d = the number of subjects whose criterion
>   scores are free to vary once each group
>   mean has been calculated.

It was previously determined that R_a^2 equals .80 and R_w^2
equals .20. df_n equals 2-1, or 1, because only one group mean
was free to vary once the grand mean was calculated. df_d is 100-2
because within each group the last (50th) score was not free to
vary since the group mean plus the other 49 scores determined the
last boy's score.

Substituting these values into Equation 2.4, one obtains:

$$\underline{F}(df_n, df_d) = \frac{R_a^2/df_n}{R_w^2/df_d}$$

$$= \frac{.80/1}{.20/98} = \frac{.80}{.00204}$$

$$\underline{F}(1,98) = 392.15$$

This $\underline{F}$ value is exactly the same as obtained by using $\hat{V}_a$ = 10,000 and $\hat{V}_w$ = 2,500.

What has been gained? Not only can one determine how likely it is that the R_a^2 is due to sampling error, but one also gets an idea of what proportion of the sample variance can be explained by the piece of information, group membership. Further, the R_w^2 indicates how much ignorance still exists with regard to the source of the observed criterion variance.

For this example, R_a^2 = 1 - R_w^2, so the two points might seem redundant. This will not always be the case. The more general equation for $\underline{F}$ will be developed later and appears as Equation 4.6.

Theoretical and heuristic benefits of R^2 in analysis of variance

The clear-cut result provided in the example of the two methods (80% of the variance explained) is seldom observed in the behavioral sciences. This is particularly true when one enters the domain of theory testing . A typical example is a recent study using a 2-by-2 factorial analysis of variance design which reports an $\underline{F}$ value of 5.08 (with df of 1 and 54) for an interaction effect. An $\underline{F}$ of 5.08 with these degrees of freedom is an unlikely chance event (p < .01). The author writes: "More important is the strong interaction [between two of the predictor variables]."

It is not reported in the study, but one can determine from the $\underline{F}$ and df values that the combined effects of A, B, and A * B account for less than 9% of the observed criterion variance. Over 91% of the observed criterion variance in the study is due to unknown sources. When one starts off with no explained variance, accounting for 9% of the variance is surely an improvement (and the $\underline{F}$ test indicates that that 9% is likely to be non-chance). This is all well and fine, but the traditional reporting of analyses using only the level of probability for a particular $\underline{F}$ ratio obscures the degree of ignorance remaining in most studies with regard to the unexplained criterion variance. And indeed, one might wonder whether accounting for approximately 8% of the observed criterion variance represents a "strong interaction."

The implication of the preceding discussion is in no way meant to downgrade present research. The intent is to show the added value that specifying the percent of accounted-for variance provides. This specification should result in a tempering of stated conclusions.

A moderating comment is called for here. The 8 percent of variance due to interaction could yield a warranted strong conclusion when the unknown variance (error variance) is small. For example, if the A and B effects accounted for, say, 80% of the variance, then an added 8% due to interaction would be very remarkable because this effect removes almost half of the remaining unknown variance.

Heuristically, the use of R^2 keeps one's ignorance in full view because $1-R^2$ is the proportion of variance yet to be accounted for. If theoretic expectations of causal relationships

are empirically found to account for less than 50% of the criterion variance, then it seems that additional "relevant" variables
must enter into the theoretical model. The determination of which
are "relevant" variables is empirically a function of the criterion variable and the other predictor variables that are already
in the predictive system. And of course such model building is in
the spirit of Chapter One. Researchers should want as much as
possible to reduce the unaccounted-for variance--the researcher's
true enemy.

Many studies reported in journals give sufficient information to allow the reader to calculate the R^2 value for himself.
A preferable procedure would be one in which the researcher reported the R^2 in his published study, but as yet this is not commonly done. The following procedure will enable the reader to
calculate R^2 values for himself.

Some journals provide Analysis of Variance source tables
such as Table 2.1 on the next page.

One can obtain the total sum of squares (SS_t) by adding all
entries under SS. The proportion of variance accounted for due to
any source can be derived by dividing the SS (sum of squares) associated with that source by SS_t. SS_t for the data provided in
Table 2.1 is 1531.72. SS for the A*B interaction is 131.44. The
proportion of variation accounted for by the A*B interaction
would be:

$$R^2_{A*B} = \frac{SS_{A*B}}{SS_t} = \frac{131.44}{1531.72} = .08$$

Table 2.1 A typical source table.

Source	df	SS	MS	F
A	1	.09	.09	<1
B	1	3.88	3.88	<1
A*B	1	131.44	131.44	5.08
Error (within)	54	1396.31	25.86	
Total	57	1531.72		

The reader might examine crucial research studies in his field and determine how much variance the significant results account for.

3
Vectors and Vector Operations

The multiple linear regression approach can be most easily understood within the framework of vector notation. If a researcher is to develop a command over the linear regression approach, he must become relatively familiar with vector notation and vector operations. Once the researcher has grasped the basic components of vector algebra, he can construct for himself the models used to test a particular research question. Thus, knowledge of vector algebra, coupled with mastery of the linear regression approach, allows the researcher to ask the research question in the particular way he wants and then to generate the statistical test. This approach may seem heretical to those researchers who were taught that they must frame their research question in a particular fashion, or that only certain research questions are, in fact, permissible.

Most people come into contact with vectors in their everyday lives; and thus, the mastery of vector algebra is not really a difficult chore. Many of the properties of vector algebra are simply generalizations from the algebra with which most readers are already familiar.

Why use vectors?

A vector is an ordered set of numbers which allows data to be represented in a very concise fashion. Some readers may be familiar with other definitions of vectors. For the purposes of this text, a vector is simply an ordered set of numbers. Once the conventions are learned and the symbols are identified, a large amount of data can be represented in a small amount of space. A familiar example of vectors is shown in Table 3.1.

Table 3.1 Example of a single vector; the same measurement for each of several entities.

City	Noon Temperature (in degrees Fahrenheit)
Chicago	20
Miami	80
Pittsburgh	45
St. Louis	30

The above type of table is familiar to everyone. The same kind of measurement has been made on each of the entities represented; i.e., the noon temperature has been recorded for each of the cities listed. The collection of temperatures is called a vector because each member of the collection is, in fact, a number; and the members (or "elements") are in some particular order. That is, the temperature of 20° was observed in Chicago, and the temperature of 30° was observed in St. Louis.

The data in Table 3.2 indicates another use of vectors, representing several measurements on a single entity. The above information could have come from a biographical data sheet. If one

Table 3.2 Example of several measurements on a
         single entity.

Height (in inches)	68
Weight (in pounds)	150
Age (in years)	25
Schooling (in years)	19
Married (yes = 1; no = 0)	1
Children (total number)	1

was given the following vector (Y_1) and told that this vector

represents the same information as presented in Table 3.2, then

one would know that this individual is 72 inches tall, weighs 300

pounds, is thirty years old, has 12 years of education, is not

married, and has no children.

$$Y_1 = \begin{bmatrix} 72 \\ 300 \\ 30 \\ 12 \\ 0 \\ 0 \end{bmatrix} = \begin{bmatrix} \text{height of subject in inches} \\ \text{weight of subject in pounds} \\ \text{age of subject in years} \\ \text{years of education of subject} \\ 1 = \text{married}; \ 0 = \text{not married} \\ \text{total number of children} \end{bmatrix}$$

The data in Table 3.3 represents a combination of the kinds

of vectors previously described. Here there are several observa-

tions on each of several entities. For the purposes of this book,

the focus will be on vectors which represent the same kind of

measurement on a number of subjects. Since there are four pieces

of information about each entity or subject in Table 3.3, there

are four vectors in Table 3.3. Someone has collected three pieces

of information and then calculated a fourth piece of information

(Batting Average) from two of the observations (dividing "Hits"

by "At Bats"). This fourth piece of information is considered to

be a new piece of information, as discussed later in the chapter.

Table 3.3 Example of several vectors for each entity
(player).

Player	At Bats	Hits	Batting Average	Home Runs
Joe	120	60	.500	3
Sam	100	30	.300	5
Dick	110	22	.200	15
Jim	100	25	.250	25
Russ	120	48	.400	20

Definition of vector

The definition of a vector that will be used here is simply:
A vector is an ordered set of numbers. The number of elements in
a vector is called the dimension of a vector. With reference to
Table 3.3, and given the order of the players, it is seen that
Joe hit three home runs because Joe is the subject corresponding
to the first element of each of the four vectors, and the numerical value of the first element of the "Home Run" vector is 3.
Likewise, since .400 is the fifth element of the "Batting Average" vector, and Russ is the subject corresponding to the fifth
element, Russ's batting average is .400.

It should be emphasized here that any number can be an element of a vector. Positive numbers, zeroes, and decimal fractions
have already been used. Negative numbers and common fractions are
also valid candidates as elements of vectors.

The other property of a vector, that there is some order,
should not be taken lightly. The order does not have to be in-

herent in the data; in fact, usually it is not. Referring again

to Table 3.3 and the "hits" vector, there seems to be no order.

The numbers are not arranged from high to low nor from low to

high. The reader should be well aware, though, that the order has

already been defined by the first column of names.

Vector notation

The general vector notation is to represent vectors with

subscripted capital letters (such as A_1) and to represent ele-

ments of a vector by double subscripted capital letters (such as

A_{1_1}). The particular subscripted letter which is associated with

a vector is purely arbitrary but will sometimes have some mne-

monic advantages. The hits vector of Table 3.3 could be repre-

sented as H_1. H_{1_1} would then be the number corresponding to Joe's

hits (60), H_{1_2} would be the number corresponding to Sam's hits

(30), and so on. The second subscript thus refers to the position

of the element in the vector. In general terms, the vector W_1

with t elements can be represented by the symbol W_1, or by the

bracketed symbols, or by the bracketed numbers;

$$W_1 = \begin{bmatrix} W_{1_1} \\ W_{1_2} \\ W_{1_3} \\ \cdot \\ \cdot \\ \cdot \\ W_{1_t} \end{bmatrix} = \begin{bmatrix} 6 \\ 10 \\ 3 \\ \cdot \\ \cdot \\ \cdot \\ 17 \end{bmatrix}$$

where the three dots mean "and so on continuing to." These three dots are necessary because the value of t is not known; there could be 50 elements of W_1, or 100, or just 4.

Example 3.1--vector elements

$$\text{If the vector} \quad \begin{bmatrix} 1 \\ 5 \\ 6 \\ 13 \\ -2 \end{bmatrix} \quad \text{is represented by } X_1,$$

then X_{1_1} is equal to 1, X_{1_2} is equal to 5, X_{1_3} is 6, X_{1_4} is 13, and X_{1_5} is -2. Note that X_{1_6}, X_{1_7}, etc., are not defined for this vector.

Categorical vectors

Many occasions will be found later on to use what some people call "categorical vectors." These vectors will usually be helpful in representing group membership. One number will be assigned to entities which belong to the group under consideration and another number to those entities which do not belong to that group. Any two numbers could be used to designate the two groups, but it is more efficient to use a "1" if an entity is a member of the group under consideration and a "0" if the entity is not a member of that group. The fact cannot be over-emphasized that the judicious use of "1's" and "0's" for representation of group membership is a crucial aspect of linear regression.

Example 3.2--group membership vectors

A researcher is interested in studying the college student population and is interested in the effect on some behavior of three variables: sex, college status, and marital status. With respect to the behavioral model introduced in Chapter One, he has reason to believe that the behavior he is investigating is a function of sex, college status, and marital status. In order to complete the example, assume that the behavior under investigation is the degree of one's political liberalism. The group membership vectors that need to be constructed for this problem are:

S_1 = 1 if person is a male, and 0 otherwise;

S_2 = 1 if person is a female, and 0 otherwise;

C_1 = 1 if person is a freshman, and 0 otherwise;

C_2 = 1 if person is a sophomore, and 0 otherwise;

C_3 = 1 if person is a junior, and 0 otherwise;

C_4 = 1 if person is a senior, and 0 otherwise;

M_1 = 1 if person is married, and 0 otherwise; and

M_2 = 1 if person is not married, and 0 otherwise.

Note that two vectors are used to represent sex and two vectors to represent marital status, while four vectors are used to represent college status. Another investigator might not be satisfied with this particular set of group designations. An additional marital status group could well be that of "divorced." This latter group could provide relevant information and thus increase the predictability of the behavior under consideration. It is up to the researcher to define the groups that he wants to include in his research. The linear regression approach is flexible in that it will handle as many groups as the researcher is willing to define and obtain data from. Even sex can be divided into more than two groups if one is willing to re-define the variable as "sex-role interests" and divide the continuum of responses to such a questionnaire into, say, three or four groups.

Given the following vectors, defined as above,

	S_1	S_2	C_1	C_2	C_3	C_4	M_1	M_2
Person 1	1	0	1	0	0	0	0	1
Person 2	0	1	0	0	0	1	1	0
Person 3	1	0	1	0	0	0	1	0
Person 4	1	0	0	0	0	1	1	0

it can be seen at once that the first person (the first element in each vector) is a male, freshman, and unmarried. Inspection of the second element in each vector indicates a female senior who is married.

Elementary Vector Algebra Operations

Addition of vectors

The addition of two vectors is defined as the addition of each element in one vector to the corresponding element in the other vector. An implicit requirement is that both vectors have the same number of elements before addition is possible. The addition of two vectors, A_2 and B_3, to produce the vector C_5 is written symbolically as:

$$C_5 = A_2 + B_3 = \begin{bmatrix} A_{2_1} \\ A_{2_2} \\ A_{2_3} \\ \cdot \\ \cdot \\ \cdot \\ A_{2_t} \end{bmatrix} + \begin{bmatrix} B_{3_1} \\ B_{3_2} \\ B_{3_3} \\ \cdot \\ \cdot \\ \cdot \\ B_{3_t} \end{bmatrix} = \begin{bmatrix} A_{2_1} + B_{3_1} \\ A_{2_2} + B_{3_2} \\ A_{2_3} + B_{3_3} \\ \cdot \\ \cdot \\ \cdot \\ A_{2_t} + B_{3_t} \end{bmatrix} = \begin{bmatrix} C_{5_1} \\ C_{5_2} \\ C_{5_3} \\ \cdot \\ \cdot \\ \cdot \\ C_{5_t} \end{bmatrix} = C_5$$

Example 3.3--addition of two vectors

$$A_4 = \begin{bmatrix} 1 \\ 0 \\ 3 \\ -4 \end{bmatrix} \quad B_4 = \begin{bmatrix} 2 \\ 1 \\ 5 \\ 7 \end{bmatrix} \quad D_4 = \begin{bmatrix} 3 \\ 0 \\ 4 \end{bmatrix} \quad C_4 = A_4 + B_4$$

$$C_4 = A_4 + B_4 = \begin{bmatrix} 1 \\ 0 \\ 3 \\ -4 \end{bmatrix} + \begin{bmatrix} 2 \\ 1 \\ 5 \\ 7 \end{bmatrix} = \begin{bmatrix} 1 + 2 \\ 0 + 1 \\ 3 + 5 \\ -4 + 7 \end{bmatrix} = \begin{bmatrix} 3 \\ 1 \\ 8 \\ 3 \end{bmatrix} = C_4$$

The addition of B_4 and D_4 is not possible because these two vectors do not have the same number of elements.

Subtraction of two vectors

The subtraction of two vectors is as straightforward as the addition of two vectors. In order to subtract B_6 from A_2 to produce the vector C_5, one simply subtracts corresponding elements of B_6 from A_2.

Example 3.4--subtraction of two vectors

$$A_2 = \begin{bmatrix} 2 \\ 0 \\ 7 \\ -4 \end{bmatrix} \qquad B_6 = \begin{bmatrix} 1 \\ 1 \\ 5 \\ 7 \end{bmatrix} \qquad C_5 = A_2 - B_6$$

$$C_5 = A_2 - B_6 = \begin{bmatrix} 2 \\ 0 \\ 7 \\ -4 \end{bmatrix} - \begin{bmatrix} 1 \\ 1 \\ 5 \\ 7 \end{bmatrix} = \begin{bmatrix} 2 - 1 \\ 0 - 1 \\ 7 - 5 \\ -4 - 7 \end{bmatrix} = \begin{bmatrix} 1 \\ -1 \\ 2 \\ -11 \end{bmatrix} = C_5$$

Multiplication of a vector by a number

Many occasions will be found later on to multiply a vector by a number; and so it is necessary to become familiar with this operation. The multiplication of a vector by a number is defined as the multiplication of each element of the vector by that number.

$C_6 = k*A_5$ (where k is a constant;and the * indicates multiplication) is computed by multiplying each element of A_5 by the constant k.

$$C_6 = k*A_5 = \begin{bmatrix} k*A_{5_1} \\ k*A_{5_2} \\ k*A_{5_3} \\ \cdot \\ \cdot \\ \cdot \\ k*A_{5_t} \end{bmatrix} = C_6$$

Example 3.5--multiplication of a vector by a number

Some weight observations have been made on a number of entities, and it is desired to change the data from the original unit of pounds to the new unit of ounces. Thus, each observation must be multiplied by the constant 16 because there are 16 ounces per pound. This operation can be represented in vector notation as:

$$Z_3 = 16*P_3$$

where the vector P_3 represents the original observations in terms of pounds, and the vector Z_3 represents the observations in the form of the new units. Suppose that the original vector looked like this:

$$P_3 = \begin{bmatrix} 2 \\ 4 \\ 6 \\ 3 \\ 0 \end{bmatrix}$$

Then the new vector Z_3 would be:

$$Z_3 = 16*P_3 = 16* \begin{bmatrix} 2 \\ 4 \\ 6 \\ 3 \\ 0 \end{bmatrix} = \begin{bmatrix} 16*2 \\ 16*4 \\ 16*6 \\ 16*3 \\ 16*0 \end{bmatrix} = \begin{bmatrix} 32 \\ 64 \\ 96 \\ 48 \\ 0 \end{bmatrix} = Z_3$$

The reader should note that it makes sense to multiply each and every element of the vector by the constant. The fourth element of vector Z_3, for instance, should represent (in ounces) the number of pounds corresponding to the fourth element of P_3. Note that the fourth element of Z_3 (48 oz.) is, conceptually, equivalent to the fourth element of P_3 (3 pounds).

Example 3.6--the conciseness of multiplying a vector by a number

Another example is introduced here to illustrate the simplicity and conciseness of vector algebra. Suppose that it is desired to change a 100-element vector of observations reported in
units of feet to units in terms of inches. This operation can be
represented with vectors in the following fashion: Where F_2 is
the vector of observations in feet, I_2 is the new vector of observation in inches, and k is the multiplication constant (12 in
this case, because there are 12 inches per foot):

$$I_2 = k*F_2 = 12*F_2$$

Again, there is not just one multiplication inferred by the above
expression; but every element of F_2 is multiplied by the constant
k to produce the corresponding element in I_2.

Useful Properties of Vectors

Combining the above knowledge of vectors with some knowledge
about the properties of ordinary numbers, the following useful
properties can be expressed:

Property 1: The subtraction of a vector from itself yields the
null vector.

$$X_3 + (-1)X_3 = X_3 - X_3 = \underline{0}$$

where $\underline{0}$ represents a vector whose every element is equal to 0.
Such a vector is often called the null vector. The property becomes useful when one has an occasion to subtract a vector from
itself. The result of such an operation is a vector whose every

element is equal to zero, or the null vector $\underline{0}$.

$$
\begin{bmatrix} X_{3_1} \\ X_{3_2} \\ X_{3_3} \\ \cdot \\ \cdot \\ \cdot \\ X_{3_t} \end{bmatrix} + -1 \begin{bmatrix} X_{3_1} \\ X_{3_2} \\ X_{3_3} \\ \cdot \\ \cdot \\ \cdot \\ X_{3_t} \end{bmatrix} = \begin{bmatrix} X_{3_1} \\ X_{3_2} \\ X_{3_3} \\ \cdot \\ \cdot \\ \cdot \\ X_{3_t} \end{bmatrix} + \begin{bmatrix} -X_{3_1} \\ -X_{3_2} \\ -X_{3_3} \\ \cdot \\ \cdot \\ \cdot \\ -X_{3_t} \end{bmatrix} = \begin{bmatrix} X_{3_1}-X_{3_1} \\ X_{3_2}-X_{3_2} \\ X_{3_3}-X_{3_3} \\ \cdot \\ \cdot \\ \cdot \\ X_{3_t}-X_{3_t} \end{bmatrix} = \begin{bmatrix} 0 \\ 0 \\ 0 \\ \cdot \\ \cdot \\ \cdot \\ 0 \end{bmatrix} = \underline{0}
$$

Property 2: Multiplication of the sum of two or more vectors by a constant is equivalent to the multiplication of each vector by the constant and then the addition of the resulting products.

$$a*(X_3+Y_4) = (a*X_3) + (a*Y_4)$$

$$
a(X_3+Y_4) = a \left(\begin{bmatrix} X_{3_1} \\ X_{3_2} \\ X_{3_3} \\ \cdot \\ \cdot \\ \cdot \\ X_{3_t} \end{bmatrix} + \begin{bmatrix} Y_{4_1} \\ Y_{4_2} \\ Y_{4_3} \\ \cdot \\ \cdot \\ \cdot \\ Y_{4_t} \end{bmatrix} \right) = a \begin{bmatrix} (X_{3_1}+Y_{4_1}) \\ (X_{3_2}+Y_{4_2}) \\ (X_{3_3}+Y_{4_3}) \\ \cdot \\ \cdot \\ \cdot \\ (X_{3_t}+Y_{4_t}) \end{bmatrix} = \begin{bmatrix} a(X_{3_1}+Y_{4_1}) \\ a(X_{3_2}+Y_{4_2}) \\ a(X_{3_3}+Y_{4_3}) \\ \cdot \\ \cdot \\ \cdot \\ a(X_{3_t}+Y_{4_t}) \end{bmatrix}
$$

$$
= \begin{bmatrix} aX_{3_1}+aY_{4_1} \\ aX_{3_2}+aY_{4_2} \\ aX_{3_3}+aY_{4_3} \\ \cdot \\ \cdot \\ \cdot \\ aX_{3_t}+aY_{4_t} \end{bmatrix} = \begin{bmatrix} aX_{3_1} \\ aX_{3_2} \\ aX_{3_3} \\ \cdot \\ \cdot \\ \cdot \\ aX_{3_t} \end{bmatrix} + \begin{bmatrix} aY_{4_1} \\ aY_{4_2} \\ aY_{4_3} \\ \cdot \\ \cdot \\ \cdot \\ aY_{4_t} \end{bmatrix} = a \begin{bmatrix} X_{3_1} \\ X_{3_2} \\ X_{3_3} \\ \cdot \\ \cdot \\ \cdot \\ X_{3_t} \end{bmatrix} + a \begin{bmatrix} Y_{4_1} \\ Y_{4_2} \\ Y_{4_3} \\ \cdot \\ \cdot \\ \cdot \\ Y_{4_t} \end{bmatrix} = aX_3+aY_4
$$

Example 3.7--simplification of two vectors multiplied by the same
                        constant

Suppose that one wanted to multiply the total GRE (Graduate

Record Exam) of five subjects by a constant of 3. The following

section indicates that one could multiply both the verbal (V_3)

and quantitative (Q_3) sections by 3 and add these products; or

one could first sum the verbal and quantitative sections and then

multiply this sum (T_3) by the constant 3.

$$V_3 = \begin{bmatrix} 200 \\ 250 \\ 300 \\ 400 \\ 500 \end{bmatrix} \qquad Q_3 = \begin{bmatrix} 400 \\ 450 \\ 500 \\ 600 \\ 700 \end{bmatrix} \qquad T_3 = V_3 + Q_3 = \begin{bmatrix} 600 \\ 700 \\ 800 \\ 1000 \\ 1200 \end{bmatrix}$$

$$3*(V_3 + Q_3) = (3*V_3) + (3*Q_3)$$

$$3* \begin{bmatrix} 600 \\ 700 \\ 800 \\ 1000 \\ 1200 \end{bmatrix} = \begin{bmatrix} 1800 \\ 2100 \\ 2400 \\ 3000 \\ 3600 \end{bmatrix} = 3* \begin{bmatrix} 200 \\ 250 \\ 300 \\ 400 \\ 500 \end{bmatrix} + 3* \begin{bmatrix} 400 \\ 450 \\ 500 \\ 600 \\ 700 \end{bmatrix} = \begin{bmatrix} 600 \\ 750 \\ 900 \\ 1200 \\ 1500 \end{bmatrix} + \begin{bmatrix} 1200 \\ 1350 \\ 1500 \\ 1800 \\ 2100 \end{bmatrix} = \begin{bmatrix} 1800 \\ 2100 \\ 2400 \\ 3000 \\ 3600 \end{bmatrix}$$

The addition of the two vectors, before multiplying by the

constant, is the most frequently used option. There is one less

mathematical operation than if one were to multiply both vectors

by the constant and then add the results.

Furthermore, it will become necessary later in this chapter

to reduce the number of vectors used to the simplest possible

form. The left-hand side of Property 2 [a*(X_3+Y_4)] is in a sim-

pler form than the right-hand side [($a*X_3$) + ($a*Y_4$)].

Property 3: Multiplication of a vector by the sum of two constants

is equivalent to the separate multiplication of that vector by

each constant, and then summed.

$$(a+b)*X_3 = (a*X_3)+(b*X_3)$$

$$(a+b)*X_3 = \begin{bmatrix} (a+b)X_{3_1} \\ (a+b)X_{3_2} \\ (a+b)X_{3_3} \\ \cdot \\ \cdot \\ \cdot \\ (a+b)X_{3_t} \end{bmatrix} = \begin{bmatrix} aX_{3_1} + bX_{3_1} \\ aX_{3_2} + bX_{3_2} \\ aX_{3_3} + bX_{3_3} \\ \cdot \\ \cdot \\ \cdot \\ aX_{3_t} + bX_{3_t} \end{bmatrix} = \begin{bmatrix} aX_{3_1} \\ aX_{3_2} \\ aX_{3_3} \\ \cdot \\ \cdot \\ \cdot \\ aX_{3_t} \end{bmatrix} + \begin{bmatrix} bX_{3_1} \\ bX_{3_2} \\ bX_{3_3} \\ \cdot \\ \cdot \\ \cdot \\ bX_{3_t} \end{bmatrix} = aX_3+bX_3$$

The above property indicates that when a vector is to be

multiplied by the sum of two constants, one can either add the

two constants together and then multiply the vector by the sum

of the two constants, or one can multiply the vector by the two

separate constants and then add the resultant products. Adding

the two constants first is much easier and quicker because fewer

operations are involved.

As with the previous property, there will be many occasions

to simplify vectors and their weighting coefficients. The left-

hand side of Property 4 is simpler than the right-hand side, even

though they are numerically equal. The left-hand side clearly in-

dicates that there is only one vector; whereas the right-hand

side contains two vectors which can be reduced, or simplified, to

one vector. More on these notions will be presented in the sec-

tion on linear dependencies.

Property 4: Multiplication of a vector by zero yields a vector with all elements equal to zero, the null vector.

$$(0)*X_4 = \underline{0}$$

$$0*X_4 = 0* \begin{bmatrix} X_{4_1} \\ X_{4_2} \\ X_{4_3} \\ \cdot \\ \cdot \\ \cdot \\ X_{4_t} \end{bmatrix} = \begin{bmatrix} 0*X_{4_1} \\ 0*X_{4_2} \\ 0*X_{4_3} \\ \cdot \\ \cdot \\ \cdot \\ 0*X_{4_t} \end{bmatrix} = \begin{bmatrix} 0 \\ 0 \\ 0 \\ \cdot \\ \cdot \\ \cdot \\ 0 \end{bmatrix} = \underline{0}$$

Multiplication of a vector by the number "0" yields the null vector, $\underline{0}$. No matter what the elements of a vector, if one multiplies the vector by 0, one will end up with the null vector as the product.

Property 5: Multiplication of a vector by one yields that same vector.

$$1*X_3 = X_3$$

$$1*X_3 = 1* \begin{bmatrix} X_{3_1} \\ X_{3_2} \\ X_{3_3} \\ \cdot \\ \cdot \\ \cdot \\ X_{3_t} \end{bmatrix} = \begin{bmatrix} (1)*X_{3_1} \\ (1)*X_{3_2} \\ (1)*X_{3_3} \\ \cdot \\ \cdot \\ \cdot \\ (1)*X_{3_t} \end{bmatrix} = \begin{bmatrix} X_{3_1} \\ X_{3_2} \\ X_{3_3} \\ \cdot \\ \cdot \\ \cdot \\ X_{3_t} \end{bmatrix} = X_3$$

Multiplication of any vector by the constant "1" yields the same vector.

These last two properties may seem trivial, and indeed they are quite trivial--although useful. The reader is encouraged to understand all of the above properties before proceeding. These properties of vector algebra, coupled with the concept to be discussed in the next section, form the structure of multiple linear regression. The more adept the reader becomes with the concepts presented in the present chapter, the more adequately he can handle the building and simplification of linear regression models.

Linear Combinations of Vectors

Linear combinations of two vectors

There will be many situations later on where vectors will be combined and also where it will be necessary to determine if certain vectors are, in fact, linear combinations of other vectors. Therefore, the important concept of linear combinations needs to be defined.

Definition of a linear combination of vectors is as follows: Vector X_3 is said to be a linear combination of vectors Y_3 and Z_3 if there exist two numbers (numerical constants referred to as "weighting coefficients"), a and b (of which at least one is not zero), such that the following relationship holds:

$$X_3 = (a*Y_3) + (b*Z_3)$$

This rather sterile definition may become understandable with the following examples.

Example 3.8--determining linear combinations

Given: $X_3 = \begin{bmatrix} 3 \\ 4 \\ 5 \end{bmatrix}$ $Y_3 = \begin{bmatrix} 1 \\ 2 \\ 3 \end{bmatrix}$ $Z_3 = \begin{bmatrix} 1 \\ 0 \\ -1 \end{bmatrix}$

X_3 is a linear combination of Y_3 and Z_3 because:

$$X_3 = (a*Y_3) + (b*Z_3) \quad \text{when } a = 2 \text{ and } b = 1$$

$$(2*Y_3) + (1*Z_3) = 2* \begin{bmatrix} 1 \\ 2 \\ 3 \end{bmatrix} + 1* \begin{bmatrix} 1 \\ 0 \\ -1 \end{bmatrix} = \begin{bmatrix} 2 \\ 4 \\ 6 \end{bmatrix} + \begin{bmatrix} 1 \\ 0 \\ -1 \end{bmatrix} = \begin{bmatrix} 3 \\ 4 \\ 5 \end{bmatrix} = X_3$$

Some special linear combinations of vectors

A total test score vector (computed by simply adding to-gether the two subtest scores) is a linear combination of the two subtest vectors. In this instance, the weighting coefficients, a and b, are both equal to 1.

Example 3.9--total score as a linear combination of subtest
                                    scores

Consider the example of the total GRE score, which is com-puted by adding together the verbal and quantitative subtest scores. Given the verbal vector (V_4) and the quantitative vector (Q_4):

$$V_4 = \begin{bmatrix} 400 \\ 500 \\ 600 \\ 350 \end{bmatrix} \qquad Q_4 = \begin{bmatrix} 400 \\ 300 \\ 400 \\ 750 \end{bmatrix}$$

The total GRE vector (T_4) is:

$$T_4 = (1*V_4) + (1*Q_4) = V_4 + Q_4$$

Therefore:
$$T_4 = \begin{bmatrix} 400 \\ 500 \\ 600 \\ 350 \end{bmatrix} + \begin{bmatrix} 400 \\ 300 \\ 400 \\ 750 \end{bmatrix} = \begin{bmatrix} 800 \\ 800 \\ 1000 \\ 1100 \end{bmatrix}$$

Another special case of linear combination occurs when a vector is multiplied by a number. The weight of the "second vector" is in this instance zero; and because it is zero, the elements of the second vector are of no consequence. That is, it does not matter what the elements of the second vector are because it is already known that multiplication of any vector by zero will yield the null vector.

Example 3.10--linear combination of two vectors

$$A_3 = \begin{bmatrix} 4 \\ 3 \\ 0 \\ 1 \end{bmatrix} \qquad B_5 = \begin{bmatrix} B5_1 \\ B5_2 \\ B5_3 \\ B5_4 \end{bmatrix} \qquad C_6 = \begin{bmatrix} 24 \\ 18 \\ 0 \\ 6 \end{bmatrix}$$

$C_6 = (6*A_3) + (0*B_5)$

$$C_6 = 6* \begin{bmatrix} 4 \\ 3 \\ 0 \\ 1 \end{bmatrix} + 0* \begin{bmatrix} B5_1 \\ B5_2 \\ B5_3 \\ B5_4 \end{bmatrix} = \begin{bmatrix} 6*4 \\ 6*3 \\ 6*0 \\ 6*1 \end{bmatrix} + \begin{bmatrix} 0*B5_1 \\ 0*B5_2 \\ 0*B5_3 \\ 0*B5_4 \end{bmatrix} = \begin{bmatrix} 24 \\ 18 \\ 0 \\ 6 \end{bmatrix} + \begin{bmatrix} 0 \\ 0 \\ 0 \\ 0 \end{bmatrix} = \begin{bmatrix} 24 \\ 18 \\ 0 \\ 6 \end{bmatrix}$$

The vector C_6 is thus a linear combination of the vector A_3.

The concept of a linear combination of vectors is not restricted to just two vectors. A vector may be a linear combination of more than two vectors. The following example will help to clarify this point.

Example 3.11--linear combination of several vectors

$$A_4 = \begin{bmatrix} 4 \\ 3 \\ 2 \\ 1 \end{bmatrix} \quad B_6 = \begin{bmatrix} 1 \\ 0 \\ 0 \\ 0 \end{bmatrix} \quad C_9 = \begin{bmatrix} 0 \\ 1 \\ 0 \\ 0 \end{bmatrix} \quad D_8 = \begin{bmatrix} 0 \\ 0 \\ 1 \\ 0 \end{bmatrix} \quad E_7 = \begin{bmatrix} 0 \\ 0 \\ 0 \\ 1 \end{bmatrix}$$

A_4 is a linear combination of B_6, C_9, D_8, and E_7 because:

$A_4 = (4*B_6) + (3*C_9) + (2*D_8) + (1*E_7)$

The reader should verify the above statement by actually carrying out the implied multiplications.

The vector B_6 is not a linear combination of vectors C_9, D_8, and E_7 because there does not exist any weighting coefficients such that

$B_6 = (a*C_9) + (b*D_8) + (c*E_7)$

Mutually exclusive group membership vectors

Another special linear combination of vectors occurs when mutually exclusive group membership vectors are added together.

Example 3.12--representation of mutually exclusive vectors

Suppose that one had occasion to deal with the variables of sex and marital status. The group membership vectors may be defined as follows:

S_1 = 1 if subject is female, 0 otherwise;

S_2 = 1 if subject is male; 0 otherwise;

M_1 = 1 if subject is married; 0 otherwise; and

M_2 = 1 if subject is not married, 0 otherwise.

	S_1	S_2	M_1	M_2
Sam	0	1	1	0
Sue	1	0	1	0
Sally	1	0	0	1
Jane	1	0	0	1
Jack	0	1	0	1
Joe	0	1	1	0

S_1 and S_2 are mutually exclusive group membership vectors; that is, all subjects belong to one or the other categories of "male" and "female." Likewise, the categories of "married" and "not married" exhaust all of the possibilities of marital status (as far as the present researcher is concerned). Other categories of marital status could have been included, but evidently the research question here was not concerned with any additional categories.

One way of checking to see if the stated categories are, in fact, mutually exclusive is to compute the linear combination (using all weights equal to 1), of the vectors under consideration. If these vectors are in fact mutually exclusive, then they will take into account each and every subject once and only once. That is, group membership vectors are represented by "1's" and "0's", and the resultant sum of the mutually exclusive group membership vectors will yield a vector with all elements equal to one (1).

The unit vector

A vector with all of its elements equal to 1 is called the unit vector and is symbolized as "U". Because of the frequent use of the unit vector, the symbol U is reserved for that vector.

Note that the unit vector is not subscripted since all of its elements are known to be equal to one.

Consider adding together the two sex vectors. Here a linear combination is being computed because the weights can be thought of as being equal to one (1) as in Example 3.10.

$$(1*S_1) + (1*S_2) = S_1 + S_2 = \begin{bmatrix} 0 \\ 1 \\ 1 \\ 1 \\ 0 \\ 0 \end{bmatrix} + \begin{bmatrix} 1 \\ 0 \\ 0 \\ 0 \\ 1 \\ 1 \end{bmatrix} = \begin{bmatrix} 0+1 \\ 1+0 \\ 1+0 \\ 1+0 \\ 0+1 \\ 0+1 \end{bmatrix} = \begin{bmatrix} 1 \\ 1 \\ 1 \\ 1 \\ 1 \\ 1 \end{bmatrix} = U$$

Thus, S_1 and S_2 are mutually exclusive because their sum is equal to the unit vector.

M_1 and M_2 are mutually exclusive vectors, as shown in the following:

$$(1*M_1) + (1*M_2) = M_1 + M_2 = \begin{bmatrix} 1 \\ 1 \\ 0 \\ 0 \\ 0 \\ 1 \end{bmatrix} + \begin{bmatrix} 0 \\ 0 \\ 1 \\ 1 \\ 1 \\ 0 \end{bmatrix} = \begin{bmatrix} 1+0 \\ 1+0 \\ 0+1 \\ 0+1 \\ 0+1 \\ 1+0 \end{bmatrix} = \begin{bmatrix} 1 \\ 1 \\ 1 \\ 1 \\ 1 \\ 1 \end{bmatrix} = U$$

The reader should now have a good feeling for the fact that the unit vector can be considered as a linear combination of mutually exclusive group membership vectors. In fact, the unit vector was relied upon to define mutually exclusive group membership vectors; so the above statement is simply a consequence of that definition--but a very important consequence, as will be shown in later chapters. The unit vector is assumed by computer programs to be in any regression model, so the relationship of the unit vector to other vectors needs to be known.

The concept of linear dependency

The concept of linear dependency is very important. It can be easily introduced at this point because it deals with linear combinations of vectors. A linear dependency occurs when one vector in a set of vectors can be expressed as a linear combination of the other vectors. Such a vector is said to be linearly dependent upon the other vectors. A linearly dependent vector, because it can be expressed in terms of other vectors, is redundant information; and, as such, it is not useful in terms of predicting behavior. There will be times, though, when one will want to include linearly dependent vectors; but one needs to know and be aware of when one has done so.

Example 3.13--linearly dependent vectors

In Example 3.9, the total GRE was expressed as the sum of the two subtests. The reader should verify for himself that the verbal subtest (V_3) can be expressed as the total GRE (T_3) minus the quantitative subtest (Q_3); or $V_3 = 1*T_3 + -1*Q_3$. It is also true that the quantitative subtest is a linear combination of the total GRE and the verbal subtest; or $Q_3 = 1*T_3 + -1*V_3$.

Any one of the three vectors in Example 3.14 is linearly dependent on the other two because it can be expressed as a linear combination of the other two. The total GRE cannot be expressed in terms of the verbal subtest alone nor the quantitative subtest alone. That is, no weight (a) can be found such that: $T_3 = a*V_3$ nor such that $T_3 = a*Q_3$.

Likewise, no weight (a) can be found such that: $V_3 = a*Q_3$.

The discussion so far has centered on linearly dependent vectors. But one will generally want to determine how many of a set of vectors are linearly <u>independent</u>, which is the opposite of "linearly dependent." If there is one linear dependency in a set of three vectors, then two vectors are linearly independent. In Example 3.14, the final result is a set of two vectors of information which are said to be linearly <u>independent</u>.

> Definition of linear independence: A vector is said to be linearly independent if that vector cannot be expressed as a linear combination of the other vectors in the set.

If a vector can be expressed as a linear combination of the other vectors in the set, then it is redundant information and as such must be thought of as being eliminated from the set of vectors when attempting to determine the number of linearly independent vectors in a set of vectors. The following example is intended to clarify the determination of the number of linearly independent vectors.

Example 3.14--determining the number of linearly independent vectors

$$V_1 = \begin{bmatrix} 1 \\ 2 \\ 3 \\ 4 \\ 5 \end{bmatrix} \quad V_2 = \begin{bmatrix} 2 \\ 4 \\ 6 \\ 8 \\ 10 \end{bmatrix} \quad V_3 = \begin{bmatrix} 4 \\ 8 \\ 12 \\ 16 \\ 20 \end{bmatrix} \quad V_4 = \begin{bmatrix} 1 \\ 2 \\ 3 \\ 4 \\ 4 \end{bmatrix}$$

V_2 is a linear combination of the other vectors.

$$V_2 = 2*V_1 + 0*V_3 + 0*V_4$$

or

$$V_2 = 2*V_1$$

Therefore, V_2 is essentially eliminated from the set of vectors for the purpose of determining the number of linearly independent vectors in the set. There is now a potential set of three linearly independent vectors, V_1, V_3, and V_4.

V_3 is a linear combination of the remaining vectors:

$$V_3 = 4*V_1 + 0*V_4$$

or

$$V_3 = 4*V_1$$

Therefore, V_3 is essentially eliminated from the set of vectors for the purpose of determining the number of linearly independent vectors in the set.

The two vectors, V_1 and V_4, remain. The problem is to find a weight (a) such that: $V_1 = a*V_4$.

Note that the first four elements of V_4 must be multiplied by a weight of 1, whereas the fifth element must be multiplied by a weight of 1.25, in order to equal the elements of V_1. Thus, there is no one weight which will suffice. Vectors V_1 and V_4 are thus linearly independent. Therefore, there are two linearly independent vectors in the set of vectors in Example 3.15.

One could have first eliminated from the set of four vectors V_1, because $V_1 = .5*V_2 + 0*V_3 + 0*V_4$.

Then V_3 could be eliminated because it is a linear combination of V_2 and V_4: $V_3 = 2*V_2 + 0*V_4$.

Again, since V_2 cannot be shown to be a linear combination of V_4, there are two linearly independent vectors in the set of vectors in Example 3.15.

It can also be shown that V_3 and V_4 are two linearly independent vectors in the set of vectors in Example 3.15. It does not matter which two vectors remain in the set; the crucial point is that in this set of four vectors, only two vectors contain "new information." The other two vectors contain redundant information, and thus would not increase predictability if used in a prediction equation.

Complexity is operationally defined here as the number of linearly independent vectors. Thus, in Example 3.15, the level of complexity is 2, since two "pieces of information" exist. In the following example the level of complexity is found to be 5.

Example 3.15--complexity and the number of linearly independent
vectors

S_1	S_2	C_1	C_2	C_3	C_4	U
1	0	1	0	0	0	1
0	1	1	0	0	0	1
1	0	0	1	0	0	1
0	1	0	1	0	0	1
1	0	0	0	1	0	1
0	1	0	0	1	0	1
1	0	0	0	0	1	1
0	1	0	0	0	1	1

The first two vectors are the two sex vectors, and the next four vectors are the four class rank vectors. Whenever dichoto-

mous vectors are being considered along with the unit vector, it
is quite beneficial to leave the unit vector in the set. Of the
seven vectors above,

$$S_1 = 1*U + -1*S_2, \text{ and}$$

$$C_1 = 1*U + -1*C_2 + -1*C_3 + -1*C_4$$

Therefore, C_1 and S_1 are two linear dependencies in the set.
No other vectors can be eliminated; therefore, there are five
pieces of information in the set of seven vectors. (One group
membership vector in each mutually exclusive group can always be
eliminated if the unit vector is in the initial set of vectors).

Most linear dependencies can be ascertained by knowing the
variables and how they are defined. Indeed, very seldom would one
want to look at the actual data and attempt to find the weighting
coefficients. Suppose the following vectors were being considered: U   = 1 for all subjects;

X_1 = 1 if male, 0 otherwise;

X_2 = 1 if female, 0 otherwise;

X_3 = 1 if freshman, 0 otherwise;

X_4 = 1 if sophomore, 0 otherwise;

X_5 = 1 if junior, 0 otherwise;

X_6 = 1 if senior, 0 otherwise;

X_7 = Math Achievement Test A;

X_8 = Math Achievement Test B;

X_9 = Verbal IQ;

X_{10} = Non-verbal IQ; and

X_{11} = Total IQ (verbal IQ plus non-verbal IQ)

One of the sex vectors (X_1 or X_2) can be eliminated. One of the class rank vectors (X_3, X_4, X_5, or X_6) can be eliminated, and one of the IQ vectors can be eliminated. One probably would eliminate the total IQ, so direct measures of both Verbal and Non-Verbal IQ would remain in the analysis. Note that although two math achievement tests are being considered, it is highly unlikely that these two tests would be providing perfectly redundant information. To do so, they would have to be perfectly correlated, which is a very unlikely state of affairs. It could be that one of the math tests, though, is empirically linearly dependent upon the whole set of vectors. Again, this is not a likely state of affairs, but a possible one. It is sometimes difficult to determine linear dependencies by inspection of the model. Indeed, the data itself determines whether or not a vector is linearly dependent. Fortunately, the computer solution will verify the actual number of linearly independent vectors, although one would want to rely on that procedure as a last resort only. In the prediction of a given criterion, the number of non-zero weighting coefficients will be the number of linearly independent pieces of information.

In later chapters the prediction of a criterion variable through the use of a set of predictor variables will be discussed. The notion of linear dependencies will be used to eliminate redundancies from the predictor set, to eliminate "excess baggage." It could be said, though, that one of the major goals of research is to find weighting coefficients such that the criterion variable is linearly dependent upon the predictor set. It is therefore desirable to have linear dependency when considering a criterion variable, whereas it is not desirable to have linear dependency when considering predictor variables.

4

Building Linear Models which Reflect
Research Hypotheses and which Minimize Errors
of Prediction

The F statistic enables one to determine how likely it is
that the amount of variance explained by a particular piece of
information is due to sampling variation. (This was discussed
briefly in Chapter Two and will be more fully developed here.)
The linear regression model is one of the most general procedures
for calculating total and partialled sums of squares (into various components such as within- or among-group) which are used to
calculate R^2, estimates of criterion variance, F values, and,
ultimately, probability values. This chapter introduces the construction of linear models in the service of answering behavioral
science research questions.

The simple linear model and its components are presented. A
simple data set is given to provide concrete meaning for these
components. The calculation of linear components in complex models is an easy process with existing computer programs. Thus, procedures for calculating weighting coefficients are discussed only
in a brief section. The emphasis of this chapter is upon (1)
stating a research question and (2) constructing linear models
designed to answer that research question. An actual problem,
with data and answers, is provided.

It is assumed that anyone studying this text has access to an appropriate regression program, such as the one provided in Appendix A. Basic Fortran transformation procedures are included in Appendix C, which easily enable the researcher to construct vectors from the original data to reflect the various linear models he may need.

Research hypotheses requiring a single straight line of best fit

Given a criterion behavior (Y_1) exhibited by a group of individuals under study, one may wish to know if the variance in these criterion scores can be accounted for. The information at hand to account for criterion behavior may take many forms (sex of the subject, previous test scores, knowledge of which treatment the subject received, etc.).

For illustrative purposes, suppose one is interested in establishing that "there is a relationship between Ability and Performance" (the research hypothesis). Now suppose that scores on the performance behavior (Y_1), and the ability (X_1) hypothesized to be relevant to that performance behavior are obtained for five subjects. Implied by the research hypothesis above is a supposition that there is a systematic relationship between the X_1 and Y_1 variables. This relationship could take many forms, but since the particular form is not specified, it is conventional to assume that the relationship being investigated is a linear one. (Chapter Seven deals with non-linear relationships). The next step is to graph the two sets of scores and describe the graph.

Figure 4.1 shows the graph of the observed performance and ability scores for the five individuals, A, B, C, D, and E.

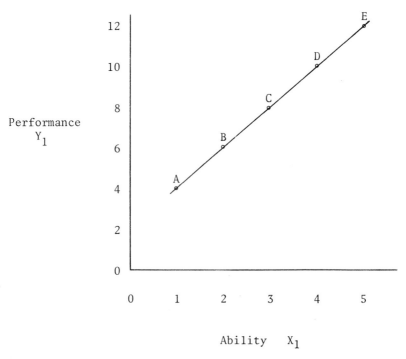

Figure 4.1. The observed X_1 and Y_1 scores for
individuals A, B, C, D, and E.

The description of this graph of scores involves using the formula for a straight line:

Equation 4.1: $Y_1 = a + bX_1$

       where: Y_1 = the scores on the Y axis variable (the per-
                formance scores);

            X_1 = the scores on the X axis variable (the ability
                scores);

            a = the point at which the straight line crosses
                the Y axis (called the "Y-intercept"); and

$\qquad$ b $\;=\;$ the increase in the Y axis for every one-unit increase in the X axis scores (called the "slope of the line").

In Figure 4.1, notice the lightly drawn line that connects the five points. This line can be specified by finding values for a and b in Equation 4.1. To get a feel for the value of the slope, b, notice that individual A scored 1 on Ability (and 4 on Performance) while individual B scored 2 (and 6). Thus, for a one unit increase in Ability from 1 to 2, Performance increased two units (from 4 to 6). For every unit increase on X_1 there is a two-unit increase on Y_1. Therefore, b (in the equation $Y_1 = a + bX_1$) must equal 2. Putting this value into Equation 4.1, the value for the intercept, "a", can also be determined.

So far, $\hat{Y}_1 = a + 2X_1$ has been obtained, where $\hat{Y}_1$ = predicted Y_1 and b = 2. In order to complete the equation, assume for the moment that a = 0. In vector form the equation would be:

$$\hat{Y}_1 \quad = \quad aU \quad + \quad 2X_1$$

$$\begin{array}{c} A \\ B \\ C \\ D \\ E \end{array} \begin{bmatrix} 2 \\ 4 \\ 6 \\ 8 \\ 10 \end{bmatrix} = 0 \begin{bmatrix} 1 \\ 1 \\ 1 \\ 1 \\ 1 \end{bmatrix} + 2 \begin{bmatrix} 1 \\ 2 \\ 3 \\ 4 \\ 5 \end{bmatrix}$$

[Note that "a" can be represented as "a" times the unit vector. Indeed, the value of "a" is added to each subject.]

Note that the first element in vector X_1 is 1 and must be the score on X_1 for individual A. Likewise, the X_1 of 5 must represent individual E's score on X_1. The vector U provides a one for every person, and when multiplied by "a", adjusts each score

by a constant amount (hence the weighting coefficient for the

unit vector is often referred to as the regression constant).

When each element in X_1 is multiplied by 2 (the value of b)

and added to the value of "0 times the unit vector," the result

is the vector $\hat{Y}_1$. In finding values for a and b, one is trying to

make $\hat{Y}_1$ equal to Y_1. If "a" really equals 0, as has been assumed

above, then Y_1 should equal $\hat{Y}_1$. Note, however, that with a = 0,

Y_1 and $\hat{Y}_1$ are not equal.

$$Y_1 \qquad \hat{Y}_1 \qquad Y_1 - \hat{Y}_1$$

$$
\begin{matrix}
A \\
B \\
C \\
D \\
E
\end{matrix}
\begin{bmatrix}
4 \\
6 \\
8 \\
10 \\
12
\end{bmatrix}
\begin{bmatrix}
2 \\
4 \\
6 \\
8 \\
10
\end{bmatrix}
\begin{bmatrix}
2 \\
2 \\
2 \\
2 \\
2
\end{bmatrix}
$$

To make $\hat{Y}_1$ equal to Y_1, a value of 2 must be added to every

score in $\hat{Y}_1$. This means that a = 2 rather than a = 0. Now with

a = 2 and b = 2, a line of perfect fit is obtained (see Figure

4.2).

None of the observed ability scores was zero, but if there

was an X_1 score of zero, the line would cross the Y axis at the

Y_1 value of 2, which is therefore the intercept for this data.

The original hypothesis for which this data was obtained

was, "There is a (linear) relationship between Ability and Per-

formance." Figure 4.1 indicated that there was indeed a linear

relationship between X_1 and Y_1; without exception, as X_1 in-

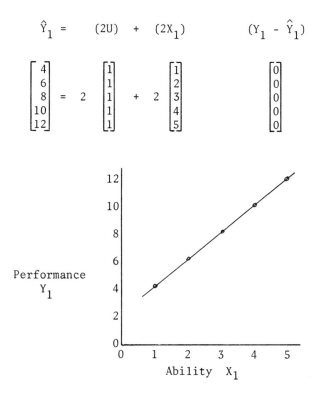

$$\hat{Y}_1 = \quad (2U) \quad + \quad (2X_1) \qquad (Y_1 - \hat{Y}_1)$$

$$\begin{bmatrix} 4 \\ 6 \\ 8 \\ 10 \\ 12 \end{bmatrix} = 2 \begin{bmatrix} 1 \\ 1 \\ 1 \\ 1 \\ 1 \end{bmatrix} + 2 \begin{bmatrix} 1 \\ 2 \\ 3 \\ 4 \\ 5 \end{bmatrix} \qquad \begin{bmatrix} 0 \\ 0 \\ 0 \\ 0 \\ 0 \end{bmatrix}$$

Figure 4.2. The vector representation of $\hat{Y}_1 = 2 + 2X_1$ and the line of best fit.

creases, so does Y_1. And since the increase has a constant value (e.g., 2), Y_1 is a linear function of X_1.

However, the observed relationship may be due to sampling error as discussed in Chapter Two. If one wishes to generalize beyond the five subjects, one wants to know, "How likely is it that the observed linear relationship between X_1 and Y_1 in the sample is due to sampling variation?" This can be determined by the F statistic, the use of which is discussed later in this chapter. In this problem, X_1 had only one person for each of the five X_1

scores. If there were more subjects and therefore more observations at each point on X_1, it would seldom happen that all subjects with a particular score on X_1 would have the same criterion
(Y_1) value. Each point on the X_1 variable has a population of
people with that score. In any particular study, the individuals
at the scale point $X_1 = 2$ are a sample of the population of people who have an X_1 score of 2. This holds true for all scale
points on X_1. When a line is fit to the data, then all observed
squared deviations on Y_1 from that line for all scale points on
X_1 can be viewed as a within-group sum of squares. Knowledge of
X_1 cannot explain these deviations. The difference between this
within-group sum of squares and the total sum of squares yields
the variation explained by knowledge of the estimated linear relationship between X_1 and Y_1.

For the data in Figure 4.1, there are no data points which
do not fall on the line. There is perfect prediction for these
data, a very unlikely situation. Figure 4.3 more clearly indicates the notion of variability about the line of best fit for
each value of the predictor variable X_1. Note that in Figure 4.3
two persons have an ability score of 6, but one is at a performance level of 5, whereas the other is at a performance level of
only 3. Figure 4.3 also portrays an unlikely situation in that
the criterion means for each ability level fall right on the line
of best fit. Further discussion of that situation is presented
later.

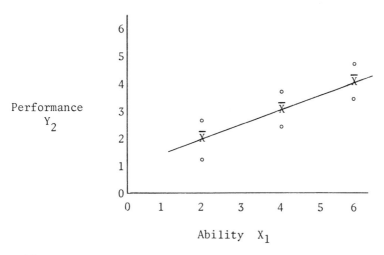

Figure 4.3. Criterion variability for subjects having
the same X_1 values.

To test a research hypothesis, one must compare the accuracy
of two linear models in fitting the same data, one model reflecting the state of affairs supposed by the research hypothesis, and
the other model reflecting the state of affairs supposed by the
statistical hypothesis. The two models will provide two estimates
of the population criterion variance accounted for by the predictor variables. An $\underline{F}$ test can then be calculated from these two
variance estimates. Since the present research hypothesis specifies a systematic straight line relationship, the one regression
model would depict a single straight line:

Model 4.1: $Y_1 = a_1U + b_1X_1 + E_1$

    where: Y_1 = a vector of Performance scores;

              U = the unit vector;

              X_1 = a vector of Ability (predictor) scores;

E_1 = a vector of deviations from the line of
best fit (when squared and summed, this
yields the within sum of squares);

a_1 = regression constant; and

b_1 = the slope of the line.

Since the model reflecting the research hypothesis will always contain more information than the model reflecting the statistical hypothesis, the model reflecting the research hypothesis
is referred to as the "full model."

Dividing the sum of squared error components (ESS_f) by the
number of data points free to vary yields a within estimate of
the criterion variance ($\hat{V}_w$):

Equation 4.2: $$\hat{V}_w = \frac{ESS_f}{df_w} \left[= \frac{\Sigma E_1{}^2}{N-2}, \text{ for the example} \right]$$

The degrees of freedom denominator for the single straight-line
model is N-2 because two weighting coefficients had to be calculated in order to find the straight line of best fit (i.e., a and
b).

The statistical hypothesis states that there is no relationship between Ability and Performance. This hypothesis implies
that the best prediction is the mean of all the criterion scores,
or, in other words, that the line of best fit is horizontal--im-
plying that the slope is equal to zero. The slope in Model 4.1
can be set equal to zero, which effectively eliminates X_1 from
the model. A simplification of Model 4.1 with the restriction
that b_1 = 0 results in the following (restricted) model:

Model 4.2: $Y_1 = a_0 U + E_2$

The above model has one piece of predictor information (the unit vector). The weighting coefficient for the unit vector is subscripted differently than in the full model because the numerical value may be different. Indeed, if the weighting coefficient for X_1 is other than zero in the full model, the numerical value for the unit vector will in general be different in the two models. Weighting coefficients will not always be subscripted differently in the full and restricted models in the remaining sections of this text. The reader must be aware that weighting coefficients simply hold the place for some numerical value. The weighting coefficients in Model 4.1 and 4.2 are raw score weights, or b weights. They are applicable for raw data. If all variables are standardized so that they have a mean of 0.0 and a standard deviation of 1.0, then the weighting coefficients are referred to as standard weights, or standard partial regression weights, or beta weights. There is very little value in standardizing the data, so raw score weights will be used throughout the text.

The error vector is subscripted differently in the full model than in the restricted model. This is done to imply that the full model (Model 4.1) is allowed to have a different degree of predictability (and hence error) than is the restricted model (Model 4.2). The criterion vector is subscripted the same in both models since the two models are reflections of the same research hypothesis. The full model simply uses more predictor information than does the restricted model, in accounting for the variance in a given criterion.

The error in prediction (the sum of the squared error, ESS
or ΣE^2) will probably be lower in Model 4.1 (which incorporated
the X_1 information) than in Model 4.2. The question again is,
"How likely is it that the decrease in error in prediction (or
increase in predictability) is due to randomness?" The increase
in predictability would be the difference between the two sums of
squared errors ($\Sigma E_2{}^2 - \Sigma E_1{}^2$). Dividing this difference by the
number of pieces of information which account for the difference
(the number of pieces of information in the full model minus the
number of pieces of information in the restricted model, or df_p)
yields another variance estimate. ($\hat{V}_p$ is a variance estimate from
a particular piece of information.)

Equation 4.3: $\hat{V}_p \;=\; \dfrac{\text{ESS}_r - \text{ESS}_f}{df_p} \left[= \dfrac{\Sigma E_2{}^2 - \Sigma E_1{}^2}{1} \text{, for the example} \right]$

Now that two variance estimates are available, the F test
can be performed. $\hat{V}_w$ is the best estimate of the criterion variance because it takes into account all of the predictor information that appears in the full model. $\hat{V}_p$ is a variance estimate
influenced by the "degree of worth" or gain in accuracy achieved
by the predictor (namely X_1) included in the full model that is
absent from the restricted model.

The F test for deciding if a statistically significant gain
has been produced by use of the full model, as compared to the
restricted model, is:

Equation 4.4: $F_{(df_p, df_w)} \;=\; \dfrac{\hat{V}_p}{\hat{V}_w} \;=\; \dfrac{(\text{ESS}_r - \text{ESS}_f)/df_p}{(\text{ESS}_f)/df_w}$

At the end of Chapter Two the $\underline{F}$ test was expressed in terms of R^2. Since that formulation is more easily generalizable to other research hypotheses, the above stated $\underline{F}$ test will be developed here in terms of R^2. The reader who becomes confused with mathematical gyrations may want to skip the following material, which is not terribly essential, and go directly to Equation 4.6.

The $\underline{F}$ test will now be developed in terms of R^2 from Equation 4.4.

R^2 is the proportion of variance accounted for:

$$R^2 = \frac{SS_p/N}{SS_t/N}, \text{ or}$$

the proportion of sum of squares accounted for:

$$R^2 = \frac{SS_p}{SS_t}$$

$1-R^2$ would then be the proportion of sum of squares unaccounted for:

$$1-R^2 = \frac{ESS}{SS_t}$$

thus

$$\frac{ESS_f}{SS_t} = 1-R_f^2$$

$$\frac{ESS_r}{SS_t} = 1-R_r^2$$

Given Equation 4.4:

$$\underline{F} = \frac{(ESS_r - ESS_f)/df_p}{(ESS_f)/df_w}$$

and dividing by SS_t yields

$$F = \frac{\left[\dfrac{ESS_r}{SS_t} - \dfrac{ESS_f}{SS_t}\right] / df_p}{\dfrac{ESS_f}{SS_t} / df_w}$$

Substituting the above derived equalities:

$$F = \frac{\left[(1-R_r^2) - (1-R_f^2)\right]/df_p}{(1-R_f^2) / df_w}$$

and by rearranging terms, the above formulation can be simplified to:

Equation 4.5: $F = \dfrac{(R_f^2 - R_r^2) / df_p}{(1-R_f^2) / df_w}$

It is the case that $(R_f^2 - R_r^2)$ yields the gain in accuracy due to the additional information in the full model. This gain is compared to the error variation shown by the denominator in Equation 4.4 as ESS_f and in Equation 4.5 as $(1-R_f^2)$. Thus, the amount of gain is compared against the measure of random error variation. If this gain is large in comparison to the measure of random variation, as determined by the magnitude of F and its associated chance probability distribution, then the statistical (or null) hypothesis may be rejected (with the realization that such a decision will be wrong a proportion of the time equal to alpha). In general, the degrees of freedom for the numerator and denominator will be referred to as df_n and df_d, respectively. Therefore, the F test that will be used throughout the remainder of this text is the same as first introduced in Equation 2.4 on page 50 and is reproduced here:

Equation 4.6: $F_{(df_n, df_d)}$ $= \dfrac{(R_f^2 - R_r^2)/df_n}{(1-R_f^2)/df_d}$

The R^2 values for the full and restricted models will now be calculated. Model 1 was the full model, and its vector representation is in Figure 4.4.

$$Y_1 \quad = \quad a_1 U \quad + \quad b_1 X_1 \quad + \quad E_1 \quad (E_1)^2$$

$$\begin{bmatrix} 4 \\ 6 \\ 8 \\ 10 \\ 12 \end{bmatrix} = 2 \begin{bmatrix} 1 \\ 1 \\ 1 \\ 1 \\ 1 \end{bmatrix} + 2 \begin{bmatrix} 1 \\ 2 \\ 3 \\ 4 \\ 5 \end{bmatrix} + \begin{bmatrix} 0 \\ 0 \\ 0 \\ 0 \\ 0 \end{bmatrix} \quad \begin{bmatrix} 0 \\ 0 \\ 0 \\ 0 \\ 0 \end{bmatrix} \quad \Sigma(E_1)^2 \ = 0 = ESS_f$$

Figure 4.4. Vector representation of the error in
prediction in the full model.

If each element in E_1 is squared and then summed, the error sum of squares associated with the predictors in that model is obtained. The total sum of squares can be calculated by subtracting the criterion mean from each criterion score, squaring this deviation, and then summing all the squared deviations. The total sum of squares for the data under consideration is 40. Therefore,

$$R_f^2 = \frac{(SS_t - ESS_f)}{SS_t} = \frac{(40 - 0)}{40} = 1.00$$

Model 4.2 was the restricted model, and its vector representation is in Figure 4.5.

In order to minimize the sum of the squared elements in E_2, the weight "a_0" will in this case be the mean of the scores in

$$Y_1 \;=\; a_0 U \;+\; E_2 \qquad (E_2)^2$$

$$\begin{bmatrix} 4 \\ 6 \\ 8 \\ 10 \\ 12 \end{bmatrix} = 8 \begin{bmatrix} 1 \\ 1 \\ 1 \\ 1 \\ 1 \end{bmatrix} + \begin{bmatrix} -4 \\ -2 \\ 0 \\ 2 \\ 4 \end{bmatrix} \quad \begin{bmatrix} 16 \\ 4 \\ 0 \\ 4 \\ 16 \end{bmatrix} \qquad \Sigma (E_2)^2 \;=\; 40 \;=\; ESS_r$$

Figure 4.5. Vector representation of the error in
prediction in the restricted model.

the criterion vector (Y_1). The mean of Y_1 is 8; therefore, $a_0 =$
8. Each element of E_2 is the difference of the individual score
from the criterion mean. The first element in vector U is multiplied by the weighting coefficient of 8 and the resultant value
is 8. The first element in Y_1 is 4; thus (4 - 8) = -4, which is
the first element in E_2. Vector E_2 contains the deviation of each
individual's score from the criterion mean. When these elements
are squared and summed, the error sum of squares for the restricted model (for this data, 40) is obtained.

No other value than 8 (the criterion mean) for "a_0" will
give a smaller error sum of squares for the restricted model. The
reader may wish to try other values for "a_0" to see what happens
to E_2.

The proportion of criterion variance accounted for by the
restricted model is

$$R_r^2 \;=\; \frac{(SS_t - ESS_r)}{SS_t} \;=\; \frac{(40 - 40)}{40} \;=\; .00$$

Now for the calculations of degrees of freedom. Degrees of

freedom was conceptualized in Chapter Two in terms of how many data points are free to vary. Degrees of freedom can also be conceptualized in terms of the number of linearly independent predictor vectors in a model. The notion of linearly independent vectors was introduced at the end of Chapter Three.

The degrees of freedom in the numerator (df_n) in Equation 4.6 is the difference between the number of linearly independent vectors in the full model (symbolized m_1 in this text) and the number of linearly independent vectors in the restricted model (symbolized m_2 in this text). One way to conceptualize the degrees of freedom for the numerator of the F test that is being constructed is to ask, "How many new, linearly independent predictors were added to the predictors in the restricted model in order to form the full model?"

The number of linearly independent vectors in a model would not include the criterion vector nor the error vector. Weighting coefficients for the predictor vectors are mathematically determined so as to make the criterion vector linearly dependent upon the weighted combination of the predictor vectors; hence the criterion vector is not a linearly independent vector. In most instances a perfect fit will not be possible; therefore, an error vector is included. But note that the values in the error vector are determined only after the other weighting coefficients are determined; hence the error vector is not a predictor vector.

The full model ($Y_1 = a_1 U + b_1 X_1 + E_1$) has two linearly independent vectors (U and X_1). The restricted model ($Y_1 = a_0 U + E_2$)

has only one linearly independent vector (U). Therefore, df_n =

(2-1) = 1. The full model has one more piece of information than

the restricted model and that piece of information accounts for

the difference between R_f^2 and R_r^2.

The R_r^2 is not always equal to .00. Often, several predictors

are under consideration, and the researcher wishes to restrict

one of the predictors from the full model to test the proportion

of unique variance that that one piece of information adds to the

others when used along with them. (Such problems will soon be

presented.) The restricted model will then contain information in

addition to the unit vector.

The F ratio for the comparison of Models 4.1 and 4.2 will

now be calculated. The general F equation has as the denominator:

$(1-R_f^2)/df_d$. The 1 represents the maximum proportion of criterion

variance that could be accounted for. When the proportion of

accounted-for variance in the full model (R_f^2) is subtracted from

1, the proportion of unaccounted or error variance is obtained.

The number of observations (N) minus the number of linearly in-

dependent vectors in the full model will always be df_d (there are

two linearly independent vectors in the full model and N = 5;

hence df_d = 5-2 = 3). In a sense, for every weighting coefficient

calculated in the full model, one less criterion score is free to

vary. This is analogous to saying that once the group mean is

calculated, one of the observed values in the group is fixed.

In the problem of the relationship of X_1 to Y_1 tested by

Models 4.1 and 4.2 the following data is obtained:

$$R_f^2 = 1.00 \qquad\qquad R_r^2 = .00$$

$$df_n = 2-1 \qquad\qquad df_d = N-2$$

$$F_{(df_n, df_d)} = \frac{(R_f^2 - R_r^2)/df_n}{(1-R_f^2)/df_d} = \frac{(1.00 - .00)/1}{(1-1.00)/(5-2)} = \frac{100/1}{0/3} = \text{infinity}$$

F is equal to infinity because of the denominator of 0. This difficulty is encountered because it was desirable to facilitate understanding of the slope and Y-intercept concepts, and so data
that yield an R_f^2 of 1.00 for the full model were employed. To
overcome the "ill effects" of the unusual data chosen for that
illustration, a little error can be introduced. Let $1-R_f^2 = .03$.
This makes $R_f^2 = .97$ rather than 1.00, and the resultant F would
be:

$$F_{(1,3)} = \frac{(.97 - .00)/1}{(1-.97)/3} = \frac{.97/1}{.03/3} = \frac{.97}{.01} = 97$$

With 1 and 3 degrees of freedom, an observed F of 34 or
larger is found less than one time in a hundred ($p<.01$) due to
sampling variation. The observed F (with a little error thrown
in) was 97, which is much larger than 34.

The next step is to ascertain how tenable is the research
hypothesis: "There is a linear relationship between Ability and
Performance." It is known that the observed F is a very rare occurrence where there is no systematic relationship and that
knowledge of Ability explains 97% of the criterion variance.
Knowing that the five subjects are a random sample of a specified population, most researchers would accept a positive answer:

"Yes, there is a linear relationship between Ability and Performance."

Yet it must be remembered that the observed data may be that
one-time-in-a-hundred chance finding. It is highly unlikely, but
possible. To base a decision upon these data depends upon the importance of the outcome. If someone's life may be lost by incor-
rectly accepting the research hypothesis, one may want several
more samples--if the decision involves only a few dollars, one
probably would accept the research hypothesis based upon the one
sample.

The point that statistics is only an aid to decision-making
must be stressed. The a priori establishment of a level of probability and percent of error variance one can tolerate are not
statistical decisions. Logical forethought, including an analysis
of the costs of the various decisions, must be accomplished by
the researcher before the collection and analysis of the data.

Summary of F and R^2

The general F formula using R^2 is:

$$F_{(df_n, df_d)} = \frac{(R_f^2 - R_r^2)/df_n}{(1-R_f^2)/df_d}$$

The R_f^2 is the proportion of observed criterion variance that
the fullest linear model accounts for.

The R_r^2 is the proportion of observed criterion variance that
the restricted linear model explains. Some information in the
full model is restricted (e.g., if $b_1 = 0$ is hypothesized, then

the variable associated with b_1, (X_1), does not appear in the restricted model).

In the numerator $(R_f^2 - R_r^2)$ is the proportion of unique variance that the deleted variable(s) explain. The degrees-of-freedom term in the numerator (df_n) is the number of linearly independent vectors used to account for the proportion of variance difference between R_f^2 and R_r^2. The difference between the number of linearly independent vectors in the full model and the number of linearly independent vectors in the restricted model is df_n.

One minus R_f^2, (i.e., $1-R_f^2$) is the proportion of variance unexplained by the full model (error variance), and df_d equals the number (N) of observations minus the number of linearly independent vectors in the full model. In essence, df_d is the number of observations which are free to vary after weights for each of the linearly independent vectors in the full model have been calculated.

A second illustrative example of a single straight line of best fit

The second example contains data that is not as systematic as in the first example and which is closer to the real-world state of affairs. The research hypothesis under consideration is the same as for the first example, and can be stated in a number of ways:

1. There is a linear relationship between Ability and Performance.

2. For every unit of increase in Ability, there is a constant change in Performance.

3. The correlation between Ability and Performance is other than zero.

4. Ability is linearly predictive of Performance.

The rival or statistical hypotheses for the above stated research hypotheses would be, respectively:

1. There is no linear relationship between Ability and Performance.

2. For every unit increase in Ability, there is no change in Performance.

3. The correlation between Ability and Performance is zero.

4. Ability is not linearly predictive of Performance.

The observed data points for six subjects' Ability and Performance is given in Figure 4.6.

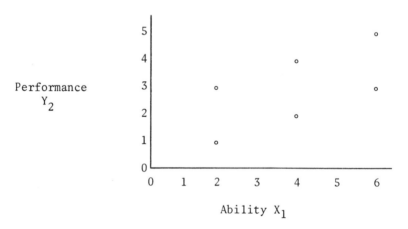

Figure 4.6. Observed Performance scores (Y_2) and Ability scores (X_1) for six individuals.

The six data points reflected in the scattergram (Figure 4.6) seem to follow a trend; however, it is apparent that a single straight line cannot be cast so as to go through all the data

points. Two people had an Ability score of 2, but their performance scores differed; one had a Performance score of 3, and the
other had a Performance score of 1. A similar type of discrepancy is noted for Ability levels 4 and 6. There is not a perfect
linear relationship since a single straight line does not fit all
the data points. A line of best fit, though, can be cast which
will minimize the sum of the squared distances from that line
(elements in the error vector).

The full model required by the research hypothesis is the
same single straight line model as for the first example.

Model 4.3: $Y_2 = a_1U + b_1X_1 + E_3$

The task is to find values for a_1 and b_1 so as to minimize
the sum of the squared elements in E_3. First, a_1 and b_1 will be
solved for intuitively, and then the weights will be derived
formally.

Figure 4.7 shows the vector representation of the full model.

$$Y_2 \quad = \quad a_1U \quad + \quad b_1X_1 \quad + \quad E_3$$

$$\begin{bmatrix} 1 \\ 3 \\ 2 \\ 4 \\ 3 \\ 5 \end{bmatrix} = a_1 \begin{bmatrix} 1 \\ 1 \\ 1 \\ 1 \\ 1 \\ 1 \end{bmatrix} + b_1 \begin{bmatrix} 2 \\ 2 \\ 4 \\ 4 \\ 6 \\ 6 \end{bmatrix} + \begin{bmatrix} ? \\ ? \\ ? \\ ? \\ ? \\ ? \end{bmatrix}$$

Figure 4.7. Vector representation of data in
scattergram of Figure 4.6.

Look at the scattergram presented in Figure 4.6 and place a
point halfway between the two scores for the individuals who

scored 2 on the ability measure. That point will be the mean cri-

terion score for these two individuals (the mean score is 2).

Likewise, if one obtains the mean criterion scores for those who

had an Ability score of 4 and an Ability score of 6 and places

them between the respective observed points, one can easily cast

a straight line which is the line of best fit (that is, the sum

of the squared elements in the error vector (E_3) is minimized).

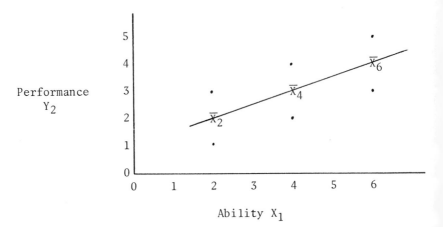

Figure 4.8. An intuitive selection of the line of
          best fit.

The line in Figure 4.8 is straight and passes through the

mean data points; thus the line of best fit has been obtained.

What is the value for a_1? The line intersects the Y axis at 1;

therefore, $a_1 = 1$. Also, a two-unit increase on X_1 from zero to

two gives a one unit increase on Y_2. Since the slope is the same

throughout, 2 to 4 and 4 to 6, b_1 should be .5. One can check

this value by noting that the value on the line at $X_1 = 2$ is 2,

and at $X_1 = 4$ is 3. So for each unit increase on X_1, a .5 in-

crease on Y_2 is observed. The next step is to calculate E_3 with
these weights, and then square the elements of E_3, as in Figure
4.9.

$$Y_2 \quad = \quad 1U \quad + \quad .5X_1 \quad + \quad E_3 \qquad (E_3)^2$$

$$\begin{bmatrix} 1 \\ 3 \\ 2 \\ 4 \\ 3 \\ 5 \end{bmatrix} = 1 \begin{bmatrix} 1 \\ 1 \\ 1 \\ 1 \\ 1 \\ 1 \end{bmatrix} + .5 \begin{bmatrix} 2 \\ 2 \\ 4 \\ 4 \\ 6 \\ 6 \end{bmatrix} + \begin{bmatrix} -1 \\ +1 \\ -1 \\ +1 \\ -1 \\ +1 \end{bmatrix} \qquad \begin{bmatrix} 1 \\ 1 \\ 1 \\ 1 \\ 1 \\ 1 \end{bmatrix} \quad \Sigma (E_3)^2 \ = \ 6$$

Figure 4.9. Vector representation of Model 4.3:
$Y_2 = a_1U + b_1X_1 + E_3$.

The first element in vector E_3 is -1. This value was obtained by multiplying a_1 times U (1*1) and adding that to b_1
times the first element of X_1 (.5*2), which gives a predicted
score of 2. Solving for E_3, the observed score (1) minus the predicted score (2) equals -1. This procedure is followed to obtain
each element in vector E_3. Squaring each element in the error
vector E_3, and summing yields the error sum of squares found with
the full model. No other values for a_1 and b_1 than 1 and .5, respectively, yield a smaller error sum of squares.

The statistical hypothesis sets $b_1 = 0$, reflecting the notion that there is no linear relationship between Ability and Performance. The sample data indicates that b_1 is not 0, but that
$b_1 = .5$. However, is this value of .5 merely a chance sampling
discrepancy from a true population value of 0? To answer this

question, the $\underline{F}$ test between a full and restricted model can be calculated and evaluated according to a predetermined alpha level. Suppose the alpha level is set at .05 (that is, if the observed $\underline{F}$ can be expected less than 5 times in 100, the statistical hypothesis will be rejected and the research hypothesis will be accepted).

The full model was Model 4.3: $(Y_2 = a_1U + b_1X_1 + E_3)$. To test the statistical hypothesis, b_1 is restricted to 0, effectively deleting X_1 from the full model. The restricted model is then:

Model 4.4: $Y_2 = a_2U + E_4$

(Note that the weighting coefficient for the unit vector is designated as a_2, rather than as a_1, just to point out that this weight will probably not be the same in the restricted model as in the full model; indeed, in this case it is not.) The value for a_2 that will satisfy the minimal sum of squared elements in E_4 is the criterion mean of Y_2. $\overline{Y}_2 = \dfrac{\Sigma Y}{N} = \dfrac{18}{6} = 3$. Therefore, $a_2 = 3$. Figure 4.10 shows the vector representation of the restricted model with E_4 and $(E_4)^2$.

$$
\begin{array}{ccccccc}
Y_2 & = & 3U & + & E_4 & & (E_4)^2
\end{array}
$$

$$
\begin{bmatrix} 1 \\ 3 \\ 2 \\ 4 \\ 3 \\ 5 \end{bmatrix} = 3 \begin{bmatrix} 1 \\ 1 \\ 1 \\ 1 \\ 1 \\ 1 \end{bmatrix} + \begin{bmatrix} -2 \\ 0 \\ -1 \\ +1 \\ 0 \\ +2 \end{bmatrix} \qquad \begin{bmatrix} 4 \\ 0 \\ 1 \\ 1 \\ 0 \\ 4 \end{bmatrix} \qquad \Sigma(E_4)^2 = 10
$$

Figure 4.10. Vector representation of the restricted model and the sum of the squared elements in E_4.

As in the case of the first example, when only the criterion mean is used to predict the criterion, the sum of the squared elements in the error vector is numerically equal to the total sum of squares. For the data under consideration, SS_t = 10. To obtain the R_f^2 and the R_r^2 one must place the sum of squares explained by that model over the total sum of squares.

The sum of squares in vector E_3 was 6 and is the sum of squares <u>not</u> accounted for by knowledge of Ability. The total sum of squares is 10. The proportion of variance accounted for in the full model is the difference between the total sum of squares (or SS_t) and $\Sigma(E_3)^2$ (or ESS_f), divided by the total sum of squares. Equation 4.6a presents this notion symbolically as well as finding the value of R_f^2 for the example data.

$$\text{Equation 4.6a: } R_f^2 = \frac{(SS_t - ESS_f)}{SS_t} = \frac{(10 - 6)}{10} = \frac{4}{10} = .40$$

The proportion of variance accounted for in the restricted model is the total sum of squares (SS_t) minus the sum of the squared elements in E_4 (or ESS_r) over the total sum of squares. In this case the sum of the squared elements in E_4 is equal to the total sum of squares. This will always be the case when the restricted model contains only the unit vector. Therefore, the R^2 for the restricted model can be found from Equation 4.7.

$$\text{Equation 4.7: } R_r^2 = \frac{(SS_t - ESS_r)}{SS_t} = \frac{(10 - 10)}{10} = .00$$

where: ESS_r = the sum of the squared elements in the error vector associated with the restricted model.

Substituting into the general $\underline{F}$ formula, Equation 4.4, the $\underline{F}$ value for comparing Model 4.3 with Model 4.4 can be found:

$$\underline{F}_{(df_n, df_d)} = \frac{(R_f^2 - R_r^2)/df_n}{(1-R_f^2)/df_d}$$

$$\underline{F}_{(1,4)} = \frac{(.40 - .00)/1}{(1-.40)/4} = \frac{.40/1}{.60/4} = \frac{.40}{.15} = 2.667$$

df_n equals the difference between the number of linearly independent vectors in the full model and the restricted model; therefore, $df_n = 1$ because the full model ($Y_2 = a_1U + b_1X_1 + E_3$) has two linearly independent vectors and the restricted model ($Y_2 = a_2U + E_4$) has one linearly independent vector.

df_d equals N minus the number of linearly independent vectors in the full model; N = 6, so $df_d = (6-2) = 4$.

With one and four degrees of freedom, an $\underline{F}$ value of 7.71 or larger is observed five times in 100 when the population value of b_1 really is zero. An $\underline{F}$ value of 2.667 is observed more than 5 times in 100 due to sampling variation; therefore, since an alpha of .05 had been adopted, the statistical hypothesis that $b_1 = 0$ cannot be rejected and it is concluded that the observed apparent linear relationship between ability and performance may only reflect a chance event.

A straight line will not often go through all of the data points as it did in the first example, nor will it often go through all of the mean data points as in the second example. Hence, mathematical formulas are usually required to find the weighting coefficients for the line of best fit, as discussed in the following section.

Mathematical calculation of the single straight line of best fit

    The data used up to this point were constructed so as to

provide intuitive solutions to obtaining the line of best fit. In

a bivariate case where the line of best fit is not intuitively

obvious, the following equations can be used to obtain the Y-

intercept weight (a) and the slope weight (b) that will minimize

the sum of the squared elements in the error vector:

Equation 4.8: $b = \dfrac{\Sigma XY - \left[\dfrac{(\Sigma X)\ (\Sigma Y)}{N}\right]}{\Sigma X^2 - \left[\dfrac{(\Sigma X)^2}{N}\right]}$

Equation 4.9: $a = \overline{Y} - b\overline{X}$

    To find the values for a and b in the example just given,

Equation 4.8 can be used to derive the weight "b", and Equation

4.9 to solve for "a". The calculations are given in Figure 4.11.

    The weights in Figure 4.11 (b_1 = .5 and a_1 = 1) are the same

values previously obtained using the intuitive approach.

    It should be quite apparent to the reader that the real

world seldom yields data which can be solved intuitively, and in-

deed real world problems extend beyond the bivariate case such

that many predictors are needed to account for the criterion vari-

ance. Nevertheless, as complex models are investigated, the read-

er should be aware of the fact that the complex models break down

into subsets of the basic linear model which has the following

form: $Y_1 = a_1U + b_1X_1 + E_1$. The mathematical solution of weights

Y	X	X^2	XY
1	2	4	2
3	2	4	6
2	4	16	8
4	4	16	16
3	6	36	18
5	6	36	30
$\Sigma Y = \overline{18}$	$\Sigma X = \overline{24}$	$\Sigma X^2 = \overline{112}$	$\Sigma XY = \overline{80}$

$$\overline{Y} = \frac{\Sigma Y}{N} = \frac{18}{6} = 3$$

$$\overline{X} = \frac{\Sigma X}{N} = \frac{24}{6} = 4$$

$$b_1 = \frac{80 - \dfrac{(24)\,(18)}{6}}{112 - \dfrac{(24)^2}{6}} = \frac{80-72}{112-96} = \frac{8}{16} = .5$$

$$a_1 = 3 - .5\,(4) = 3-2 = 1$$

Figure 4.11. Mathematical solution for Y-intercept (a_1)
and slope (b_1) for the second data set.

for the multiple predictor set becomes involved and has been
treated extensively elsewhere (Mendenhall, 1968). The intent of
this book is to explicate conceptual research problems and the
linear models required to answer the resulting hypotheses. The
models can then be executed mathematically by the numerous computer programs such as the one provided in Appendix A.

The use of the Applied Research Hypotheses: Single straight-line
computer problem

Now for an introduction to computer processing of data. Rather than calculating weighting coefficients by hand or by calcu-

lators, most researchers are now turning to the computer. Appendix C contains data which illustrates all example problems provided in this text. These data can be punched on cards to be used with a linear regression program to help solve the problems given (Applied Research Hypotheses). Appendix A gives one computer program which has been found to be very useful.

A number of "Applied Research Hypotheses" are discussed in the remainder of this text as a means of providing examples of research problems and appropriate ways of testing research hypotheses with linear regression analysis. Some of these Applied Research Hypotheses have an associated "Generalized Research Hypothesis" which gives a more general example of that hypothesis with generalized full and restricted models. The Applied Research Hypotheses are based on data in the data set in Appendix C. Each provides: (1) a research and a statistical hypothesis, (2) a full model and a restricted model, (3) the information which leads to an interpretation (R^2_f, R^2_r, $\underline{F}$, and probability) and which is found by computer solution, and (4) the permissible interpretation. The reader is urged to use the given hypotheses and full and restricted models to get computer solutions and to check those solutions against the ones provided here.

Applied Research Hypothesis 4.1 is designed to fit the single straight-line notions just presented. The vectors in the problem are the same as those in the data set in Appendix C. That is, X_2 and X_3 refer to variables two and three, respectively, in the data set. At this point, it is strongly recommended that time be spent using a linear regression program to obtain the informa-

tion (R_f^2, R_r^2, $\underline{F}$, and p) provided in Applied Research Hypothesis

4.1.

The first computer run is relatively simple but illustrates

one widespread application of the general linear model. The re-

search hypothesis stated in Applied Research Hypothesis 4.1 re-

quires a single straight line of best fit. Most regression pro-

grams automatically provide the Pearson Product Moment Correla-

tion between each variable and every other variable in the analy-

sis. Note that the correlation between the predictor variable and

the criterion variable is provided on the computer output; if one

would square this correlation, one would obtain the R_f^2 listed on

the output. Thus, the single straight-line model provides the

same information as does the correlation coefficient. Note: the

letters used to symbolically designate predictors and criteria

are arbitrarily chosen. In many cases in this text, both pre-

dictors and criteria are referred to as "X" variables (i.e.,

X_2, X_3, X_{27}, etc.). This corresponds to the use of X's in the

Applied and Generalized Research Hypotheses and helps to indicate

that any variable may serve as a predictor in one hypothesis but

as a criterion in another hypothesis.

A special note regarding the unit vector in computer solutions

Both the full model and restricted model require the unit

vector when testing the single straight-line hypothesis. Most

computerized regression programs provide the unit vector auto-

---

Applied Research Hypothesis 4.1

Research Hypothesis: For a given population, X_3 is linearly predictive of X_2.

Statistical Hypothesis: For a given population, X_3 is not linearly predictive of X_2.

Full Model: $X_2 = a_0U + a_3X_3 + E_1$

Restricted Model: $X_2 = a_0U + E_2$

Note: The actual mechanics of getting the necessary information into the computer program will vary depending upon the program used and the computer installation. One should check with the computer center staff for those details.

alpha = .001

$R_f^2 = .92$ $R_r^2 = .00$

$F_{(1,58)} = 747$ nondirectional probability < .0000001

Interpretation: Since the calculated probability is less than the predetermined alpha, the following conclusion can be made: X_3 is linearly predictive of X_2.

---

matically because of its widespread use in linear models. Since the unit vector is automatically provided, the researcher does not have to punch that vector onto the data cards. The unit vector must, though, be considered in the determination of linearly independent vectors. Because each element of the unit vector is a "1" and therefore the weighting coefficient of the unit vector will be added to every individual's predicted score, the weight of the unit vector is often referred to as the "regression constant." The computer program in Appendix A refers to the unit vector weight as "CONSTANT".

Research hypotheses requiring both continuous and dichotomous
variables in the predictor set

Ambiguous results regarding the effects of specific treatments are found in a number of research domains. For example, in
the field of physical conditioning and strength improvement, some
studies report that isometric procedures are more effective in
increasing strength than isotonic procedures, while others report
the opposite outcome. (The following material dealing with isotonic and isometric conditioning is adapted from an article by
Bender, Kelly, Pierson, and Kaplan (1968). Isometrics are exercises in which the individual tenses the muscle to be strength-
ened against a static object; isotonic exercises use the muscle
in such activities as weight lifting, push-ups, etc.) When studies consistently provide contradictory results, one may suspect
that some other undetected variable is operating to cause one
method to work best for one type of individual, while another
method may work best for another type of individual. (It is assumed that the researchers are competent and conscientious; thus,
hopefully, experimenter bias is eliminated and the data are good.)
The subjects that have been previously researched may represent
different populations with respect to this undetected variable.

As an example, Figures 4.12 a and b may represent the data
regarding two such contradictory studies on improving arm
strength. Study I in Figure 4.12a shows that isotonic exercises
resulted in a higher mean Post-test arm strength, while Study II
in Figure 4.12b shows that isometric exercises resulted in a

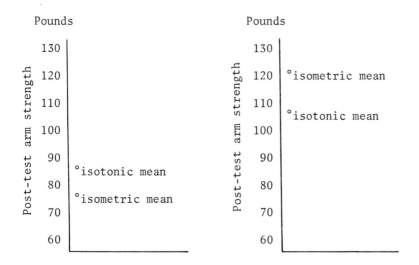

Figure 4.12a. Study I mean differences on Post-test arm strength for isotonic and iso-metric procedures, favoring isotonic.

Figure 4.12b. Study II mean differences on Post-test arm strength for isotonic and iso-metric procedures, favoring isometric.

higher mean Post-test arm strength.

The search for the underlying variable which can clarify the issue would most likely start with an examination of the details of the contradictory studies to determine if they represent dif-ferent populations on some variable. Suppose that in the litera-ture review it was found that most of the studies which indicated superiority for isometric procedures used skilled athletes who were relatively strong initially, and that most of the studies supporting the superiority of isotonic procedures used subjects with average Pre-strength levels. Some studies may have available the initial arm strength of the subjects. A subsequent researcher could incorporate that information and modify Figures 4.12 a and b to result in Figures 4.13 a and b. Note that Figure 4.13a shows that isotonic procedures result in higher Post-test arm strength for all Pre-test scores studied except for a few at the higher end; whereas Figure 4.13b shows that isometric procedures result in higher Post-test arm strength for all Pre-test scores studied except for a few at the lower end.

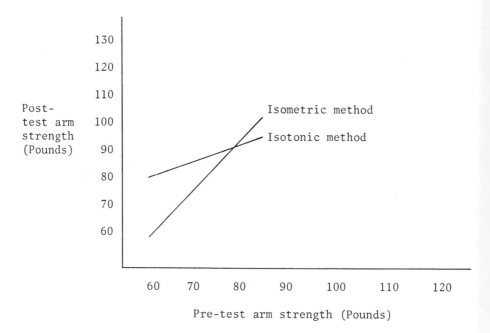

Figure 4.13a. Study I: lines of best fit for data from isotonic and isometric procedures, when viewed across levels of Pre-test arm strength.

The hypothetical data represented in Figures 4.13 a and b reveal a number of important considerations. The first thing one should note is that Study I (Figure 4.13a) included subjects with

a range of Pre-test scores from 60 pounds to 85 pounds, whereas

the Pre-test strength range for subjects in Study II (Figure

4.13b) was from 75 pounds to 120 pounds. It appears, then, that

these two studies each included subjects from different popula-

tions with respect to Pre-test arm strength.

In addition, one might note another consideration by looking

at the lines for the isotonic method in both Figures 4.13 a and

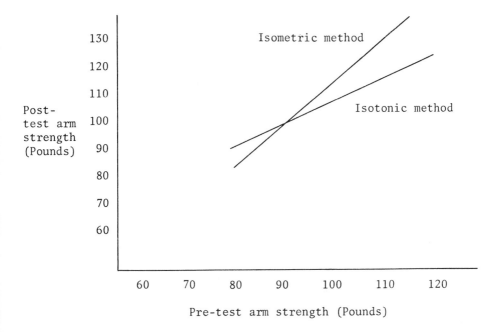

Figure 4.13b. Study II: lines of best fit for data from
            isotonic and isometric procedures, when
            viewed across levels of Pre-test arm strength.

b. In Figure 4.13a, at all levels of observed Pre-test scores,

the Post-test score is higher for the isotonic method, indicating

that the individual's strength improved. But in Figure 4.13b, the
line for the isotonic method shows less and less difference between Pre- and Post-test scores until at a Pre-test strength of
120 pounds the Post-test score is also 120 pounds, indicating no
improvement. It seems that if a person is originally relatively
weak, isotonic procedures are better than isometric for improvement; and if a person is already relatively strong, isometric
procedures are more effective than isotonic. Suppose one looks
through more literature and the observed trend is found repeatedly. It appears that the variable which explains the contradictory
results reported in the literature has been isolated. These review findings lead to the following research hypothesis:

> Isometric exercises are superior to isotonic exercises in
> terms of Post-test arm strength for men who are high on
> the Pre-test strength continuum, and isotonic exercises
> are superior to isometric exercises for men who are ini-
> tially low on the Pre-test strength continuum.

The rival hypothesis is:

> The differential effects (if any) of both isotonic and
> isometric methods are the same across the range of ob-
> served Pre-test scores.

Another way of stating the statistical hypothesis is:

> The slopes of the lines for the two treatments are the
> same across Pre-test scores, although the Y-intercepts are
> allowed to be different.

Since very weak men (below a Pre-test score of 75) are not
included in Study II, and very strong men (above 85) are not included in Study I, a new study is required to test the research
hypothesis against the statistical hypothesis to determine that
the two studies differed only in sampling.

Suppose the researcher has the resources to select and measure 60 men who range in strength from 60 pounds Pre-test arm
strength through 110 pounds arm strength. He wants to select two
groups, one to be given isometric treatment and the other isotonic treatment. Random selection is needed, but he also wants to
be sure that all levels of strength are included in both treatments. He may take strength intervals of 10 pounds (e.g., from
60 to 70 pounds) and randomly assign half of the individuals
within that range to each treatment group. He can do this for all
10-pound intervals to be sure the two treatment groups represent
the population fairly.

Testing for interaction

The research hypothesis requires that the data for both
treatments be compared over the same range of interest on Pre-
test arm strength. Thus it would not be appropriate to find and
look at a line of best fit for only the isotonic procedure or to
look at a line of best fit for only the isometric procedure. Two
lines of best fit (one for each method) must be found simultaneously, and then their slopes must be compared. Figure 4.14 represents the data obtained from the 60 subjects.

An inspection of Figure 4.14 suggests that the research hypothesis is supported because weak men improve more with isotonic
exercise and strong men improve more with isometric exercise. But
how likely is it that the observed interaction is due to sampling
error? To answer the research hypothesis, the researcher must

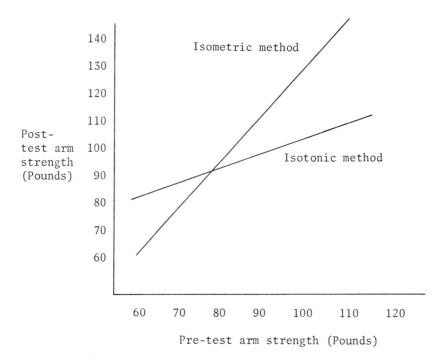

Figure 4.14. The two lines of best fit between Pre- and
            Post-test strength measures for isometric
            and isotonic exercise groups.

analyze all of the data simultaneously.

One linear model must be constructed to reflect the two

lines with differing slopes in the research hypothesis. That mod-

el is Model 4.5 below.

Model 4.5: $Y_1 = a_1U_1 + b_1X_1 + a_2U_2 + b_2X_2 + E_5$

where: Y_1 = Post-test arm strength scores;

        U_1 = 1 if the score on the criterion is from a man
              given _isotonic_ exercises, zero otherwise;

        X_1 = Pre-test arm strength if the score on the cri-
              terion is from a man given _isotonic_ exercises,

zero otherwise;

U_2 = 1 if the score on the criterion is from a man given <u>isometric</u> exercise, zero otherwise;

X_2 = Pre-test arm strength if the score on the criterion is from a man given <u>isometric</u> exercise, zero otherwise; and

a_1, a_2, b_1, and b_2 are least squares weights calculated to minimize the sum of the squared elements in E_5.

Model 4.5 yields the vectors in Figure 4.15.

Individual	Y_1	=	a_1U_1	+	b_1X_1	+	a_2U_2	+	b_2X_2	+	E_5
1	78		1		60		0		0		?
2	99		1		100		0		0		?
3	91		1		80		0		0		?
.	.		.		.		.		.		.
.	.	= a_1	.	+ b_1	.	+ a_2	.	+ b_2	.	+	.
30	120		1		120		0		0		?
31	84		0		0		1		80		?
.	.		.		.		.		.		.
.	.		.		.		.		.		.
60	120		0		0		1		122		?

Figure 4.15. Schematic representation of a model allowing for two lines with different slopes (linear interaction).

Figure 4.15 has subject identification numbers at the left for each of the 60 subjects. Individual 1 has a criterion score of 78 (the first element in vector Y_1). His corresponding value in U_1 is 1 because he participated in the isotonic exercise group. His Pre-test score (X_1) was 60 and his values for U_2 and

X_2 are zero because he was <u>not</u> in the isometric group.

Individual 31 has a Y_1 score of 84; and he has zeroes in vectors U_1 and X_1, which means he was <u>not</u> given isotonic exercise. The value for person 31 in U_2 is 1 and in X_2 is 80, thus he was given isometric exercise and his Pre-test score was 80 pounds.

Model 4.5 has four linearly independent predictor vectors. These vectors allow for two lines because they allow for two slopes (b_1 and b_2), and for two Y-intercepts (a_1 and a_2). If only the top half of each of the vectors (individuals 1 through 30) in Figure 4.15 is considered, the vectors U_2 and X_2 would contain only zeroes, so the model reduces to $Y_1 = a_1U_1 + b_1X_1 + E_5$, a single straight line for the isotonic group. If only the bottom half of each of the vectors in Figure 4.15 is considered, the model reduces to $Y_1 = a_2U_2 + b_2X_2 + E_5$, a single straight line for the isometric group.

To support the research hypothesis under study, the plot of the two lines of best fit must show that the isometric treatment yields higher performance than isotonic for the strong, and isotonic yields higher performance than isometric for the weak. Figure 4.14 shows that this condition is met. Another algebraically equivalent way to express this hypothesis is to expect that the slope b_2 is greater than the slope b_1.

Since Model 4.5 fits the research hypothesis (by allowing the lines to have separate slopes and intercepts), it will be considered the "full" model. (This would also be the full model

if the hypothesis were that b_1 is greater than b_2. An inspection
of the actual slopes is necessary to verify that the difference
(if any) in slopes is in the direction hypothesized.)

Solving for weights a_1, a_2, b_1, and b_2, one obtains an R_f^2
value which specifies the proportion of criterion variance the
four predictor pieces of information account for. The hypothesis
that b_2 is greater than b_1 can be tested by casting a regression
model which forces $b_1 = b_2$. This restriction implies two straight
lines which have a single common slope for both treatments--
reflected by the vector X_3, which contains Pre-test scores for
all subjects, regardless of which treatment they received.

Model 4.6: $Y_1 = a_1U_1 + a_2U_2 + b_3X_3 + E_6$

Model 4.6 gives up one piece of information from the full
model, which is the knowledge of the treatment group associated
with Pre-test scores.

Model 4.6 reduces to $Y_1 = a_1U_1 + b_3X_3 + E_6$ when considering
only subjects 1-30 (the isotonic subjects). Model 4.6 reduces to
$Y_1 = a_2U_2 + b_3X_3 + E_6$ when considering only the isometric students. The model results in two lines, with possibly different
intercepts, but the same slope.

The algebraic restriction on Model 4.5 is to set the two
slopes equal, $b_1 = b_2$, or both equal to a common weight b_3.
Model 4.5, rewritten with the restriction, is:

$Y_1 = a_1U_1 + a_2U_2 + b_3X_1 + b_3X_2 + E_6$

From the properties in Chapter Three, the above can be simplified to:

$$Y_1 = a_1U_1 + a_2U_2 + b_3(X_1 + X_2) + E_6$$

But $(X_1 + X_2)$ is a new vector of Pre-test scores, and if represented by X_3, results in Model 4.6.

If there is a difference in slopes, then the sum of the squared values in the error vector E_6 will be larger than in E_5 and the R_r^2 obtained using only three pieces of information will be smaller than R_f^2. The F test indicates how likely it is that the difference between R_f^2 (using Model 4.5) and R_r^2 (using Model 4.6) is due to sampling error. Thus, the same general F test can be used to answer this research hypothesis.

$$F_{(df_n, df_d)} = \frac{(R_f^2 - R_r^2)/df_n}{(1-R_f^2)/df_d} = \frac{(R_f^2 - R_r^2)/(4-3)}{(1-R_f^2)/(60-4)}$$

where: df_n = the number of linearly independent vectors in the full model [which is four: (U_1, U_2, X_1, and X_2)] minus the number of linearly independent vectors in the restricted model [three: (U_1, U_2, and X_3)];

df_d = N minus the number of linearly independent vectors in the full model.

It should be recalled that $[(R_f^2 - R_r^2)/df_n]$ is a proportional estimate of the population variance and in this application of the general F test, rather than being an among group estimate, it is an estimate of the population variance using interaction. Likewise, $[(1-R_f^2)/df_d]$ is the best proportional estimate of the population variance using all the information in the full model.

With 1 and 56 degrees of freedom, a F of 7.12 or larger is observed less than one time in 100. Therefore, if an alpha of .01 has been adopted and the resultant F is 7.12 or larger, one would reject the statistical hypothesis that $b_1 = b_2$. The research hypothesis would be accepted only if b_2 is indeed greater than b_1. A large F can be obtained if b_1 is greater than b_2, but this result would contradict the research hypothesis. This is a problem regarding probability levels associated with directional and non-directional research hypotheses which will be dealt with extensively in a later section of this chapter.

Even though a large F (and low probability) is obtained, the plotting of the lines of best fit as illustrated in Figure 4.14 must be inspected closely before the research hypothesis can be accepted. When $b_2 > b_1$, the directional interaction exists; however, before the research hypothesis that was specified can be accepted, the lines of best fit must cross within the range of observed Pre-test scores.

Looking back to Figure 4.14 one notes that the lines cross at about 80 pounds on the Pre-test, thus isotonic seems to be superior to bring about strength increase for men who initially have an arm strength of less than 80 pounds and isometric is superior for those who initially have an arm strength of 80 pounds or more.

Inspection of the lines of best fit is very crucial because a number of data configurations can generate a large F value. Figure 4.16 is one such situation. It is possible that the data could have turned out this way. The research hypothesis indicated that, among other things, initially weak men will benefit more from isotonic exercise. Yet an inspection of Figure 4.16 shows

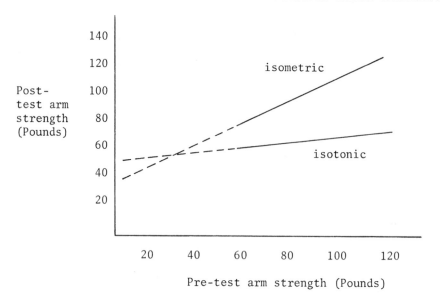

Figure 4.16. A treatment by Pre-test interaction where
the lines cross beyond the range of interest.

that the lines cross at 40 pounds on the Pre-test. The weakest

men in the study had a Pre-test strength of 60 pounds. If results

such as in Figure 4.16 occur, then the research hypothesis must

be rejected. The results in Figure 4.16 show that isometric is

best for all strength levels sampled, although the stronger the

man, the relatively more effective is the isometric treatment.

Such a finding provides data which leaves the original problem

unresolved; the relevant variable which accounts for the previ-

ously reported contradictory findings would not yet have been

found. The reader may be interested in knowing that statisticians

have coined the term ordinal interaction for data wherein the

lines of best fit do not cross within the range of interest--

disordinal interaction for data wherein the lines of best fit do cross within the range of interest.

If the data are like Figure 4.14 (the lines crossing in the range of interest), the research hypothesis can be accepted, and the data are worth reporting since it looks like one source of "contamination" contributing to contradictory results has been found. If the researcher had done a good job in analyzing past research and the research hypothesis was based upon past findings, it would be surprising if the results did not conform to the situation as shown in Figure 4.14.

There is a procedure available (the Johnson-Neyman technique) which allows one to determine the regions where the differences between groups are significant (Cahen and Linn, 1971). The approach taken in this text is to pay attention to the whole range. We acknowledge that, with respect to Figure 4.14, the isometric procedure may not be significantly better than the isotonic procedure for some subjects with Pre-test arm strength greater than 80. But if the two procedures are equally costly, and one has the option of giving either treatment, then one would assign the isometric treatment to subjects who have a Pre-test arm strength greater than 80, whereas the isotonic treatment would be assigned to those whose Pre-test arm strength was less than 80.

Review of the directional interaction example

Contradictory results were reported in the literature re-

garding the effectiveness of two treatment methods. The examination of the literature led to the expectation that isometric
treatment would yield a steeper slope across Pre-test arm
strength levels when compared with the slope associated with men
given isotonic exercise. Furthermore, the researcher was led to
expect that the starting point (the Y-intercept) would be higher
for isotonic because previous data indicated that weaker men performed better under isotonic exercise treatment.

(1) A research hypothesis was stated which expressed the expectation that two straight lines with differing slopes
would be needed to reflect the data. From this research
hypothesis, a full model was constructed which allowed
for two Y-intercepts and two slopes.

(2) A statistical hypothesis was stated which expressed the
expectation that two straight lines with a single common
slope would reflect the data. From the statistical hypothesis a restricted model was constructed which allowed for two Y-intercepts, and only one common slope.

(3) Upon observing that it was unlikely that the two slopes
were the same, the lines of best fit from the full model
were plotted to determine whether the lines crossed
within the range of interest. Given the data expressed
in Figure 4.14, along with the statistical rejection of
the rival hypothesis, the research hypothesis can then
be accepted.

The generalization of the differential effect of two treatments over the range of a continuous predictor

Generalized Research Hypothesis 4.1 (GRH 4.1) depicted below summarizes the regression approach to the solving of an "interaction" kind of research hypothesis. Several Generalized Research Hypotheses will be presented in this text. The intent of presenting a GRH is not to limit the kinds of questions asked by the researcher, but to assist in the asking of a question if the question fits the same structure as a certain Generalized Research Hypothesis. Generalized Research Hypothesis 4.1 presents a way of testing interaction between the treatment an individual receives and some initial score on a single continuous variable. The notion of interaction is extended to more than two groups and more than one continuous predictor variable in Chapter Six.

The reader familiar with analysis of variance may realize that this approach is equivalent to testing the assumption of homogeneity of regression coefficients underlying the analysis of covariance F test. Furthermore, if the pre-treatment strength were represented qualitatively with group membership vectors, the approach is equivalent to the usual test of no interaction in a two-factor design.

Many regression programs (the one in Appendix A is an exception) will not run if linear dependencies are left in the predictor set. If such a program is being used, either U_1 or U_2 in GRH 4.1 must be omitted. Which vector is omitted doesn't matter.

Generalized Research Hypothesis 4.1

Directional Research Hypothesis: For a given population, as X_1 increases, the relative superiority of method A over method B on X_4 will linearly increase.

Non-directional Research Hypothesis: For a given population, as X_1 increases, the relative superiority of method A over method B on X_4 will linearly change.

Statistical Hypothesis: For a given population, as X_1 increases, the difference between method A and method B on X_4 will remain the same.

Full Model: $X_4 = a_1U_1 + b_1X_2 + a_2U_2 + b_2X_3 + E_7$

Restriction: $b_1 = b_2$

Restricted Model: $X_4 = a_1U_1 + a_2U_2 + b_3X_1 + E_8$

where:

X_4 = the criterion;
U_1 = 1 if the score on the criterion is from a subject in method A, zero otherwise;
X_1 = the continuous predictor variable;
$X_2 = (U_1 * X_1)$ = the continuous predictor variable if the criterion is from a subject in method A, zero otherwise;
U_2 = 1 if the score on the criterion is from a subject in method B, zero otherwise;
$X_3 = (U_2 * X_1)$ = the continuous predictor variable if the criterion is from a subject in method B, zero otherwise;
a_1, a_2, b_1, b_2, and b_3 are least squares weighting coefficients calculated so as to minimize the sum of the squared values in the error vectors.

Degrees of freedom numerator = (4-3) = 1

Degrees of freedom denominator = (N-4)

where: N = number of subjects

The R^2 resulting from such a model will be equivalent to the Full Model in GRH 4.1.

Directional and non-directional research hypotheses

Most textbooks in statistics place emphasis upon analysis of variance models which primarily use non-directional research hypotheses, and most F tables are designed to assist the reader in finding two-tailed (non-directional) probability values.

The first example in this chapter dealt with testing the relationship between an ability measure and performance on some task. The research question was: "Does knowledge of Ability account for variance in Performance?" The question was cast into a research hypothesis:

> "There is a linear relationship between Ability
> and Performance."

A rival or competing hypothesis was cast which was called the statistical hypothesis:

> "There is no linear relationship between Ability
> and Performance."

A linear model was cast $(Y_2 = a_1U + b_1X_1 + E_3)$ to reflect the research hypothesis, and a restricted model was cast $(Y_2 = a_2U + E_4)$ to reflect the statistical hypothesis. With the obtained R_f^2 and R_r^2 values, an F value was calculated; and it was found from an F table that an F value of 7.71 or larger for degrees of freedom 1 and 4 is expected to be found due to sampling error 5 times in 100. An F value of 2.67 was observed; therefore, with an alpha level of .05, one fails to reject the sta-

tistical hypothesis, and by implication doesn't accept the research hypothesis.

The research hypothesis in this first example was a non-
directional hypothesis; therefore, the non-directional tabled $\underline{F}$
was appropriate for the specified level (alpha = .05).

But really, does that research hypothesis reflect what the
researcher expects based on prior research and knowledge? For
some naive researcher somewhere, looking for any kind of relationship between variables may be appropriate because he is work-
ing in the dark. On the other hand, many times a researcher does
have a basis for predicting the relationships between variables,
thereby enabling the formulation of directional hypotheses. Furthermore, one usually desires to draw a directional conclusion--
something that cannot be done with a non-directional research hypothesis.

Consider the Ability-Performance relationship. If all past
data strongly suggest that high ability seems to be associated
with high performance, and the researcher therefore wants to con-
firm (technically, to provide empirical support for) this hypothesis which emanates from past research, exactly what question
does he want to ask? He really wants to know if there is a non-
chance positive relationship between Ability and Performance. If
this is the case, the research hypothesis should be:

> "Ability is positively and linearly related
> to Performance."

The linear model which reflects the research hypothesis is

the same as before: $Y_2 = a_1 U + b_1 X_1 + E_3$. The <u>statistical</u> (or

rival) <u>hypothesis</u> is that:

> "Ability is <u>not</u> linearly related to Per-
> formance."

Again, the statistical hypothesis implies that $b_1 = 0$ and

thus the resulting restricted model is: $Y_2 = a_2 U + E_4$.

Suppose an <u>F</u> of 7.72 is observed when R_f^2 is compared with

R_r^2. Can the statistical hypothesis be rejected? No, not yet, be-

cause the percent of criterion variance for which Ability ac-

counts can be due to either a positive or a negative relation-

ship! The research hypothesis demands b_1 to be positive, but the

same <u>F</u> value can be obtained when b_1 is the same magnitude, but

negative. If and only if the weight for b_1 is positive can the

research hypothesis be accepted.

What can be said about the level of probability associated

with the observed <u>F</u> of 7.72 when b_1 is positive? On the basis of

sampling error when $b_1 = 0$, half of the time an observed <u>F</u> would

be due to a negative relationship and half of the time due to a

positive relationship.

If the researcher expects a positive relationship and will

accept only a significant positive relationship when interpreting

the observed <u>F</u> value, then one can say an <u>F</u> of 7.72 with a posi-

tive b_1 value is observed less than 2.5 times in one hundred

(p < .025). (The non-directional probability level is divided in

half to get the directional probability.)

In most research, the investigator goes to a great deal of

effort to review the literature and to develop a strong rationale for his expectation that a particular state exists, which logically requires the use of a directional hypothesis. Numerous examples exist in educational research where new methods are laboriously developed to insure that the new method is better than the existing method; yet when the <u>research hypothesis</u> is cast, it states: "The two methods are different." The hypothetical expectations are <u>directional</u>, yet the test is cast into <u>non-directional</u> terms!

There are at least two harmful outcomes as a consequence of using <u>non-directional</u> tests when the real question is directional. First, some researchers who use <u>non-directional</u> hypotheses find that the F is larger than the tabled F for their stated alpha level and cite that the findings are, to paraphrase, "significant but in the opposite direction than hypothesized." This is a serious error because the original research hypothesis did <u>not</u> state directionality. Indeed, the findings may be contrary to the theoretic rationale, but not the research hypothesis as stated. If the question was directional, then the only finding which can support the hypothesis is a significant F value with the weighting coefficient in the hypothesized direction. If the F is large but the weight is not in the direction hypothesized, then the research hypothesis is untenable. However, one cannot say "The results are significant but in the opposite direction" because that question was not under investigation. Such a finding may be very important since it may suggest something is seriously wrong with the theoretic background of the methods of treatment. These data

can be used for reconsideration of the theoretic position under

investigation, but no inferential statement should be made be-

cause these are findings after the fact, rather than findings

that were theoretically sensible and consequently predicted.

The second harmful outcome from inappropriately using the

non-directional hypothesis relates to the level of probability

for a specified F value. A non-directional test of, say, A and

B treatments yields an observed F that can be due to: A is better

than B; or B is better than A. If expectations are that A is

better than B, and the test shows A to be better than B, what is

the probability that the observed F is due to sampling error? If

the observed F in the two-tailed test (which most tables report)

is due to sampling error only 5 times in 100 ($p = .05$), then the

probability for A better than B in a directional test is .025 for

the same F value. In a directional test, the probability of

chance variation is 1/2 the two-tailed table value.

Given a directional expectation but using a non-directional

test and non-directional F table can result in erroneously fail-

ing to reject the statistical hypothesis and not accepting the

correct directional hypothesis because the reported probability

is twice as large as it should be. If the F is observed 8 times

in 100 with usual tables, then the correct probability for a sup-

porting finding (assuming the weighting coefficient is in the hy-

pothesized direction) is 4 times in 100. Good results may be re-

jected because they apparently do not quite reach the predetermined alpha level (as reported in tables and computer output) when indeed the chance probability of the test results is erroneously indicated to be twice as large as it should be. Such attention to what may appear as an academic point is really quite important when sample sizes are limited. On the other hand, with large sample sizes, the size of R^2 should be considered as well as statistical significance.

For the specific research hypothesis in the isotonic-isometric example, the slope question was directional ($b_2 > b_1$), thus if the results are in the hypothesized direction, the reported probability from typical tables or typical computer output should be divided by 2. If the linear regression program in Appendix A is used, the probability statements are given both for directional and non-directional hypotheses. One must state a directional hypothesis and check to make sure the hypothesized weights are in the hypothesized direction before reporting the "directional probability."

Recommendations concerning directional hypothesis testing

The authors strongly believe that past findings and theoretic expectations should dictate the research hypothesis. If one is groping in the dark regarding the field of investigation, then the non-directional test might yield a set of variables which may be relevant for understanding a particular criterion. Following hunches is certainly one of the inductive phases of inquiry and directionality may not be called for. Indeed, if one is

searching for relevant variables, inferential statistics in itself may not be needed. Bivariate correlations and sample means
are fine descriptors to express hunches.

On the other hand, if prior knowledge is available in a researcher's domain, the authors firmly believe that the directional hypothesis is not only to be preferred but that it is the appropriate hypothesis to test. The continuing emphasis upon directional hypotheses will appear throughout the remainder of this
text. It should be noted that the regression approach can be applied to either directional or non-directional research hypotheses. The choice of either directional or non-directional hypotheses is not a statistical choice; it depends on what the researcher wants to conclude from his analysis.

Present statistical developments limit the directional hypothesis to a single restriction on a full model. Thus, there
will only be one degree of freedom for the numerator of the F
ratio. The F ratio is equal to t^2 when there is only one degree
of freedom in the numerator. Hence the directional research hypotheses are sometimes referred to as "one degree of freedom
questions."

If the degrees of freedom in the numerator of the F ratio
is greater than one, then a directional research hypothesis has
not been tested.

Directional interaction computer problem

Applied Research Hypothesis 4.2 is designed to fit the directional interaction notions just presented. As with all computer problems in this text, the vectors are in Appendix C. Appendix B contains a number of basic FORTRAN statements which will enable the computer user to collect and transform vectors, rather than punching all of the vectors onto the data cards.

Neither the full model nor the restricted model calls for the unit vector when testing for directional interaction. As discussed earlier, most computerized regression programs provide the unit vector automatically. Since the unit vector is automatically provided, the researcher must consider it as being in his models. For the directional interaction hypothesis (ARH 4.2), the unit vector is linearly dependent upon the two treatment vectors. Therefore, even though the computer program automatically includes the unit vector, the R_f^2 and R_r^2 obtained by the computer program will be the same R_f^2 and R_r^2 as desired.

Many computer programs use a matrix inversion solution and will not run if linear dependencies are included in the model. If such a program is being used, one of the two treatment vectors should not be included in the model. Computer programs using an iterative solution will simply assign a zero weight to one of the treatment vectors. The preference of the authors is to use an iterative program, as the specification of all vectors in the model seems to make the model more communicative. For the two-group directional interaction hypothesis, the Y-intercepts can be found by adding the unit vector weight to the weight for the respective treatment group.

Applied Research Hypothesis 4.2

Research Hypothesis: For a given population, as X_8 increases, the relative effectiveness of method A (X_{10}) as compared to method B (X_{11}) on the criterion of interest (X_9) will linearly increase.

Statistical Hypothesis: For a given population, as X_8 increases, the difference between method A (X_{10}) and method B (X_{11}) on the criterion of interest (X_9) will remain the same.

Full Model: $X_9 = a_0U + a_{10}X_{10} + a_{11}X_{11} + a_{15}X_{15} + a_{16}X_{16} + E_1$

where: $X_{15} = X_{10}*X_8$, and
$X_{16} = X_{11}*X_8$

Restricted Model: $X_9 = a_0U + a_{10}X_{10} + a_{11}X_{11} + a_8X_8 + E_2$

alpha = .001

$R^2_f = .98$ $R^2_r = .87$

$F_{(1,56)} = 319$ directional $p < .0000001$

Interpretation: Since the weighting coefficient a_{15} is numerically larger than a_{16}, the directional probability is appropriate and the following conclusion can be made: As X_8 increases, the relative effectiveness of method A (X_{10}) as compared to method B (X_{11}) on the criterion of interest (X_9) will linearly increase.

The generalized single straight-line model

As a means of summarizing the "single straight line" kind of research hypothesis presented in this chapter, another Generalized Research Hypothesis is now presented. If a researcher has an hypothesis which includes only one continuous predictor variable, then Generalized Research Hypothesis 4.2 should be referred to.

Note that both directional and non-directional research hypotheses are included, although the position of this text is that most researchers will be stating directional research hypotheses. Also note that the Research Hypothesis is stated somewhat differently than earlier in this chapter. There are a number of different ways to state a research hypothesis, and a researcher should state his research hypothesis in the way that makes the most sense to him and his intended audience. The crucial matter is that the full and restricted models should be unambiguously implied by the research hypothesis.

Review of hypothesis testing via multiple regression

The process of hypothesis testing via multiple regression analysis is essentially no different than with any other kind of statistical analysis. One does find more of an emphasis on the research hypothesis, that state of affairs which the researcher is desiring to demonstrate. In most present research, the research hypothesis is too often summarily ignored, resulting in a situation in which the researcher, once he has his results, does not know what they mean. Figure 2.1 is repeated here as Figure 4.17, showing the steps in hypothesis testing advocated in this text. The initial specification of the research hypothesis, statistical hypothesis, and full and restricted models may seem tedious, but the reader will find that they are of much benefit in the final analysis.

The research hypothesis contains the statement of what the

Generalized Research Hypothesis 4.2

Directional Research Hypothesis: For a given population, as X_1
increases, X_2 linearly in-
creases.

Non-directional Research Hypothesis: For a given population,
as X_1 increases, X_2 lin-
early changes.

Statistical Hypothesis: For a given population, as X_1 increases
there is no systematic linear change in
X_2.

Full Model: $X_2 = a_0U + a_1X_1 + E_1$

Restriction: $a_1 = 0$

Restricted Model: $X_2 = a_0U + E_2$

where: X_2 = the criterion;
U = the unit vector;
X_1 = the continuous predictor variable; and
a_0 and a_1 are least squares weighting coefficients calculated so
as to minimize the sum of
squared values in the error
vectors.

Degrees of freedom numerator = (m_1-m_2) = (2-1) = 1

Degrees of freedom denominator = $(N-m_1)$ = (N-2)

where: N = number of subjects

---

researcher is interested in finding out, what he would like to

demonstrate or establish. Usually the research hypothesis will

contain some notions about directionality. The authors are be-

coming more and more convinced that a directional research hy-

pothesis is the only meaningful kind of research hypothesis.

1. Statement of the research hypothesis

2. Statement of the statistical hypothesis

3. Statement of alpha

4. Collection of data

5. Calculation of the test statistic:

    a. Formulation of the <u>full</u> <u>model</u> - all variables must be implied unambiguously by the research hypothesis

    b. Statement of the restriction implied by the statistical hypothesis (may also state the desired direction of the weighting coefficient--i.e., restriction: $a_1 = 0$; desire: $a_1 > 0$)

    c. Formulation of the <u>restricted</u> <u>model</u> - reflecting the statistical hypothesis

    d. Calculation of <u>the number of linearly independent vectors</u> - in the full model (m_1) and in the restricted model (m_2), along with the degree of predictability of the full model (R_f^2) and of the restricted model (R_r^2)

    e. Calculation of <u>F</u> from general <u>F</u> formula:

$$\underline{F}_{(m_1 - m_2,\ N - m_1)} = \frac{(R_f^2 - R_r^2)\ /\ (m_1 - m_2)}{(1-R_f^2)\ \ /\ \ (N - m_1)}$$

6. Comparison of probability (p) of calculated <u>F</u> occurring by chance alone with the preset alpha level, in order for the researcher to make a <u>decision</u>:

    a. If $p \leq$ alpha, then reject statistical hypothesis and accept research hypothesis (the full model represents the data better)

    b. If $p >$ alpha, fail to reject statistical hypothesis and fail to accept research hypothesis (the restricted model represents the data as well as the full model, given that there are fewer pieces of information in the restricted model)

Figure 4.17. Steps in hypothesis testing via multiple regression.

The research hypothesis is then reflected in a full model. Every predictor variable mentioned in the research hypothesis must be depicted in the full model. Beginning regression students often mention variables in the research hypothesis which are not contained in the full model, or fail to mention variables in the research hypothesis which they intend to use in the full model. These, of course, are mistakes and consequently any statistical testing will not be illustrative of the variables under consideration in the research hypothesis.

The statistical hypothesis is the antithesis of the research hypothesis. It must state a specific population value, whereas the research hypothesis allows for a multitude of population values. The statistical hypothesis states a fact which the researcher does not want to demonstrate. Indeed, one cannot "prove the statistical hypothesis," as many researchers try to do. The statistical hypothesis suggests restrictions on the weighting coefficients of the full model. The restrictions take the form of $a_1 = 0$; $a_2 = k$; $a_3 = a_4$; $a_5 = [(a_6 + a_7)/2]$, to give a few examples.

The restrictions on the weighting coefficients are then placed on the full model. The question being asked is, "Do these restrictions reduce my predictability more than I can tolerate?" The full model indicates the degree of predictability of the predictor variables described in the research hypothesis. The predictability of the full model is symbolized by R_f^2. The restricted model describes the degree of predictability of the predictor variables after the restrictions have been forced on the full

model. The restrictions must be placed on the full model and then

the full model must be algebraically simplified so as to convey

to the researcher the number and the nature of the linearly in-

dependent vectors on the restricted model. The predictability of

the restricted model is symbolized by R_r^2.

The simplification of the restricted model is important in

order to determine the appropriate degrees of freedom for the

statistical test. The degrees of freedom for the numerator of the

$\underline{F}$ test will be the difference between the number of linearly in-

dependent vectors in the full and restricted models. This will

also correspond to the number of restrictions made on the full

model, thus giving a partial check to the researcher on the cor-

rectness of the restricted model. The calculation of predicted

scores and/or simplification of the model, a process which will

be described later, yields a better check on the correctness of

the restricted model.

The degrees of freedom in the denominator of the $\underline{F}$ test will

be equal to the number of entities being analyzed minus the num-

ber of linearly independent vectors in the full model. The number

of linearly independent vectors can also be thought of as pieces

of information. When information is given up, the loss of some

predictability is expected. The extent of lost predictability is

the crucial question. The difference between R_f^2 and R_r^2 is that

unique variance which the variables deleted account for.

The models and research and statistical hypotheses deal with population values. When a model is applied to sample data, one is finding out how well that model reflects the data. The numerical values of the weighting coefficients are estimates of the population values. A model doesn't force a certain state of affairs, it allows for it. The resulting R^2 indicates how well the model fits the data. The data will always dictate which is the best model, as the model cannot dictate the data.

One advantage of using computers is that the exact probability associated with the F ratio can be provided. Thus more flexibility in the choice of alpha is afforded. Most statistical books present tables of F values for only selected alpha levels-- .05 or .01. This text presents no F tables, as it is assumed that a computer program is available which displays the probability level. Given the probability value, the decision rule is simple: if the probability is less than or equal to alpha, then reject the statistical hypothesis and accept the research hypothesis (prefer the full model); if the probability is greater than alpha, then fail to reject the statistical hypothesis and fail to accept the research hypothesis (prefer the restricted model).

5

Research Control of Possible Confounding Variables

One of the major goals of research is to isolate the unique effects of variables. To show that a particular set of variables is influencing the criterion in a particular way, other competing explainers of the data must be eliminated. Competing explainers can be eliminated, to various degrees, by logical arguments, by research design, or by statistical control of those variables which constitute the competing explainers. Logical argument is, of the three, the weakest defense, whereas research design is the strongest. Unfortunately, it is not always possible in the behavioral sciences to "design away" the competing explainers. Statistical control, through the analysis of covariance, can be a very useful tool for isolating the effects of variables.

Control of Confounding Variables Through Research Design

Randomization

Research conducted in the laboratory is directed toward determining the effects of a particular independent variable upon some outcome state. Some examples of laboratory research questions are:

1. Is titanium a better luminescent filament than carbon-
   ized cotton?

2. Are pigs fed daily 6 grams of complete protein heavier
   than pigs fed daily 2.5 grams of complete protein after
   60 days of feeding?

3. Do dogs raised in isolation for 80 days after birth re-
   spond to noxious stimuli less effectively than dogs
   raised in a natural kennel environment?

To answer these questions, the researcher attempts to con-

trol for all known contaminants in his research design. In the

case of the dog question, breed differences surely may contami-

nate the results. If the isolated dogs were Doberman Pincers and

the control dogs were Mongrels picked out of the pound, the ob-

served differences in sensitivity to noxious stimuli might be

contaminated by breed differences in sensory responsivity.

The researcher may control for the effects of a contamina-

ting variable (in this case, breed) by: (1) choosing only one

"level" of that variable (one breed), or (2) choosing several

levels (breeds); and in either case randomly assigning half of

the members of each breed to each treatment. If only one level

(breed) is selected, that is the only level to which the results

can be generalized. In the dog question, the researcher may se-

lect a specific breed of dogs, and randomly split litters of

puppies into experimental and control groups. The analysis of

variance on the responsivity criterion would then yield an F

value which gives the researcher an estimate of how likely it is

that the observed differences are due to sample variation. By

using the split litter of a specific breed, the population is

defined as that specific breed, and thus large response deviations between the two treatment samples would not be contaminated by the possibility that the two groups originally came from
different breed populations.

In the case of laboratory research, the researcher may have
most of the relevant (potentially contaminating) variables under
his physical control. Such control may also be achieved in the
laboratory for complex manipulations of more than one variable.
Applied research in the behavioral sciences, though, is very often conducted in natural settings where physical control of contaminants is either costly or impossible.

Matching

Consider the "simple" case where a curriculum specialist
wants to test the effectiveness of a new program designed to improve reading ability. She expects students exposed to her material to perform better than students exposed to the old method.
Suppose she has a sample of 200 students (a sample which represents some specific population) available for random assignment.
Past research may indicate that in relation to her criterion
score of reading, a number of variables are known to be related
to performance: (1) girls score better than boys; (2) children
from middle-class homes tend to perform better; (3) high I.Q.
children score higher; and (4) past reading ability is positively
related to the criterion of interest. These four variables are
possible sources of contamination if subjects are not assigned

randomly to treatments, or if the random assignments, by chance, place more of one kind of student in one of the treatment groups. To avoid such contaminating effects (i.e., non-equivalent random samples), one may attempt a matched assignment procedure, such as getting: (1) two girls, (2) from middle-class homes, (3) with high I.Q., and (4) high initial reading ability, and then randomly assigning one to the new treatment and the other to the old treatment. If followed rigorously for all ranges of the possible contaminants, the researcher will have two groups of matched pairs. With the limited original sample of 200 children, however, one would typically find after obtaining 30 or 40 pairs there are no "real" pairs left. Some of the rest may pair up on one or two of the variables, but not on the others. The mind boggles at the effort to form matched pairs for more than one or two variables, and successful matching would surely be unlikely.

Given that matching is successfully accomplished, one may conduct the study, but then the results would be generalizable only to the peculiar population that the 30 or 40 sample pairs represent. Experimental control such as this parallels the rigor of the laboratory, but does it really solve this curriculum specialist's question? Of her 200 subjects only 60 to 80 subjects were used, and they are not necessarily the same on the other relevant variables as those non-matched children. The population to which the results can be generalized is not readily apparent. If the 200 students represent the population to which she wants to generalize, then the selected matched sample does not really represent the population to which she wants to generalize and hence cannot answer her question.

Control of Confounding Variables Through Statistical Control

Covariance: the over and above question

An alternate approach to the control difficulty in the proposed curriculum study would be to match the two groups the best one can and then statistically control for the contamination that was not under experimental control. Given a rough match between the experimental and control groups, one may find that one group has a few more high I.Q. students, a few less girls, fewer middle-class children, and a slight difference in initial reading ability when compared with the other group. These differences are contaminants whose magnitude would be unknown if not accounted for in the research question. The researcher really wants to know, "Over and above the influence of I.Q., Sex, Entering reading ability, and Social class, is the innovative procedure superior to the traditional procedure as measured by Post-test reading ability scores?"

The question can be cast into the following Research Hypothesis:

"Over and above the influence of I.Q., Sex, Entering reading ability, and Social class, the innovative procedure is superior to the traditional procedure as measured by the Post-test reading ability scores."

The competing hypothesis would be the following Statistical Hypothesis:

"Over and above the influence of I.Q., Sex, Entering
reading ability, Social class, and the regression
constant, the innovative procedure is equal to the
traditional procedure as measured by the Post-test
reading ability scores."

The Full Model which reflects the research question has

Post-test reading ability as the criterion, and the predictor set

includes four covariates (I.Q., Sex, Entering reading ability,

and Social class scores) and two treatment vectors.

The Full Model would be:

Model 5.1: $Y_1 = a_1U_1 + a_2U_2 + c_1I_1 + c_2S_2 + c_3R_3 + c_4C_4 + E_1$

where: Y_1 = Post-test reading ability;
U_1 = 1 if the element in Y_1 is from a member of
the innovative treatment, zero otherwise;
U_2 = 1 if the element in Y_1 is from a member of
the traditional treatment, zero otherwise;
I_1 = I.Q. score of each individual represented
in Y_1;
S_2 = 1 if the subject is female, zero otherwise;
R_3 = Entering reading ability score of each
individual represented in Y_1;
C_4 = numerical value on Social class index for
each individual represented in Y_1;
E_1 = the error vector, the difference between
the observed criterion and the predicted
criterion $(Y_1 - \hat{Y}_1)$; and
a_1, a_2, c_1, c_2, c_3, and c_4 are regression weights
calculated to minimize the sum of the squared
elements in vector E_1.

Note that the weights for the covariates are labeled c_1,

c_2, c_3, and c_4. One could use any letter to represent these

weights, yet some researchers may find that for mnemonic pur-

poses the "c" helps to suggest that these weights are associated

with the covariates.

In order to show how the Full Model might look in vector

form, Table 5.1 is presented.

Table 5.1 Illustration of the vectors representing the Full
Model with possible observed scores.

Subject Y_1 $=$ $a_1 U_1$ $+$ $a_2 U_2$ $+$ $c_1 I_1$ $+$ $c_2 S_2$ $+$ $c_3 R_3$ $+$ $c_4 C_4$ $+$ E_1

$$
\begin{array}{cccccccccc}
1 & \begin{bmatrix}9\\7\\5\\8\\7\\4\\\cdot\\\cdot\\\cdot\\6\end{bmatrix} & = a_1 & \begin{bmatrix}1\\1\\1\\0\\0\\0\\\cdot\\\cdot\\\cdot\\0\end{bmatrix} & +a_2 & \begin{bmatrix}0\\0\\0\\1\\1\\1\\\cdot\\\cdot\\\cdot\\1\end{bmatrix} & +c_1 & \begin{bmatrix}120\\115\\100\\120\\120\\105\\\cdot\\\cdot\\\cdot\\107\end{bmatrix} & +c_2 & \begin{bmatrix}1\\0\\1\\0\\1\\1\\\cdot\\\cdot\\\cdot\\0\end{bmatrix} & +c_3 & \begin{bmatrix}7\\6\\3\\7\\6\\3\\\cdot\\\cdot\\\cdot\\4\end{bmatrix} & +c_4 & \begin{bmatrix}2\\1\\3\\1\\3\\2\\\cdot\\\cdot\\\cdot\\2\end{bmatrix} & + & \begin{bmatrix}?\\?\\?\\?\\?\\?\\\cdot\\\cdot\\\cdot\\?\end{bmatrix}
\end{array}
$$

(Subject numbers: 1, 2, 3, 4, 5, 6, ... 200)

To read Table 5.1, look at the row of scores for subject number one: She has a Post-test score of 9, was a member of the innovative treatment (U_1 = 1), was not a member of the traditional
treatment (U_2 = 0), I.Q. of 120, is a female (S_2 = 1), pre-tested
at 7, and was in the second level of Social class.

Subject number 4 had a Post-test score of 8, was not in the
innovative treatment (U_1 = 0), but was in the traditional treatment (U_2 = 1), had an I.Q. of 120, was a boy (S_2 = 0), pre-tested
at 7 (R_3), and was from a home of the highest Social class.

The restriction implied by the statistical (or competing)
hypothesis of interest is a_1 = a_2 (the research hypothesis implies a_1 is greater than a_2). Setting a_1 = a_2, both equal to a
common weight a_0, one obtains the Restricted Model:

Model 5.2: $Y_1 = a_0 U_1 + a_0 U_2 + c_1 I_1 + c_2 S_2 + c_3 R_3 + c_4 C_4 + E_2$

where: a_0 is the common weight for both groups which
reflect the equality of a_1 and a_2; and the
vector E_2 is the new error vector which is the

> difference between the observed criterion score
> and the predicted criterion, predicted from the
> variables in the Restricted Model $(Y_1 - \hat{Y}_1)$.

Since U_1 and U_2 are mutually exclusive vectors whose ele-

ments are 1's and 0's, and since they share a common weight, one

can simplify the Restricted Model to:

Model 5.3: $Y_1 = a_0U + c_1I_1 + c_2S_2 + c_3R_3 + c_4C_4 + E_2$

> where: $U = U_1 + U_2$ and yields the unit vector (U)
> with 1's for everyone in the study.

The Restricted Model forces both groups to have the same

constant and thus any predictability is due solely to the co-

variates and the regression constant, a_0. Since almost all scales

in the social sciences are arbitrarily scaled, the regression

constant is almost always employed to adjust all of the predicted

scores up or down so as to have the same mean as the criterion.

Strictly speaking, the unit vector in the restricted model is a

covariate and should be specified in the research hypothesis.

Customary usage has led researchers to always place the unit vec-

tor in the full and restricted models. As with most customs, the

custom can be ignored as is discussed in Chapters Eight and Elev-

en. Unless otherwise indicated, the unit vector will be assumed

to be in both the full and restricted models.

The R_f^2 is the proportion of the observed sample criterion

variance accounted for by group membership and the covariates.

The R_r^2 is the proportion of the observed sample criterion vari-

ance accounted for by the covariates alone. Any loss in R^2 be-

tween the Full and Restricted Models will be the proportion of

<u>unique</u> sample criterion variance accounted for by knowledge of which treatment the subject received (over and above the effects of the covariates).

The $\underline{F}$ test equation is, as always, the general $\underline{F}$ formula:

$$\underline{F}_{(m_1-m_2,N-m_1)} = \frac{(R_f^2 - R_r^2)/(m_1 - m_2)}{(1-R_f^2) / (N-m_1)}$$

The number of linearly independent vectors in the Full Model (m_1) is six. Since there are five linearly independent vectors in the Restricted Model (the regression constant and the four co-variates), m_2 is five. With 200 subjects, the degrees of freedom are 1 and 194. The resulting $\underline{F}$ is:

$$\underline{F}_{(m_1-m_2,N-m_1)} = \frac{(R_f^2 - R_r^2)/(6-5)}{(1-R_f^2)/(200-6)} = \frac{(R_f^2 - R_r^2)/1}{(1-R_f^2)/194}$$

Assume $R_f^2 = .53$ and $R_r^2 = .45$.

Substituting, one obtains:

$$\underline{F}_{(1,N-6)} = \frac{(.53 - .45)/1}{(1-.53)/194} = \frac{.08/1}{.47/194}$$

The value of .08 is the proportion of the <u>sample</u> criterion variance uniquely due to knowledge of group membership, over and above the covariate knowledge. When this is divided by the numerator degrees of freedom, the result is a proportional <u>estimate</u> of the <u>population</u> criterion variance using the unique knowledge of group membership. One may want to label this $\hat{V}_u$, where the u indicates the proportionally estimated population variance is due to <u>unique group knowledge</u>.

The value .47 is the proportion of the <u>sample</u> criterion
variance <u>unaccounted</u> for by the variables in the Full Model, and
is called error variance. When this value is divided by the denominator degrees of freedom, one obtains a proportional <u>estimate</u>
of the <u>population</u> criterion variance which is the most stable
estimate (see Chapter Two, page 34 for review).

$$\underline{F}_{(1,194)} = \frac{.08/1}{.47/194} = \frac{\hat{V}_u}{\hat{V}_w} = \frac{.08}{.0024} = 33.33$$

With an F of 33.33, one can determine how often a value this
large or larger is observed due to sampling variation. To answer
the research question, the researcher should first set the alpha
level before she starts her analysis. The value one selects depends upon cost, inconvenience, and other matters. In the case of
the innovative curriculum, the curriculum specialist may decide
that conversion to the new procedure is a bother, but that if the
observed adjusted criterion mean is due to sampling variation
less than 25 times in a thousand (alpha = .025), she would be
willing to convert to the new program. The "tabled F value" is
3.88. That is, an F value of 3.88 or larger (with 1 and 194 degrees of freedom) will be observed 2.5% of the time when the statistical hypothesis is really true. The observed F of 33.33 meets
her decision point criterion; and if upon inspection of the
weights a_1 and a_2 her research hypothesis is supported, that is,
if $a_1 > a_2$, then she accepts the research hypothesis and rejects
the competing hypothesis. Given these findings, she is on <u>fairly</u>
safe grounds to convert to the innovative reading program. Fur-

thermore, the curriculum specialist may report her findings in an appropriate journal so her research community can benefit from her study. However, with a demonstrated R^2 gain of 8%, cost conscious educators may be less enthusiastic. One might do well to set an R^2 criterion, as well as an alpha level.

Note should be made here that the authors of this text take a somewhat different position than do others regarding the legitimacy of the covariance analysis. Some statisticians would take the position that lack of random assignment disallows a meaningful conclusion. Our position is that research and decisions must be made in the real world. Random assignment is ideal, but insight can be gained when random assignment is not possible. Our emphasis on replication (discussed in full in Chapter Eleven) is a check on any bias which might occur from not having random assignment. Analysis of covariance cannot save a badly designed study. But controlling for confounding variables is better than ignoring them.

The reader should realize that if $a_2 > a_1$ (i.e., the traditional procedure yielded the higher Post-test mean score), she should not report the results as "significant in the opposite direction" because that was not the question under investigation. Indeed, given the apparent debilitating influence of the innovative program, the researcher should be surprised because her careful planning based upon past knowledge was not supported. A number of questions may be worth pursuing, such as:

1. Is the criterion measure appropriate?

2. Did the teachers sabotage the program?

3. Is the method interacting with one or more of the covariates?

4. Is the innovative treatment really not that good after all?

If upon tracking down possible contaminants, she finds no explanation, the researcher may then replicate the study and, given that the research hypothesis cannot again be accepted, she should publish her results to alert the reading research community to a possible flaw in the theoretical knowledge in her field.

What does the "over and above" analysis do?

The problem just presented in detail is rather complex, but it is just such complexity of applied research which makes the analysis of covariance useful. For a conceptual understanding of covariance, consider the following simple problem:

Suppose one wanted to test the influence of the innovative reading program discussed, and 50 boys and 50 girls were randomly selected for each method. (Sex cannot be a competing explainer with this way of selecting subjects because there are the same proportion of boys to girls in each method.) The researcher also knows initial reading ability will be related to the criterion. If the two groups have moderately different means on the Pre-test of reading, any observed Post-test difference between groups is likely to be influenced by those Pre-test mean differences.

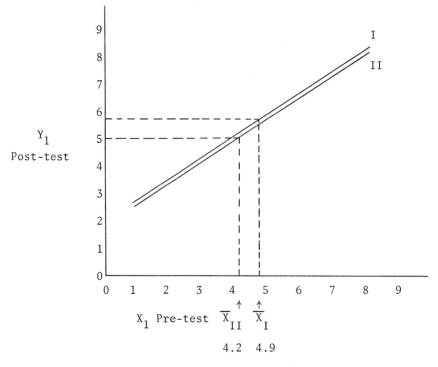

Figure 5.1. The lines of best fit representing Group I and Group II across the range of Pre-test scores on reading as they relate to Post-test scores.

An inspection of Figure 5.1 shows that on the Pre-test, Group I had a mean of 4.9 and Group II had a Pre-test mean of 4.2. Group I is also superior to Group II on the Post-test ($\overline{X}_I$ = 5.7; $\overline{X}_{II}$ = 4.9), but both groups improved. If Pre-test differences were not statistically taken into account, one may conclude that Group I is .8 of a point better than Group II; however, Group I was initially .7 of a point better on the Pre-test. On inspection of the two lines of best fit, one should note that across the range of scores where the two groups overlap, there is only about a .1 difference in favor of Group I.

These Pre-test group differences should be statistically controlled for to determine if Method I is superior to Method II over and above the criterion variance that the Pre-test accounts for.

The Research Hypothesis is:

> Method I is superior to Method II on Post-test reading over and above Pre-test scores.

This hypothesis can be stated in a number of ways; the following are two alternative wordings which say the same thing: (1) Method I is superior to Method II on Post-test reading when the Pre-test differences are statistically controlled; (2) Method I is superior to Method II on Post-test reading when the Pre-test differences are covaried out.

The Statistical (competing) Hypothesis is:

> Method I is not superior to (is as equally effective as) Method II on Post-test reading over and above the Pre-test scores.

The Full Model is:

Model 5.4: $Y_1 = a_1U_1 + a_2U_2 + c_1X_1 + E_4$

> where: $U_1 = 1$ if the criterion score is from an individual in Method I, zero otherwise;
> $U_2 = 1$ if the criterion score is from an individual in Method II, zero otherwise;
> $Y_1 =$ Post-test (criterion) score;
> $X_1 =$ Pre-test score, and
> a_1, a_2, c_1 are regression weights calculated so as to minimize the sum of the squared elements in vector E_4.

The research hypothesis implies $a_1 > a_2$, and the statistical hypothesis demands $a_1 = a_2$. Thus, the Restricted Model is:

Model 5.5: $Y_1 = a_0U + c_1X_1 + E_5$

where: a_0 is the common (no difference) weight; and $U =$ the unit vector $(U_1 + U_2)$.

There are three linearly independent vectors in the Full Model $(U_1, U_2, \text{ and } X_1)$, and two linearly independent vectors in the Restricted Model $(U \text{ and } X_1)$.

Suppose the $R_f^2 = .60$ and $R_r^2 = .595$, the $\underline{F}$ ratio would then be:

$$\underline{F}_{(1,197)} = \frac{(60-.595)/(3-2)}{(1-60)/(200-3)} = \frac{.005/1}{.40/197} = \frac{\hat{V}_u}{\hat{V}_w} \simeq \frac{.005}{.002} \simeq 2.5$$

With 1 and 197 degrees of freedom, a directional $\underline{F}$ of 3.8 or greater is observed less than 25 times in 1000. The observed $\underline{F}$ of 2.5 is smaller than 3.8, thus with an alpha of .025, one would fail to reject the statistical hypothesis that over and above Pre-test differences, Method I and Method II are equal on Post-test reading. The researcher might still maintain a belief (appropriately) that the research hypothesis is true--only that the relatively small treatment effect examined with a sample of only modest size led to a relatively small $\underline{F}$. Indeed, the research hypothesis may yet be true, but based upon the study, the research hypothesis cannot be accepted.

Figure 5.2 shows the relationship of the Pre-test scores to the Post-test scores without knowledge of group membership. The dotted lines are taken from Figure 5.1. It should be obvious that the moderate Pre-test differences are accounting for the observed Post-test group differences and therefore knowledge of group mem-

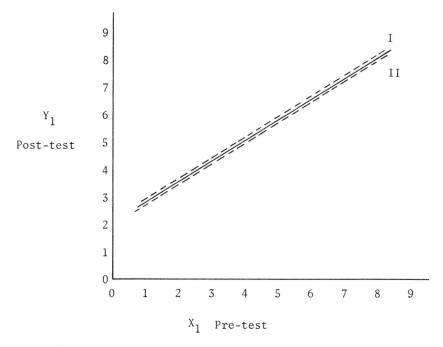

Figure 5.2. The line of best fit between Pre- and
          Post-test reading scores without knowledge
          of group membership. The dashed lines are
          the superimposed lines of best fit shown in
          Figure 5.1.

bership is almost totally redundant with knowledge of Pre-test

scores. When this is observed, one can conclude that over and

above Pre-test scores, observed group differences are chance dif-

ferences due to sampling variation.

    The analysis of the difference between two treatments, sta-

tistically controlling for the effects of one confounding vari-

able, is presented as Generalized Research Hypothesis 5.1, which

follows. An actual data analysis for this kind of question is

presented in Applied Research Hypothesis 5.1, following Generalized Research Hypothesis 5.1. The specific variables discussed in the Applied Research Hypothesis refer as always to the data in Appendix C. The notion of confounding variables will be expanded to more than one confounding variable in later chapters.

Generalized Research Hypothesis 5.1

Directional Research Hypothesis: For a given population, Method A is better than Method B on the criterion over and above the covariable.

Non-directional Research Hypothesis: For a given population, Method A and Method B are differentially effective on the criterion, over and above the covariable.

Statistical Hypothesis: For a given population, Methods A and B are not differentially effective, over and above the covariable.

Full Model: $Y_1 = a_1G_1 + a_2G_2 + c_1C_1 + E_1$

Restrictions: $a_1 = a_2$

Restricted Model: $Y_1 = a_0U + c_1C_1 + E_2$

where: Y_1 = criterion;
U = 1 for each subject;
G_1 = 1 if subject received Method A, zero otherwise;
G_2 = 1 if subject received Method B, zero otherwise;
C_1 = covariable score; and
a_0, a_1, a_2, and c_1 are least squares weighting coefficients calculated so as to minimize the sum of the squared values in the error vectors.

Degrees of freedom numerator = (3-2) = 1

Degrees of freedom denominator = (N-3)

where: N = number of subjects

Applied Research Hypothesis 5.1

Research Hypothesis: For a given population, X_{12} is better than X_{13} on the criterion X_2, over and above the covariable X_{14}.

Statistical Hypothesis: For a given population, X_{12} is equally as effective as X_{13} on the criterion X_2, over and above the covariable X_{14}.

Full Model: $X_2 = a_{12}X_{12} + a_{13}X_{13} + a_{14}X_{14} + E_1$

Restricted Model: $X_2 = a_0U + a_{14}X_{14} + E_2$

alpha = .05

$R_f^2 = .99$ $\qquad$ $R_r^2 = .95$

$F_{(1,57)} = 222$ $\qquad$ directional probability < .0000001

Interpretation: Since the weight for X_{12} is larger than for X_{13} in the full model, the directional probability can be referred to. Since the calculated probability is less than alpha, the statistical hypothesis can be rejected and the research hypothesis accepted.

Assumptions of the analysis of covariance

Analysis of covariance (the "over and above" question) makes one assumption in addition to those for the analysis of variance. This added assumption is that the slope of the line for the k groups is the same across the range of the covariate. The assumption is imposed by the linear model and can be easily seen in the linear equation: $Y_1 = a_1U_1 + a_2U_2 + a_3U_3 + \ldots + a_kU_k + c_1X_1 + E_1$. There are k groups and k weights for the groups. Each group has its own Y-intercept, yet there is only one weight (c_1) associated with the covariate (X_1). Since a one-unit increase on X_1

will yield a c_1 increase on Y_1 regardless of which group the score is associated with, all lines by necessity are parallel, as in Figure 5.3.

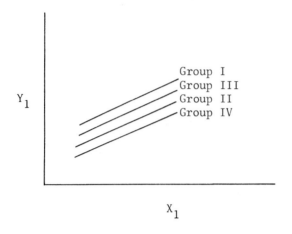

Figure 5.3. Four parallel lines which represent homogeneous slopes across X_1 for the four groups.

The assumption just presented is referred to in statistical texts as "homogeneity of regression lines." If this assumption is violated, then non-random error will be introduced into the statistical analysis and thus the interpretation of the results may be in serious error. Consider Figures 5.4a and 5.4b.

Suppose Group I received one treatment, and Group II received another treatment, and X_1 was a potential covariate of interest. If one obtained the lines of best fit for each group independently ($Y_1 = a_1 U_1 + b_1 X_1 + E_1$ [for Group I] and $Y_1 = a_2 U_2 + b_2 X_1 + E_2$ [for Group II]), and the plots looked like Figure 5.4a, it would be apparent that $b_1 \neq b_2$. The two groups do not have a

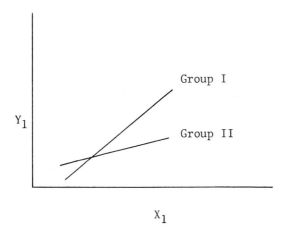

Figure 5.4a. A possible data configuration.

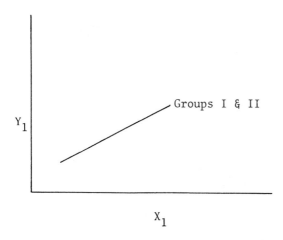

Figure 5.4b. The two lines of best fit for data in
            Figure 5.4a are superimposed when forcing
            the assumption that the slopes are common.

common slope across the potential covariate.

If the assumption of the common slope is made, however, the
Research Hypothesis might be:

> Treatment I is superior to Treatment II on the criterion
> Y_1, over and above the influence of variable X_1.

The Full Model would be:

Model 5.6: $Y_1 = a_1U_1 + a_2U_2 + b_1X_1 + E_6$

The Full Model would yield weights (a_1, a_2, and b_1) which
would yield lines of best fit as depicted in Figure 5.4b. The
lines would be superimposed or very close to it.

Thus, when the competing hypothesis for the above Research
Hypothesis is tested, the results would suggest the Research Hypothesis is untenable, and the conclusion would be the failure to
reject the Statistical Hypothesis, which is:

> Treatment I is equal to Treatment II on the criterion,
> over and above the influence of variable X_1.

The restriction implied by the Statistical Hypothesis on Model
5.6 is: $a_1 = a_2$, both equal to a common weight a_0. Therefore,
the Restricted Model is:

Model 5.7: $Y_1 = a_0U + b_1X_1 + E_7$

Holding tenable the Statistical Hypothesis is reasonable if
one is willing to consider the treatment effects averaged over
the whole range of the covariate. But averaging the effects means
ignoring a very systematic state of affairs as pictured in Figure 5.4a. The assumption of homogeneity of regression slopes is
not tenable, and hence the data should be looked at differently.
A number of statisticians shun the analysis of covariance for the

very reason just presented. The covariate, however, is not necessarily the problem. One may also fail to reject a statistical hypothesis if an important covariate is overlooked or, in fact, any
time the full model does not describe the data well (does not
achieve a high R^2). Indeed, no analysis--including an analysis of
covariance--should be made without research forethought. The
reader should note that Figure 5.4a represents a state of affairs
that was discussed extensively in Chapter Four as "interaction,"
and the test for interaction is just the test one needs to check
the assumption of homogeneity of regression slopes.

Testing for non-parallel slopes

Suppose the researcher has a new method of treatment and he
wants to test the superiority of the new method over the old.
Furthermore, he "knows" initial ability influences Post-test performance and he wants to ask the question: "Over and above the
influence of Pre-test ability, is Treatment One superior to
Treatment Two in relation to Post-test performance?"

First he must consider, "Is this really the best question?"
Is there anything about Treatment One and Two which makes one of
the treatments superior for a certain range of the Pre-test
scores, but not for another range? If upon examination of the two
treatment procedures, he has a suspicion that the new method will
"really" work better for the Pre-tested high scoring individuals
but not so well for the low Pre-test individuals; and furthermore,
if his suspicion is incorrect (when tested), he may then want to

know, "Is the new treatment (Treatment One) superior to the old (Treatment Two) over and above Pre-test ability measures?" In the case just described, the researcher would have a two-stage analysis:

(1) Stage one: test for directional interaction. If interaction is found to exist, the researcher would then plot the lines of best fit to determine the tenability of the <u>directional</u> hypothesis.

(2) Stage two: if <u>no</u> directional interaction is found, then the researcher would test the directional hypothesis that the new treatment is superior to the old treatment when Pre-test ability is covaried.

The sequence of hypothesis-testing for the two-group case would be:

<u>Research Hypothesis</u> (Stage I):

Treatment I will be increasingly more effective in producing positive Post-test performance across the levels of Pre-test ability than will Treatment II.

<u>Statistical Hypothesis</u> (Stage I):

Differential treatment effects on Post-test performance will be constant across the range of Pre-test ability.

In order to test the Research Hypothesis for Stage I, the full model must have two weights associated with Pre-test ability (one for each treatment).

<u>Full Model</u>, Stage I:

Model 5.8: $Y_1 = a_1 T_1 + b_1 (T_1 * X_1) + a_2 T_2 + b_2 (T_2 * X_1) + E_8$

where: Y_1 = the criterion of Post-test performance;

T_1 = 1 if the score on the criterion is from a subject given Treatment I, zero otherwise;

X_1 = the Pre-test ability score;

T_2 = 1 if the score on the criterion is from a subject given Treatment II, zero otherwise;

$(T_1 * X_1)$ = the Pre-test score if the individual is from Treatment I, zero otherwise;

$(T_2 * X_1)$ = the Pre-test score if the individual is from Treatment II, zero otherwise;

E_8 = the error vector; and

a_1, a_2, b_1, and b_2 are regression weights calculated to minimize the sum of the squared values in E_8.

In order to reflect the Statistical Hypothesis, the two lines of best fit must be forced to be parallel. This can be done by setting b_1 = b_2, both equal to a common weight, b_3.

Restricted Model, Stage I:

Model 5.9: $Y_1 = a_1T_1 + a_2T_2 + b_3X_1 + E_9$

where: all variables are specified as for Model 5.8 and a_1, a_2, and b_3 are new regression weights calculated to minimize the sum of the squared elements in vector E_9.

If difficulty is encountered in understanding this brief section, the reader should review Chapter Four regarding interaction. The $\underline{F}$ test will compare R_f^2 and R_r^2. Degrees of freedom numerator is the number of linearly independent vectors in the full model (m_1) minus the number of linearly independent vectors in the restricted model (m_2). Denominator degrees of freedom is $N-m_1$. Model 5.8 has four linearly independent vectors and Model 5.9 has three linearly independent vectors (thus $[m_1-m_2] = 1$).

Graphically, if the research hypothesis associated with Stage I is accepted, the plots may look like either Figure 5.5a or Figure 5.5b.

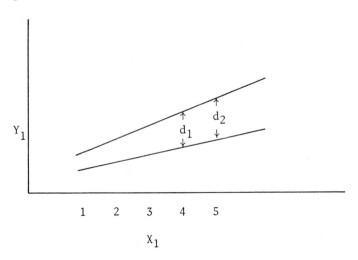

Figure 5.5a. Data supportive of directional interaction, with Treatment I always being superior to Treatment II.

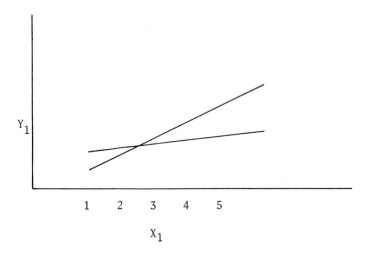

Figure 5.5b. Data supportive of directional interaction, with Treatment I not always being superior to Treatment II.

In Figure 5.5a, the lines of best fit do not cross in the range of interest, but an interaction exists because the higher the score on X_1, the greater the difference between Treatment I and Treatment II on the criterion, favoring Treatment I ($d_2>d_1$). Figure 5.5b shows the case where the lines of best fit cross within the range of interest. If the subject had an X_1 score of 3 or better, Treatment I yields superior criterion performance; but if an individual's X_1 score is below 3, Treatment II yields superior criterion performance.

Stage I tested for directional interaction. If the researcher had wished to test for homogeneity of regression slopes, the non-directional interaction would have been tested. Statisticians recommending such a test (of homogeneity of slopes) are doing so with the intent of wanting not to find significance and therefore hope the data support the restricted model over the full model. If this is the researcher's intent, then the present authors suggest making sure that the restricted model is a preferred model by setting an alpha level such as .60 or .70. When the research hypothesis describes the expected state of affairs, the researcher generally sets his alpha level at somewhere around .01 or .05. The situation is conceptually inverted when the statistical hypothesis describes the expected state of affairs, so the appropriate alpha level would be closer to .60 or .70 (Cohen, 1970).

In Stage I as presented (a directional hypothesis for interaction) the researcher's expectation was that Treatment I would have a greater slope than Treatment II. If the data indicated, on the other hand, that Treatment II has a much larger slope than Treatment I, the research hypothesis of Stage I cannot be accepted, but one certainly would not want to act as though the slopes were homogeneous. Whenever the results turn out to be in the direction opposite to that hypothesized, the researcher must go back to the theory "drawing board" to attempt to discover the anomaly. Therefore, if the slopes were quite discrepant in the direction opposite to that hypothesized in Stage I, the following Stage II analysis would not be appropriate. However, if in Stage I it was found that there was no interaction (alpha = .60), then Stage II would proceed as follows:

Research Hypothesis (Stage II):

Treatment I is superior to Treatment II in producing positive Post-test performance when the influence of Pre-test ability is covaried.

Statistical Hypothesis (Stage II):

The effects of treatment on Post-test performance are equal when the influence of Pre-test ability is covaried.

To test the Stage II Research Hypothesis, the Full Model must have two weights associated with treatment (a_1 and a_2) plus a common weight (b_3) for the covariate X_1. Thus, Full Model, Stage II:

Model 5.10: $Y_1 = a_1T_1 + a_2T_2 + b_3X_1 + E_{10}$

The reader should note that this model is the same as the model associated with the statistical hypothesis in Stage I.

To reflect the statistical hypothesis for Stage II, the weights for treatments must be set equal ($a_1 = a_2$, both equal to a common weight, a_3). Since T_1 and T_2 are mutually exclusive vec-

tors which when added together yield the unit vector, the re-

stricted model can be expressed as, Restricted Model, Stage II:

Model 5.11: $Y_1 = a_3U + b_3X_1 + E_{11}$

      where: a_3 and b_3 are regression weights calculated to
              minimize the sum of the squared elements in the
              new error vector, E_{11}.

The Full Model allows for two parallel lines, and the Re-

stricted Model forces the two lines into one common line for both

treatments. If statistical significance is obtained when Model

5.10 is compared with Model 5.11, and if a_1 is greater than a_2,

then the Stage II Research Hypothesis is supported. The two lines

of best fit will be parallel, and Treatment I will be more ef-

fective than Treatment II (a_1-a_2 units more effective). If a_1 =

a_2 or $a_1 < a_2$, the Research Hypothesis is untenable and Treatment

I is not more effective than Treatment II when Pre-test scores

are covaried.

The F test used to test the Stage II Research Hypothesis is

the same generalized F test:

$$F_{(m_1-m_2,N-m_1)} = \frac{(R_f^2 - R_r^2)/(m_1 - m_2)}{(1-R_f^2)/(N-m_1)}$$

R_f^2 is given by Model 5.10 and R_r^2 is provided by Model 5.11.

There are three pieces of information in Model 5.10 and the one

restriction on that model results in the two pieces of informa-

tion in Model 5.11; therefore $m_1 = 3$ and $m_2 = 2$. That one re-

striction results in the one degree of freedom in the numerator.

Homogeneity of regression slopes with more than one covariate

When the researcher deals with complexities such as the first covariate example in this chapter, a number of covariates may be used and tests for homogeneity of regression slopes might not be made. If the researcher suspects interaction with any of the covariates, then he should make the test because ignoring that interaction inflates the unknown variance. On the other hand, if the researcher has good reason not to expect interaction he can assume parallel slopes (or test for homogeneity) across the several covariates and test the over and above question. Some group bias on the covariates of interest will be adjusted by using the over and above test. If treatment is interacting as shown in Figure 5.5a, the assumed parallel lines in the analysis of covariance will give a larger error term, which may or may not mask the superiority of the treatment. The major risk is that one may discard a new treatment which is very good beyond a certain point on the covariate, but not much superior (and possibly inferior) at another level. In other situations, one would accept one treatment as better when the statistical hypothesis is rejected because of a difference in covariate means. The approach presented here is the best procedure at this time for dealing with groups that are not initially equivalent. The procedure is not entirely satisfactory, but it is better than assuming that the groups are equivalent.

Given the interaction illustrated in Figure 5.5b, forcing parallel lines will almost completely mask the effect that Treat-

ment I is better for people beyond Pre-test level 3 and inferior
to Treatment II for individuals below Pre-test level 3.

Whether or not one tests for homogeneity in an analysis of
covariance is a judgmental matter and depends upon expectations
of how the treatments will function for individuals at different
points across the covariate.

Analysis of covariance to obtain increased power

Huck (1972) urges researchers to use analysis of covariance
even when random assignment is made and even when the group means
of the covariate are identical. In cases where Pre-test ability
is known, the inclusion of these data in an over and above analysis will usually provide a better estimate of within group variance $(\hat{V}_w)$. Usually people who score high on the Pre-test will
score relatively high on the Post-test, and those who score low
on the Pre-test will usually score fairly low on the Post-test.
The correlation between Pre- and Post-tests is often greater than
zero. Therefore, the R^2 of a model containing knowledge of both
treatment and Pre-test scores will be larger than an R^2 of a
model using only knowledge of treatment. The over and above test
still tests the unique contribution of the independent variable
(e.g., treatment), but the proportional estimate of the population variance within $[(1-R_f^2)/(N-m_1)]$ will be smaller at the ex-
pense of only one degree of freedom (due to including the covariate) and at the cost of the collection of the covariate score.
Essentially, the reasoning is, "Why throw away knowledge regard-

ing the sample when one has it?" If the task of the researcher

is to attempt to account for all of the variance ($R^2 = 1.0$), then

one might go further and recommend that the researcher should in-

clude all information which accounts for non-random criterion

variance. Chapter One was written with this viewpoint.

6

Dichotomous Variables in Regression Analysis

Definition of dichotomous vectors

Any variable in a regression model can be either a continuous or a dichotomous variable; which it is depends on how the researcher wants to consider the variables in his research hypothesis. A dichotomous variable is one which can obtain only two values. For several reasons to be indicated shortly, the values of one and zero are almost always used. Any other two values could be used, but the usage of ones and zeros simplifies several notions. A dichotomous variable can be a real dichotomy such as: 1 if alive; 0 if dead; or the dichotomy can be artificial, such as: 1 if tall; 0 if short. Most dichotomies are artificial, the dividing line being the result of some arbitrary decision. One must keep in mind that most phenomena in the real world are of a continuous nature, and that imposing arbitrary boundaries will probably decrease predictability. More will be presented on this notion at the end of the chapter.

Dichotomous Predictors

A widely used case for dichotomous variables is in an experimental situation involving treatments. Each of the experimental

179

treatments can be identified by a dichotomous "group membership vector." First, the simplest and most widely used situation involving dichotomous predictors--the two-group problem--will be considered.

Two-group research hypotheses

Two groups of subjects are randomly selected and assigned to two different treatments. The research hypothesis of concern might be: "The experimental treatment is more effective than the control treatment with respect to the criterion under consideration of Math Achievement." After a reasonable amount of time "treating" the two groups, the criterion measure of Math Achievement is obtained. In order for the two methods to show a difference in their effectiveness, a full model must be developed to allow each group to have its own mean. That model would have as predictors a group membership vector for the experimental group and a group membership vector for the control group. These are two dichotomous vectors. The full model reflecting the Research Hypothesis would then be:

Model 6.1: $Y_1 = a_1G_1 + a_2G_2 + E_1$

where: Y_1 = the criterion variable of Math Achievement;
G_1 = 1 if criterion score is from a student in experimental method, 0 otherwise;
G_2 = 1 if criterion score is from a student in control method, 0 otherwise; and
a_1 and a_2 are least squares weighting coefficients calculated to minimize the sum of the squared values in the error vector E_1.

(There are two [m_1=2] linearly independent pieces of information in this full model.)

After formulating the full model, the task then becomes one of finding weighting coefficients a_1 and a_2 such that the sum of the squared values in E_1 is minimized. This task is generally accomplished by a computer solution, but several values for the weighting coefficients are used in Table 6.1 to illustrate the task. It should be clear that, if all three students in G_1 score 5 on the criterion test, the weighting coefficient a_1 ought to be 5. Note that the value of 5 produces no error in prediction for any of the students in G_1. The next concern is to find the numerical value of the weighting coefficient a_2. One could use the most frequently occurring score, a value of 6. Using six, one obtains perfect prediction for two of the three students, but overpredicts by three units for the first student in G_2. By using the mean score for students in G_2 as the weighting coefficient, one does not perfectly predict for any subject. But the objective for a weighting coefficient in most statistical analyses--that of least squares--is satisfied when the group mean is used. Note that in Table 6.1, the sum of the squared error components is lowest when the group mean is used as the weighting coefficient. No other value for the weighting coefficient will produce any lower sum of squared error components.

Ones and zeros are used to reflect group membership information for two reasons. First, the mean of such a vector represents the proportion of subjects in that group. The means of a set of mutually exclusive vectors should add to one; and if this is not the case, then an error has been made somewhere. More importantly,

Table 6.1 Errors produced by various values for weighting
coefficients.

$$Y_1 \;=\; a_1G_1 \;+\; a_2G_2 \qquad E_1 \qquad E_1 \qquad E_1$$

$$
\begin{bmatrix} 5 \\ 5 \\ 5 \\ 3 \\ 6 \\ 6 \end{bmatrix}
=
5\begin{bmatrix} 1 \\ 1 \\ 1 \\ 0 \\ 0 \\ 0 \end{bmatrix}
+
?\begin{bmatrix} 0 \\ 0 \\ 0 \\ 1 \\ 1 \\ 1 \end{bmatrix}
\qquad
\begin{bmatrix} 0 \\ 0 \\ 0 \\ -3 \\ 0 \\ 0 \end{bmatrix}
\qquad
\begin{bmatrix} 0 \\ 0 \\ 0 \\ -2 \\ 1 \\ 1 \end{bmatrix}
\qquad
\begin{bmatrix} 0 \\ 0 \\ 0 \\ -1 \\ 2 \\ 2 \end{bmatrix}
$$

	Using 6 for a_2	Using 5 for a_2	Using 4 for a_2
	$\Sigma E_1^2 = 9$	$\Sigma E_1^2 = 6$	$\Sigma E_1^2 = 9$

the sample means can be easily calculated when (1,0) vectors are
used.

Most computer solutions to regression models will automatically insert a unit vector into the predictor set. The unit vector may or may not be linearly dependent, depending upon what
other predictor vectors are in the system. For most research hypotheses, this automatic inclusion of the unit vector will cause
no concern as the unit vector will be desired (an hypothesis not
requiring the unit vector is presented in Chapter Eight). A complete set of dichotomous vectors reflecting all group member-
ships, when used with the unit vector, will contain a linear dependency. Those readers using a matrix inversion solution must
remember to omit linear dependencies from the model, generally
by omitting one of the group membership vectors. An iterative
solution, with the linear dependencies in the model or omitted
from the model, will give the same R^2 value as the matrix inver-

sion solution which cannot tolerate linear dependencies. The concept of linear dependency is introduced towards the end of Chapter Three.

Given that the computer solution includes a unit vector, the mean for the experimental method in Model 6.1 can be found by adding the weight for the experimental vector (a_1) to the weight for the unit vector. The mean for the control method in Model 6.1 can be found by adding the weight for the control vector (a_2) to the weight for the unit vector. A quick glance below at the Restricted Model, Model 6.2, indicates that the unit vector is in the <u>Restricted</u> Model. Therefore, the unit vector must be in the <u>Full</u> Model, either explicitly or implicitly as a linear combination of other predictors. It should always be the case that all vectors in a restricted model will be in the full model, either explicitly or implicitly as a linear combination.

If the two treatments are not differentially effective, then they must be equally effective; thus the Statistical Hypothesis: "The two treatments are equally effective with respect to the criterion of Math Achievement." This Statistical Hypothesis implies that the two treatments could be considered to be a common treatment, or that predictability is almost as good <u>without</u> knowledge as to which treatment a student received as it was <u>with</u> knowledge. The predictor variables in the Full Model give knowledge as to which treatment a student received; but knowing only that the student was in the total sample would be reflected by the following model:

Model 6.2: $Y_1 = a_0U + E_2$

The unit vector model above, wherein U has a one for all students, simply allows an overall mean to be calculated for a_0. Again, no other weighting coefficient will yield a smaller sum of squared error values for this model than will the overall mean. For the data in Table 6.1, the error vector in the Full Model will have the same values as will the error vector in the Restricted Model because the overall mean is the same as each of the group means. For the data in Table 6.1, the additional knowledge as to which treatment the student was in did not help in predicting his criterion score. Another way of saying this is that the two treatments were not differentially effective.

The method used above for finding the restricted model was to reflect the statistical hypothesis directly. Another way of finding the restricted model is to make restrictions on the full model implied by the statistical hypothesis. Some students have found that making algebraic restrictions is easier initially than trying to conceptualize the restricted model directly. If the two treatments are not differentially effective as stated by the statistical hypothesis, then the means of the populations will be equal. The weighting coefficients a_1 and a_2 in the Full Model are estimates of those population parameters. Hence, restricting a_1 to equal a_2 would reflect in the sample what is being hypothesized about the population. All one does then is to rewrite the Full Model with the restriction: $a_1 = a_2$.

Step 1: Recall Full Model: $Y_1 = a_1G_1 + a_2G_2 + E_1$

Step 2: Impose restrictions: $Y_1 = a_1G_1 + a_1G_2 + E_2$

(notice that when restrictions are imposed,
a potentially different error vector is
expected)

Step 3: Simplify by either collecting vectors being multi-

plied by like weighting coefficients, as in this

case: $Y_1 = a_1(G_1 + G_2) + E_2$ or by collecting

weighting coefficients which multiply like vectors

(use operations described in Chapter Three).

Step 4: Redefine vectors in the simplest form:

$Y_1 = a_1(U) + E_2$. Note that the numerical value of

a_1 will likely be different in the Full Model than

in the Restricted Model. The weighting coefficients

simply hold the place for numerical values. Note

also that Steps 2, 3, and 4 could have been as

follows:

Step 2: $Y_1 = a_2G_1 + a_2G_2 + E_2$

(Here a_1 has been restricted to, or

replaced by, a_2.)

Step 3: $Y_1 = a_2(G_1 + G_2) + E_2$

Step 4: $Y_1 = a_2U + E_2$

The point is that it does not matter which way the restric-

tion is imposed on the data; the same Restricted Model results.

The general $\underline{F}$ ratio can be applied to the above Research Hypoth-

esis.

If the R^2 for the full model is <u>not</u> significantly higher
than the R^2 for the restricted model, then the research hypothesis <u>cannot</u> be accepted. This does not imply that the statistical hypothesis is true for the population, just that enough of a
discrepancy from $a_1 = a_2$ has <u>not</u> been observed to support the hypothesis that in the population the two means are unequal.

If the R^2 for the full model is found to be significantly
higher than the R^2 for the restricted model, then the research
hypothesis can be accepted <u>if</u> the results show that the experimental method was <u>more</u> effective than the control method, as
stated in the directional research hypothesis. If the research
hypothesis had been non-directional, finding a significant difference between the R^2 values for the full and restricted models
would allow only the conclusion that the two treatments were differentially effective.

It is believed that most researchers would like to make some
directional interpretations, and should therefore state a directional research hypothesis. That is, before the data are collected, the researcher should state which treatment is expected to be
better. There are often logical reasons why one treatment may be
expected (or at least hoped) to be better than the other treatment. Any experimental treatment should carry an expectation of
whether it is to be better or worse than a control treatment on a
particular criterion measure. The full and restricted models for
directional research hypotheses are the same as for non-directional ones. Before interpreting "significant" results, though,

one must be sure to determine whether the results were in the hypothesized direction. Thus if one had made the following research
hypothesis: "The Experimental treatment is better than the Control treatment with respect to the criterion of Math Achievement"
then one would have to determine whether or not the mean of the
Experimental treatment ($a_0 + a_1$) was indeed larger than the mean
of the Control treatment ($a_0 + a_2$). If the weighting coefficients
are in the hypothesized direction, then the (non-directional)
probability of the resultant $\underline{F}$ ratio should rightfully be divided
by two. If the means turn out opposite to the situation which was
hypothesized, then the probability of the resultant $\underline{F}$ ratio must
be divided by two and then subtracted from one. In this latter
case, significance would not be obtained (unless one had some un-
realistic alpha level above .50).

The previous material leads to Generalized Research Hypothesis 6.1. Note that the two groups are defined as "Methods," although if a researcher had two non-manipulated groups such as
boys and girls, the same models could still apply. Many researchers analyze such non-manipulated predictor variables. One should
note, though, that if one sex is better than the other, one does
not know why. The position taken in this text is that it is more
profitable to discover why individuals differ than to describe
which subpopulations differ.

One should turn to Applied Research Hypothesis 6.1 to apply
the general notions in Generalized Research Hypothesis 6.1 to the
specific data in Appendix C.

Generalized Research Hypothesis 6.1

Directional Research Hypothesis: For some population, Method A is better than Method B on the criterion Y_1.

Non-directional Research Hypothesis: For some population, Method A and Method B are not equally effective on the criterion Y_1.

Statistical Hypothesis: For some population, Method A and Method B are equally effective on the criterion Y_1.

Full Model: $Y_1 = a_0U + a_1G_1 + a_2G_2 + E_1$

Restrictions: $a_1 = a_2$

Restricted Model: $Y_1 = a_0U + E_2$

where:

Y_1 = criterion;
U = 1 for all subjects;
G_1 = 1 if subject in Method A, zero otherwise;
G_2 = 1 if subject in Method B, zero otherwise; and
a_0, a_1, and a_2 are least squares weighting coefficients calculated so as to minimize the sum of the squared values in the error vectors.

Degrees of freedom numerator = $(k-1) = (2-1) = 1$

Degrees of freedom denominator = $(N-k) = (N-2)$

where:

N = number of subjects

k = number of groups or methods (See GRH 6.2 for more than two.)

Applied Research Hypothesis 6.1

Research Hypothesis: For a given population, X_{12} is more
effective than X_{13} on the criterion X_2.

Statistical Hypothesis: For a given population, X_{12} is equally
effective as X_{13} on the criterion X_2.

Full Model: $X_2 = a_0U + a_{12}X_{12} + a_{13}X_{13} + E_1$

Restriction: $a_{12} = a_{13}$

Restricted Model: $X_2 = a_0U + E_2$

alpha = .05

$R_f^2 = .00025$ $R_r^2 = .00$

$F_{(1,58)} = .0149$ directional p = .45

Interpretation: Since the weighting coefficients are in the desired direction, the directional probability
can be reported. But since it is not less than
alpha, the statistical hypothesis cannot be rejected.

NOTE: The sample means for X_{12} and X_{13} can be found in the
Full Model output by adding the regression constant to
the weight of each respective vector. The sample mean
for X_{12} would be: $a_0 + a_{12} = 23.81 + .36 = 24.17$.
The sample mean for X_{13} would be: $a_0 + a_{13} = 23.81 +
0 = 23.81$. If a matrix inversion computer program is
being used, either X_{12} or X_{13} must be omitted from the
Full Model. Sample means can still be calculated as
above, the only difference is that a_{12} or a_{13} would
not exist, depending upon whether X_{12} or X_{13} was
omitted.

Equivalency of point-biserial correlation and the test for the
difference between two means

The non-directional research hypothesis in GRH 6.1 could

also have been phrased in correlational terminology: "Treatment is correlated with the criterion." The directional hypothesis stated in correlational terminology is: "The correlation between G_1 and the criterion is positive," or alternatively, "The correlation between G_2 and the criterion is negative." A dichotomous variable (G_1) is being correlated with a continuous variable (Y_1). The computational formula is referred to as a point-biserial correlation. A Pearson correlation between these two variables would yield the same numerical value, and the square of either correlation would be equal to the R^2 of the Full Model. "Testing a correlation coefficient for significance" implies testing the Statistical Hypothesis that the population correlation is equal to zero (hence $R^2 = 0$). Notice that the statistically hypothesized population correlation, when squared, is the R^2 value of the Restricted Model. A specific computational formula exists for testing this hypothesis; and since it is a restricted case of the general F test formula, the general F test can be used as well.

Most statistical texts separate correlational procedures from analysis of variance procedures. The multiple linear regression approach does not make the distinction between these two procedures. In fact, the multiple linear regression approach underscores the underlying similarity of the two procedures and instead emphasizes the importance of the researcher asking the question that he wants to ask, in the terminology that he wants to use.

Advantage of multiple linear regression over correlation and analysis of variance

If one had asked a bivariate correlational question, relating only one predictor variable (and the unit vector) to the criterion, one could find the correlational value in the correlation matrix automatically calculated and printed by most computerized regression programs. The advantage of constructing a regression model would be in finding the weighting coefficients (in order to find group means or to plot the lines of best fit) and to be able to ascertain the probability of that high a correlation, or one higher, occurring by chance alone. Most statistics books have tables of necessary correlational values for selected alpha levels. But it just might be the case that an investigator has an alpha level which is not one of the "selected ones". Furthermore, the position of this text is consistent with the emerging notion of reporting the probability of the sample data occurring under the statistical hypothesis, rather than imposing your (arbitrarily chosen, probably without much foresight) alpha level on your readers. The position does not support the selection of alpha after the results are computed--but rather treats research readers as thinkers rather than as blind followers. A given journal author may have an alpha of .05 and find that the results are significant for him. Since the actual probability is not reported, a more conservative reader cannot ascertain whether the results are significant at .01 or not. Now if the author reports

the actual probability of .007, the conservative reader knows
that the research hypothesis is tenable for him as well as for
the author. On the other hand, if the actual probability is reported as .030, then the research hypothesis is not tenable for
the conservative reader, even though he realizes that the research hypothesis is tenable for the author.

Transforming t and $\underline{F}$ to R^2

Many researchers lament the fact that statistical significance can be obtained, yet practical significance may really not
be obtained. Statistical significance is indeed a necessary but
not sufficient condition for practical significance. Statistical
significance simply indicates that something other than chance is
operating, but statistical significance doesn't indicate to what
degree that "non-chance" phenomenon is operating. Research studies using extremely large sample sizes often do find statistical
significance, but little practical significance is obtained. The
larger the sample size, the smaller the difference between two
means needs to be before statistical significance is obtained.
One treatment mean could be 14.0001 and the other 14.0000, and if
enough subjects were in the sample, statistical significance
could be obtained. In most practical applications, a mean difference of .0001 would not be of any importance.

When the statistical hypothesis is not true, increasing the
sample size lowers the probability that a result of a given magnitude occurs by chance alone. The R^2 value, though, is not arti-

ficially inflated by the increase in sample size. Increasing sample size merely produces a more stable and closer estimate of the population R^2 value. This is one of the reasons that some researchers have been reporting R^2 values along with their indices of statistical significance. Friedman (1968) provides a table which allows a quick converting of published t, z, χ^2, and $\underline{F}$ values to correlational values. Squaring the correlational values would produce the desired R^2 values. Persons who are writing a literature review should be encouraged to compute and include the R^2 value of each finding reviewed. When an $\underline{F}$ value is reported in the literature (when R^2 of the restricted model is .00), the R^2 value of the full model is:

$$\text{Equation 6.1:} \quad R^2 = \frac{\underline{F}(\frac{df_n}{df_d})}{1 + \underline{F}(\frac{df_n}{df_d})} = \frac{\underline{F}(df_n)}{df_d + \underline{F}(df_n)}$$

When a t is reported, the R^2 value is:

$$\text{Equation 6.2:} \quad R^2 = \frac{t^2}{df + t^2}$$

For instance, a t of 2.4 with 116 subjects would be "highly significant" (p < .01), yet using Equation 6.2, it is found that the R^2 is .05, indicating that the researcher is accounting for only 5% of his phenomena. One might interpret this as "knowing something for sure about very little." In the past, probability values have been used to support theories. It is hoped that, in the future, R^2 values instead of probability values will be used to indicate the satisfaction one has with his theoretical framework.

Research hypotheses involving k groups

In order to test additional hypotheses, researchers often investigate multiple treatments, rather than just two. (The analysis of variance analog would be the one-factor or one-way analysis of variance.) Group membership vectors need to be constructed for each of the k groups. The most commonly used hypothesis is stated in Generalized Research Hypothesis 6.2.

Notice that only a non-directional research hypothesis can be tested. One way of noting this would be with the multiple restrictions being made. If significance is found, all that can be said is that "the k treatments are not equally effective on the criterion Y_2." One can accept the research hypothesis, but in reality it does not say much. For practical purposes, one would like to know which treatment is best. Indeed, one may have some reason to believe that a given treatment will be better than one other, or a combination of several others. If one has no idea whatsoever as to which treatment is best, then it is suggested that one use the means from the data to develop directional hypotheses for future verification.

Suppose that Treatment j is suspected of being better than Treatment k. The research hypothesis would be: "Treatment j is better than Treatment k on the criterion Y_2." This is a two-group

Generalized Research Hypothesis 6.2

Directional Research Hypothesis: Not appropriate because more
than one restriction (see
GRH 6.3).

Non-directional Research Hypothesis: The k treatments are not
equally effective on the
criterion Y_2.

Statistical Hypothesis: The k treatments are equally effective
on the criterion Y_2.

Full Model: $Y_2 = a_0U + a_1G_1 + a_2G_2 \ldots + a_kG_k + E_1$

Restrictions: $a_1 = a_2 = \ldots = a_k$

Restricted Model: $Y_2 = a_0U + E_2$

where:

Y_2 = criterion;
U = 1 for all subjects;
G_1 = 1 if criterion from subject in
group 1, zero otherwise;
G_2 = 1 if criterion from subject in
group 2, zero otherwise;
.
.
.
G_k = 1 if criterion from subject in
group k, zero otherwise; and
a_0, a_1, ... a_k are least squares weighting
coefficients calculated so as to minimize the sum of the squared values in
the error vectors.

Degrees of freedom numerator: $(m_1-m_2) = (k-1)$

Degrees of freedom denominator: $(N-m_1) = (N-k)$

where:

N = number of subjects

k = number of groups

hypothesis which could be tested by the method outlined in Generalized Research Hypothesis 6.1. Now suppose that data has been
collected on other groups and if the treatments applied to the
other groups produce about the same variance about the group
means, then the data from the other groups can be used with the
two groups to obtain a better estimate of the population criterion variance than with the two groups above. Notice in Generalized
Research Hypothesis 6.3 that the denominator degrees of freedom
is a function of the subjects in all groups, not just those in
the two groups being compared. Indeed, the denominator degrees of
freedom is the same as in Generalized Research Hypothesis 6.2.
But since there is only one restriction being made, the decrease
in R^2 is entirely attributable to that one restriction. Statistical significance allows one to make the definitive statement,
"Treatment j is better than Treatment k on the criterion Y_1, considering the variance in (i) similar treatments." (The verification that the weighting coefficient for the Treatment j vector is
larger than the weighting coefficient for the Treatment k vector
must be made here as in all directional hypothesis situations.)

With respect to GRH 6.3, there are only k linearly independent vectors in the Full Model. One of the group vectors could
have been eliminated. The unit vector in the Restricted Model
will have a weight which is equal to the mean of the subjects in
groups j and k.

Perhaps the research design included three slightly different control groups, and two slightly different discussion treat-

Generalized Research Hypothesis 6.3

Directional Research Hypothesis: Treatment j is better than Treatment k on the criterion Y_1, considering the variance in (i) similar treatments.

Non-directional Research Hypothesis: Treatment j is different than Treatment k on the criterion Y_1, considering the variance in (i) similar treatments.

Statistical Hypothesis: Treatment j is equal to Treatment k on the criterion Y_1, considering the variance in (i) similar treatments.

Full Model: $Y_1 = a_0U + a_1G_1 + a_2G_2 ... + a_iG_i + a_jG_j + a_kG_k + E_1$

Restrictions: $a_j = a_k$

Restricted Model: $Y_1 = a_0U + a_1G_1 + a_2G_2 ... + a_iG_i + E_2$

where:

Y_1 = criterion;
G_1 = 1 if criterion from subject in Treatment 1, zero otherwise;
G_2 = 1 if criterion from subject in Treatment 2, zero otherwise;

.
.
.

G_i = 1 if criterion from subject in Treatment i, zero otherwise;
G_j = 1 if criterion from subject in Treatment j, zero otherwise;
G_k = 1 if criterion from subject in Treatment k, zero otherwise; and
a_1, a_2, ... a_i, a_j, and a_k are least squares weighting coefficients calculated so as to minimize the sum of the squared values in the error vector.

Degrees of freedom numerator = $(m_1 - m_2) = [k - (k-1)] = 1$

Degrees of freedom denominator = $(N - m_1) = (N - k)$

where:

$$N = \text{number of subjects in all treatments}$$

$$k = \text{number of treatments}$$

ment groups, along with four other kinds of treatment groups. A research hypothesis of interest might be: "The average effect, on the criterion, of the two Discussion treatments is better than the average effect of the three Control groups, considering the variance of the four other kinds of Treatment groups."

The antithesis of the Research Hypothesis would be the following Statistical Hypothesis: "The average effect, on the criterion, of the two Discussion treatments is equal to the average of the three Control groups, considering the variance of the four other kinds of Treatment groups." The Full Model reflecting the Research Hypothesis contains a dichotomous vector for each group, and is:

Model 6.3: $Y_2 = a_0U + c_1C_1 + c_2C_2 + c_3C_3 + t_1T_1 + t_2T_2 + t_3T_3$

$$t_4T_4 + d_1D_1 + d_2D_2 + E_3$$

where: Y_2 = the criterion variable;
U = the unit vector containing a 1 for each subject;
C_1 = 1 if criterion from Control group one, zero otherwise;
C_2 = 1 if criterion from Control group two, zero otherwise;
C_3 = 1 if criterion from Control group three, zero otherwise;
T_1 = 1 if criterion from Treatment group one, zero otherwise;
T_2 = 1 if criterion from Treatment group two, zero otherwise;

T_3 = 1 if criterion from Treatment group three, zero otherwise;

T_4 = 1 if criterion from Treatment group four, zero otherwise;

D_1 = 1 if criterion from Discussion group one, zero otherwise;

D_2 = 1 if criterion from Discussion group two, zero otherwise; and

a_0, c_1, c_2, c_3, t_1, t_2, t_3, t_4, d_1, and d_2 are least squares weighting coefficients calculated so as to minimize the sum of the squared values in the error vector, E_3.

Restrictions implied by the Research and Statistical Hypothesis are:

$$\frac{c_1 + c_2 + c_3}{3} = \frac{d_1 + d_2}{2}$$

The restrictions can be solved for any of the five weighting coefficients--c_1 is solved for here.

Multiplying both sides by 3 yields:

$$c_1 + c_2 + c_3 = 3 \left[\frac{d_1 + d_2}{2} \right]$$

Subtracting c_2 and c_3 from both sides yields:

$$c_1 = \tfrac{3}{2}d_1 + \tfrac{3}{2}d_2 - c_2 - c_3$$

Now inserting the right-hand expression for c_1 into the Full Model yields:

$$Y_2 = a_0U + (\tfrac{3}{2}d_1 + \tfrac{3}{2}d_2 - c_2 - c_3)C_1 + c_2C_2 + c_3C_3 + t_1T_1$$
$$+ t_2T_2 + t_3T_3 + t_4T_4 + d_1D_1 + d_2D_2 + E_4$$
$$= a_0U + \left[\tfrac{3}{2}d_1\right] C_1 + \left[\tfrac{3}{2}d_2\right] C_1 - c_2C_1 - c_3C_1 + c_2C_2 + c_3C_3$$
$$+ t_1T_1 + t_2T_2 + t_3T_3 + t_4T_4 + d_1D_1 + d_2D_2 + E_4$$

Expanding and collecting terms results in the following Restricted Model:

Model 6.4: $Y_2 = a_0U + c_2(C_2 - C_1) + c_3(C_3 - C_1) + t_1T_1 + t_2T_2$

$$+ t_3T_3 + t_4T_4 + d_1(D_1 + \tfrac{3}{2}C_1) + d_2(D_2 + \tfrac{3}{2}C_1) + E_4$$

There are (m_1 = 9) linearly independent pieces of information in
the Full Model (Model 6.3) since one of the nine group vectors is
linearly dependent upon the other eight group vectors plus the
unit vector. There are (m_2 = 8) good pieces of information in the
Restricted Model (Model 6.4). Therefore, df_n = 1 and df_d = N-9.
If the F ratio in comparing the Full and Restricted Models produced a probability value lower than the predetermined alpha, one
could accept the Research Hypothesis, if the average of the two
Discussion weighting coefficients was greater than the average of
the three Control weighting coefficients. Some of the vectors appearing in the Restricted Model look strange, but could easily be
generated through a data generating computer subroutine. The unfortunate researcher lacking such niceties could punch those vectors onto the data cards.

The statistically sophisticated reader may see the similarities between what has just been presented and the myriad of techniques referred to as "multiple comparisons" or "post-hoc comparisons."

The above procedure is similar to the "planned comparison"
techniques. The authors of this text see three basic differences
between what has just been presented and the body of traditional
literature referred to as "post-hoc" procedures.

First, in the multiple linear regression approach as presented above, the researcher is forced to state the research hypothesis. The research hypothesis should be directional in nature allowing the researcher to make a conclusive statement. (The "planned comparison" literature apparently does not consider the notion of directional hypothesis testing.)

Second, because the research hypothesis is stated by the researcher and is a well-thought-out analysis of the design, no adjustment to the resultant probability is necessary as is the case in "post-hoc comparisons." It should be noted that a priori (as contrasted to post-hoc) comparisons do not differ from the regression presentation on this point, nor on the next point.

Thirdly, the researcher does not have to initially compare the full model to the "unit vector model." (The post-hoc comparisons require the one-way F to be significant before any specific comparisons can be made.) The "one-way F" hypothesis does not have to be run for two reasons--the "one-way F" question is not of interest to the researcher, and secondly, global statistical significance might not be obtained even though some specific questions of interest might produce significance. This could occur, for instance, when ten control groups and only one experimental group are examined. The 11 groups might statistically be considered to not be differentially effective, whereas the one experimental group might well be better than the average of the ten control groups. A more extensive discussion of multiple comparisons can be found in Williams (1974). Kerlinger and Pedhazur (1973) discuss the topic in terms of contrast and effect coding.

"Main effects" research hypotheses

The research literature is filled with analyses of variance investigating "main effects" and "interaction." It will now be shown that the "main effects" hypothesis is nothing more than one kind of multiple comparison. Suppose a researcher is interested in the effects on some criterion, Y_4, of three levels of Familiarity (high, medium, and low) and two levels of Affectivity (neutral and unpleasant). To obtain all possible combinations, the researcher would need 3 times 2, or 6, treatment groups, as indicated in Figure 6.1.

	Neutral Affectivity	Unpleasant Affectivity
High Familiarity	a_1	a_2
Medium Familiarity	a_3	a_4
Low Familiarity	a_5	a_6

$$\frac{a_1 + a_3 + a_5}{3} \qquad \frac{a_2 + a_4 + a_6}{3}$$

average "Neutral Affectivity" effect, from the sample

average "Unpleasant Affectivity" effect, from the sample

Figure 6.1. Treatment combinations represented by Model 6.5.

The regression model which would allow each of these six treatments to have its own mean would contain six dichotomous vectors and would be:

Model 6.5: $Y_4 = a_0U + a_1X_1 + a_2X_2 + a_3X_3 + a_4X_4 + a_5X_5$

$$+ a_6X_6 + E_5$$

where: Y_4 = the criterion variable;
   U = the unit vector containing a one for each subject;
   X_1 = 1 if criterion is from a subject in High Familiarity and Neutral Affectivity, zero otherwise;
   X_2 = 1 if criterion is from a subject in High Familiarity and Unpleasant Affectivity, zero otherwise;
   X_3 = 1 if criterion is from a subject in Medium Familiarity and Neutral Affectivity, zero otherwise;
   X_4 = 1 if criterion is from a subject in Medium Familiarity and Unpleasant Affectivity, zero otherwise;
   X_5 = 1 if criterion is from a subject in Low Familiarity and Neutral Affectivity, zero otherwise;
   X_6 = 1 if criterion is from a subject in Low Familiarity and Unpleasant Affectivity, zero otherwise; and
   a_0, a_1, a_2, ... a_6 are least squares weighting coefficients calculated to minimize the sum of the squared values in the error vector, E_5.

There are (m_1 = 6) linearly independent pieces of information in the Full Model. (Of six group vectors and the unit vector, six are linearly independent.)

The comparison of the average of the three Neutral Familiarity groups with the average of the three Unpleasant Familiarity groups is what statisticians call the "main effect" of Affectivity. The research and statistical hypotheses (very seldom clearly stated in statistical texts) are:

Research Hypothesis: The average effect of the three Neutral Familiarity treatments is different than the average effect of the three Unpleasant Familiarity treatments.

Statistical Hypothesis: The average effect of the three Neutral Familiarity treatments is not different than the average effect of the three Unpleasant Familiarity treatments.

The Full Model reflecting the Research Hypothesis is stated above
as Model 6.5. The restriction implied by the Research and Statistical Hypotheses is:

$$\frac{a_1 + a_3 + a_5}{3} = \frac{a_2 + a_4 + a_6}{3}$$

Multiplying both sides by 3 and then solving for a_1:

$$a_1 = a_2 + a_4 + a_6 - a_3 - a_5$$

Substituting the right-hand side for a_1 in the Full Model results in:

$$Y_4 = a_0U + (a_2 + a_4 + a_6 - a_3 - a_5) X_1 + a_2X_2 + a_4X_4$$
$$+ a_5X_5 + a_6X_6 + E_2$$

Expanding and collecting terms results in the following Restricted Model:

Model 6.6: $Y_4 = a_0U + a_2(X_2 + X_1) + a_3(X_3 - X_1) + a_4(X_4 + X_1)$
$$+ a_5(X_5 - X_1) + a_6(X_6 + X_1) + E_6$$

There are five linearly independent vectors in the above
Restricted Model and this corresponds with the fact that one restriction was made on six linearly independent vectors. Six vectors minus the one restriction equals the five linearly independent vectors in the Restricted Model.

Notice that the Research Hypothesis was non-directional.
Statistical texts present the "main effects" hypothesis as a non-
directional one, and hence computer programs and applied analyses
have stayed within that framework. (Although many illegitimate
directional conclusions are made, what is even sadder is when a
researcher presents empirical evidence for a directional hypoth-

esis and yet concludes the non-directional hypothesis.) The in-sightful reader will realize that a directional "main effects" hypothesis could have been made, and indeed should have been made, if anything conclusive is desired from the data.

Categorical interaction

Another of the infinite number of possible research hypoth-eses relevant to the design discussed in the previous section is another multiple comparison:

> Research Hypothesis: The difference between the Neutral and Unpleasant treatments given the High Familiarity treatment, and the difference between the Neutral and Unpleasant treatments given the Medium Famil-iarity treatment, and the difference between the Neutral and Unpleasant treatments given the Low Familiarity treatment are not all equal.

> Statistical Hypothesis: The difference between the Neutral and Unpleasant treatments given the High Famil-iarity treatment, and the difference between the Neutral and Unpleasant treatments given the Medium Familiarity treatment, and the difference between the Neutral and Unpleasant treatments given the Low Familiarity treatment are all equal.

The Full Model reflecting the Research Hypothesis is the same as used for the "main effects" hypothesis.

Model 6.7: $Y_4 = a_0U + a_1X_1 + a_2X_2 + a_3X_3 + a_4X_4 + a_5X_5 + a_6X_6$

$$+ E_5$$

Restrictions implied by the Research and Statistical Hypotheses are:

$$(a_1 - a_2) = (a_3 - a_4) = (a_5 - a_6)$$

One may choose to solve for a_1 and a_5:

$$a_1 = a_3 + a_2 - a_4$$

$$a_5 = a_3 - a_4 + a_6$$

Forcing these restrictions onto the Full Model results in:

$$Y_4 = a_0U + (a_3 + a_2 - a_4)X_1 + a_2X_2 + a_3X_3 + a_4X_4 + (a_3 - a_4$$
$$+ a_6)X_5 + a_6X_6 + E_8$$

By collecting terms, one obtains the Restricted Model:

Model 6.8: $Y_4 = a_0U + a_2(X_1 + X_2) + a_3(X_3 + X_1 + X_5) + a_4(X_4$
$$- X_1 - X_5) + a_6(X_6 + X_5) + E_8$$

where all vectors are defined as for Model 6.5. Notice that there are four weights to be calculated in the above Restricted Model; hence four linearly independent vectors (of the weights--a_0, a_2, a_3, a_4, and a_6--only four will be non-zero). The Full Model had six linearly independent vectors, therefore (6-4) or two restrictions were made on the Full Model. Note that there are two equal signs in the restrictions, and that in going from the Full Model to the Restricted Model, two weighting coefficients (a_1 and a_5) were restricted to a specific function of the others.

Statistical interaction viewed within the multiple linear regression approach

Interaction was discussed previously in this text when the isotonic and isometric example was presented in Chapter Four, and when heterogeneity of regression slopes was discussed in Chapter Five. The following pages attempt to bring various "interaction" notions together in one source. The following material is a paper presented by Keith McNeil and Judy McNeil at the 1973 annual convention of the American Educational Research Association. The paper appears in its entirety to retain the flow. Some of the early material may seem redundant, although the material at the end of the paper discussing curvilinear functions may require the reading of Chapter Seven. Throughout the entire paper, the notion that "interaction" is just another predictor variable is emphasized.

Statistical Interaction Viewed Within the Multiple Linear Regression Approach

Statistical interaction has traditionally been viewed as a complex phenomenon that if found would probably cause problems in the interpretation of results and is therefore hopefully not found in the data. Interaction should be viewed, on the other hand, as simply another variable or set of variables that may need to be included in a model in order to obtain acceptable predictability. The approach that researchers should take is to acknowledge that interactions do exist rather than to summarily ignore the phenomena, acting as though the criterion is more simply predicted than it actually is.

The present paper will cover the following topics: (1) traditional analysis of variance interaction, including main and simple effects as well as ordinal and disordinal notions, (2) pooling of a nonsignificant interaction term with the within variance, (3) categorical vs. continuous variables, (4) curvilinear interaction, (5) interaction in only part of the design, (6) notions about directionality vs. nondirectionality, (7) using interaction terms as covariates in analysis of covariance, (8) fitting various three-dimensional models, and (9) interpretation of complex interactions vs. predictability.

Traditional analysis of variance interaction

Consider a two-way design with four levels of IQ and three levels of drugs as in Table 1. The "interaction question" is only one of the infinite number of possible research hypotheses relevant to this design. There are an infinite number of multiple comparisons possible with this design, and traditional interaction is a part of some of those comparisons. The research and statistical hypotheses could be stated as:

> Research Hypothesis: The differences between the four IQ levels are not constant across the three drug treatments. (More explicitly: The differences between the four IQ levels given Drug 1, and the differences between the four IQ levels given Drug 2, and the differences between the four IQ levels given Drug 3 are not all equal.)

> Statistical Hypothesis: The differences between the four IQ levels are constant across the three drug treatments.

The regression formulation for testing the significance of the interaction has been well documented (Bottenberg and Ward (1963); Kelly, Beggs, McNeil, Eichelberger, & Lyon, 1969; Jennings, 1967). (Note that references cited in this paper are at the end of the paper and will not necessarily appear at the end of the text.) A Full Model, allowing for interaction, is needed to allow the 12 cell means to manifest themselves. Then a Restricted Model, allowing only the row (IQ) and column (Drug) means, to manifest them-

Table 1. Design, source table, and comparison of pooling and not
pooling for a 4x3 design.

Source of Variation	Case 1 Not Pooling		Case 2 Pooling	
	df	F	df	F
IQ	3	$\dfrac{MS_{IQ}}{MS_{Within}}$	3	$\dfrac{MS_{IQ}}{MS_{Within+Interaction}}$
Drug	2	$\dfrac{MS_{Drug}}{MS_{Within}}$	2	$\dfrac{MS_{Drug}}{MS_{Within+Interaction}}$
IQ*Drug	6	$\dfrac{MS_{IQ*Drug}}{MS_{Within}}$	(N-12+6)	
Within	N-12			
Total	N-1			

	Drug 1	Drug 2	Drug 3
IQ level 1	a_1	a_2	a_3
IQ level 2	a_4	a_5	a_6
IQ level 3	a_7	a_8	a_9
IQ level 4	a_{10}	a_{11}	a_{12}

selves can be compared to the Full Model via the generalized $\underline{F}$-ratio. The Full Model allowing the 12 means to manifest themselves would have 12 linearly independent vectors, or 12 pieces of information:

Model 1: $Y_1 = a_1C_1 + a_2C_2 + a_3C_3 + a_4C_4 + a_5C_5 + a_6C_6 + a_7C_7$

$$+ a_8C_8 + a_9C_9 + a_{10}C_{10} + a_{11}C_{11} + a_{12}C_{12} + E_1$$

where: Y_1 = the criterion score;
C_1 = 1 if IQ level 1 and Drug 1, 0 otherwise;
C_2 = 1 if IQ level 1 and Drug 2, 0 otherwise;
C_3 = 1 if IQ level 1 and Drug 3, 0 otherwise;
C_4 = 1 if IQ level 2 and Drug 1, 0 otherwise;
C_5 = 1 if IQ level 2 and Drug 2, 0 otherwise;
C_6 = 1 if IQ level 2 and Drug 3, 0 otherwise;
C_7 = 1 if IQ level 3 and Drug 1, 0 otherwise;
C_8 = 1 if IQ level 3 and Drug 2, 0 otherwise;
C_9 = 1 if IQ level 3 and Drug 3, 0 otherwise;
C_{10} = 1 if IQ level 4 and Drug 1, 0 otherwise;
C_{11} = 1 if IQ level 4 and Drug 2, 0 otherwise;
C_{12} = 1 if IQ level 4 and Drug 3, 0 otherwise; and
a_1, a_2, a_3, ... a_{12} are least squares weighting coefficients calculated to minimize the sum of the squared values in the error vector, E_1.

Model 1 allows each of the twelve means to manifest themselves. Model 1 could be written another equivalent way, which will have more value for later discussion. To some readers, the following model more clearly indicates the allowance for interaction between IQ and Drugs.

Alternate Model 1: $Y_1 = a_1(Q_1{}^*D_1) + a_2(Q_1{}^*D_2) + a_3(Q_1{}^*D_3) +$

$$a_4(Q_2{}^*D_1) + a_5(Q_2{}^*D_2) + a_6(Q_2{}^*D_3) +$$

$$a_7(Q_3{}^*D_1) + a_8(Q_3{}^*D_2) + a_9(Q_3{}^*D_3) +$$

$$a_{10}(Q_4{}^*D_1) + a_{11}(Q_4{}^*D_2) + a_{12}(Q_4{}^*D_3)$$

$$+ E_1$$

The multiplication signs in this model indicate the allowance for interaction effects, as they will do whenever they appear in subsequent models. Since Model 1 and Alternate Model 1 are equivalent models, forcing the restrictions of no interaction on either Model will result in the same restricted model. The restrictions specifying no interaction are as follows:

$$(a_1 - a_2) = (a_4 - a_5) = (a_7 - a_8) = (a_{10} - a_{11})$$

$$(a_2 - a_3) = (a_5 - a_6) = (a_8 - a_9) = (a_{11} - a_{12})$$

Forcing these restrictions onto either full model would result in the same Restricted Model:

Model 2: $Y_1 = q_1Q_1 + q_2Q_2 + q_3Q_3 + q_4Q_4 + d_1D_1 + d_2D_2 + d_3D_3 + E_2$

where: Q_1 = 1 if IQ level 1, 0 otherwise;
Q_2 = 1 if IQ level 2, 0 otherwise;
Q_3 = 1 if IQ level 3, 0 otherwise;
Q_4 = 1 if IQ level 4, 0 otherwise;
D_1 = 1 if Drug 1, 0 otherwise;
D_2 = 1 if Drug 2, 0 otherwise; and
D_3 = 1 if Drug 3, 0 otherwise.

One must remember that most computer programs will automatically provide the unit vector. In most applications the unit vector is desired, but when dealing with categorical data it can cause trouble unless appropriately considered. By this we mean the consideration of linearly dependent vectors. In Model 1, if all interaction vectors are added together, the unit vector is the result. Therefore, one of those 13 vectors (the 12 interaction vectors plus the unit vector) is not a new piece of information. For heuristic purposes, one might want to ignore the unit vector. But for computer solution and for being less likely to make an error in determining the number of linearly independent vectors, the unit vector should be retained. Therefore, one of the interaction vectors, is not a new piece of information; and hence, there are 12 pieces of information in Model 1. In Model 2, the unit vector also needs to be considered. Here, the four IQ vectors added together yield the unit vector, so if the unit vector is considered as a piece of information, then one of the IQ vectors must be considered as redundant. The same argument holds for the three drug vectors. If the three drug vectors are added together, the unit vector results. Hence, one of the drug vectors must be considered as redundant. Which IQ vector, and which drug vector is considered redundant is of no concern, indeed all three drug vectors, four IQ vectors and the unit vector can remain in the analysis. The important point to remember is that there are six pieces of information in Model 2: the unit vector, three IQ vectors, and two drug vectors. Model 2 thus has 6 linearly independent pieces of information.

The degrees of freedom can be easily determined, with the degrees of freedom numerator in this case being $(m_1 - m_2)$, or $(12 - 6 = 6)$. The degrees of freedom denominator are $(N - m_1)$, in this case $(N - 12)$.

If the interaction is significant, most statistical authors indicate that the main effects question is not appropriate, and that simple effects should be investigated. The simple effects is simply a one-way analysis of variance at each level. The regression formulation for the IQ level 1 simple effect would be:

Model 3: $Y_1 = c_1 \ (Q_1 * D_1) + c_2 \ (Q_1 * D_2) + c_3 \ (Q_1 * D_3) + E_3$

Model 4: $Y_1 = a_0 U + E_4$

Since there are three pieces of information in the Full

Model, and one in the Restricted Model, the degrees of freedom

would be (3-1) and (N-3). Since the Research Hypothesis implied

by Models 3 and 4 deals only with subjects at IQ level 1, then

only those subjects at IQ level 1 would be analyzed. N in this

case then is the number of subjects in IQ level 1. Other models

could be used to test the other IQ simple effects, as well as to

test the drug simple effects.

Even when there is significant interaction, one might still

want to look at the main effects, and this would be appropriate

if the interaction were ordinal. Ordinal interaction occurs when

the rank order of the categories of one variable on the basis of

their criterion scores is the same within each category of the

second predictor variable. When significant interaction exists in

the data, the magnitude of the main effects are not indicative of

the effects at each level, but it is the case that the one level

is uniformly superior. (Note should be made here that a signifi-

cant "main effect" indicates a non-directional difference, and

does not allow one to make a directional interpretation (McNeil,

1971).)

Testing for main effects can basically be done in two ways,

one where the within term is used as the best estimate of ex-

pected variance, and the other wherein the within source is

pooled with the nonsignificant interaction and used as a new es-

timate of expected variance. These notions are more fully de-

lineated by Jennings (1967) and Kelly et al. (1969). Jennings

(1967) presents the regression models appropriate for the case where the sources are not pooled. If the main effect for IQ were to be tested, the following restriction would be made on Model 1, if the cell Ns (N_1, N_2, etc.) were not proportional:

$$\frac{N_1a_1 + N_2a_2 + N_3a_3}{N_1 + N_2 + N_3} = \frac{N_4a_4 + N_5a_5 + N_6a_6}{N_4 + N_5 + N_6} = \frac{N_7a_7 + N_8a_8 + N_9a_9}{N_7 + N_8 + N_9} =$$

$$\frac{N_{10}a_{10} + N_{11}a_{11} + N_{12}a_{12}}{N_{10} + N_{11} + N_{12}}$$

resulting in an extremely complicated Restricted Model, which has little conceptual value. The pooling analog is much more conceptually pleasing and is discussed below.

Pooling of the interaction term

When the interaction and within sources of variance are pooled the interaction source of variance is being considered as error variance. The argument for this procedure (when there is little sample interaction) is that the result is a more stable estimate of the expected chance variance. Model 2 depicted a state of affairs wherein interaction was not allowed. If Model 2 is not significantly less predictable than Model 1, then Model 2 can be treated as a more meaningful state of affairs. Testing for IQ main effect on this model can be accomplished by the restrictions on Model 2: $q_1 = q_2 = q_3 = q_4$, resulting in the following Restricted Model:

Model 5: $Y_1 = d_1D_1 + d_2D_2 + d_3D_3 + E_5$

Since there are 6 pieces of information in the Full Model

(Model 2) and 3 pieces of information in the Restricted Model
(Model 5), then the test of significance would be $F_{(6-3,\ N-6)}$,
similar to case 2 in Table 1. One can think of the IQ main effect
question here as having information about IQ and drugs, and simply giving up information about IQ. Alternatively, one can ask,
"Does knowledge of IQ account for criterion variance, over and
above knowledge of drugs?"

The drug main effects question would be tested by forcing
the restrictions: $d_1 = d_2 = d_3$ on Model 2, resulting in the following Restricted Model:

Model 6: $Y_1 = q_1Q_1 + q_2Q_2 + q_3Q_3 + q_4Q_4 + E_6$

resulting in an F with 2 and N-6 degrees of freedom.

Categorical vs. continuous variables

One of the major weaknesses of applied analysis of variance
analyses is the artificial categorization of continuous variables. Researchers usually can obtain continuous data, and usually want to infer along some continuum, but usually analyze
categorical data. Indeed, phenomena in the real world usually
follow systematic functions, rather than discrete leaps and
bounds. The multiple linear regression approach readily allows
one to investigate continuous variables, and specifically to investigate the interaction between categorical variables and continuous variables.

Suppose that one wanted to treat IQ as a continuous variable, rather than artificially categorizing it into four levels.

Suppose also that Figure 1a depicts the pictorial representation of the suspected interaction, and Figure 1b the state of affairs allowing no interaction between drugs and IQ as they affect the criterion. The regression formulation allowing for interaction would be:

Model 7: $Y_1 = a_1D_1 + a_2D_2 + a_3D_3 + b_1S_1 + b_2S_2 + b_3S_3 + E_7$

> where: D_1, D_2, and D_3 are defined as above;
> $S_1 = D_1*Q_5 =$ IQ if have Drug 1, 0 otherwise;
> $S_2 = D_2*Q_5 =$ IQ if have Drug 2, 0 otherwise;
> $S_3 = D_3*Q_5 =$ IQ if have Drug 3, 0 otherwise; and
> $Q_5 =$ IQ for each subject.

The slope of the Drug 1 line would be b_1 and the Y-intercept would be a_1. If these three lines are not interacting, they will be parallel, which means that the three slopes must all be equal, resulting in the restriction: $b_1 = b_2 = b_3$. Setting all three slopes equal to a common slope, b_5, results in the following Restricted Model:

Model 8: $Y_1 = a_1D_1 + a_2D_2 + a_3D_3 + b_5Q_5 + E_8$

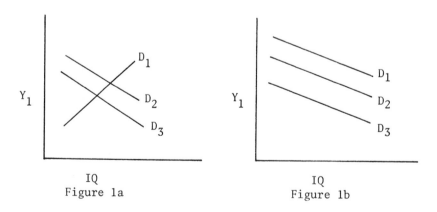

Figure 1. States of affairs allowing linear interaction (1a) and not allowing linear interaction (1b).

For the $\underline{F}$ test of significance between Models 7 and 8, there would be an $\underline{F}_{(6-4, \ N-6)}$. Note that the vectors which allow for interaction can be found by multiplying the drug levels by the continuous IQ variable.

Now suppose that the three drugs were actually three different dosage levels of the same drug, and we now desire to treat the drug variable as a continuous variable, rather than artificially categorizing it into levels. The "interaction" Research Hypothesis now would be:

> Research Hypothesis: There is an interaction between IQ and Drug dosage level, over and above the separate linear effects of IQ and Drug dosage level.

A more precise and insightful way of stating the research hypothesis would be:

> Research Hypothesis: The combined effect of IQ and Drug dosage level is needed to predict the criterion, over and above the separate linear effects of IQ and Drug dosage level.

Testing the interaction between the continuous variable of IQ and that of Drug dosage level would be accomplished by constructing Models 9 and 10.

Model 9: $Y_1 = a_0 U + a_4 D_4 + b_5 Q_5 + c_4 (D_4 * Q_5) + E_9$

> where: Y_1 = criterion;
> U = unit vector;
> D_4 = Drug dosage level for each subject;
> Q_5 = IQ for each subject; and
> $D_4 * Q_5$ = product of Drug dosage level and IQ for each subject.

The product term $(D_4 * Q_5)$ allows for the linear interaction between Drug dosage level and IQ. The restriction on Model 9 which would not allow interaction to occur would be to set c_4

equal to zero, having the effect of dropping (D_4*Q_5) out of the model:

Model 10: $Y_1 = a_0U + a_4D_4 + b_5Q_5 + E_{10}$

Note should be made that the unit vector appears in both Models 9 and 10. When all variables in the regression model are continuous, the unit vector is not generally a redundant piece of information, and hence must be counted as one of the linearly in-dependent vectors. Furthermore, the interaction term in Model 9 (D_4*Q_5) is not <u>linearly</u> dependent upon the other vectors in the model. The interaction term is a function of these variables, but it is not a linear function, hence it is linearly independent. There are four linearly independent vectors in Model 9 and three in Model 10, hence the test of significance for this comparison is: $F_{(4-3, N-4)}$. (The denominator degrees of freedom is incor-rectly calculated by Saunders (1956).)

The product of two continuous variables has been labeled in the literature as a "moderator variable" (Saunders, 1956). Since the multiplication of two continuous variables is simply an ex-tension of the multiplication of two categorical variables (al-ready referred to in the literature as "interaction"), it is unfortunate that the new terminology, "moderator variable," was introduced. For heuristic purposes the single concept of inter-action should suffice, although any variable which increases predictability has a legitimate place in a model whether it has a special name or not.

Some authors (Goldberg, 1972) indicate that few interactions

Table 2. Comparison of interaction possibilities.

	Models from text	Specific		General	
		df_n	df_d	df_n	df_d
Both predictor variables categorical	1 vs. 2	6	N-12	$(r-1)(c-1)$	$N-(c*r)$
One predictor variable continuous and one predictor variable categorical	7 vs. 8	2	N-6	$(k-1)$	$N-(k*2)$
Both predictor variables continuous	9 vs. 10	1	N-4	1	N-4

N = number of subjects
r = number of categories of the one predictor variable
c = number of categories of the other independent variable
k = number of categories of the one categorical variable

between continuous variables will be found. Goldberg's work (he calls them "configural" variables) is with highly correlated test items which one would not necessarily expect to interact with one another. Others (DuCette and Wolk, 1972; Wood and Langevin, 1972) have found limited selected instances wherein a "moderator variable" increases prediction. Whether or not any variable, whether interaction or not, is necessary is an empirical question.

Table 2 indicates the advantages of treating variables as continuous, rather than artificially categorizing them. When variables are treated as continuous, more degrees of freedom are

obtained in the denominator, and concurrently fewer degrees of
freedom exist in the numerator. The extreme situation appears
when both variables are treated as continuous, resulting in only
one degree of freedom in the numerator, corresponding to the one
restriction made on the Full Model (Model 9). Consequently, if
there is statistical significance in comparing Model 9 to Model
10, one knows which restriction is "generating" the significance.
Inspection of the weight(s) for that source will then indicate if
the results are in the hypothesized direction. But when more than
one restriction is made, as in comparing Models 7 and 8, or Models 1 and 2, and significance is found, the source of this significance cannot be pinpointed. These comments are general to any
test wherein more than one restriction is made, and not solely a
problem with interaction hypotheses.

Curvilinear interaction

The careful reader will note that there are some things that
are being assumed when continuous variables are used. The assumption is that there is a systematic (in the previous example,
linear) relationship over the range of the continuous variable
(in the one case of IQ, and in the last case of IQ and Drug dosage level). One probably should act as if there is a systematic
relationship between variables--it is hard to believe that the
world is constructed of leaps and bounds and discontinuities. The
extent to which the assumption of a systematic linear relationship is not valid determines whether or not the above interaction

is appropriate. If a researcher feels that the relationship is not linear, he should not immediately turn to categorical representation of the data; he should try to express the systematic functional relationship. Perhaps the relationship is a second degree curve, or perhaps a logarithmic function. It is difficult to believe that the world was made rectilinear and it's about time that researchers quit trying to force all of their phenomena into straight-line relationships. On the other hand, blind allegiance to maximum curvilinearity indices, such as the eta coefficient is not the answer either (McNeil, 1970a).

Suppose that a second degree interaction such as that depicted in Figure 2a is suspected. If one wanted to treat drug dosage level as a categorical variable, the research hypothesis would be:

Research Hypothesis: The three Drug dosage levels may have not only different Y-intercepts and different linear slopes over the range of IQ in the prediction of the criterion, but also different second degree slopes, rather than a common second degree slope.

The Full Model for testing this Research Hypothesis would be:

Model 11: $Y_1 = a_1D_1 + a_2D_2 + a_3D_3 + b_1F_1 + b_2F_2 + b_3F_3 + c_1S_1 + c_2S_2 + c_3S_3 + E_{11}$

where: D_1 = 1 if Drug dosage level 1, 0 otherwise;
D_2 = 1 if Drug dosage level 2, 0 otherwise;
D_3 = 1 if Drug dosage level 3, 0 otherwise;
F_1 = IQ if Drug dosage level 1; 0 otherwise;
F_2 = IQ if Drug dosage level 2; 0 otherwise;
F_3 = IQ if Drug dosage level 3; 0 otherwise;
S_1 = IQ^2 if Drug dosage level 1; 0 otherwise;
S_2 = IQ^2 if Drug dosage level 2; 0 otherwise; and
S_3 = IQ^2 if Drug dosage level 3; 0 otherwise.

If one suspected that the three curves actually had a common

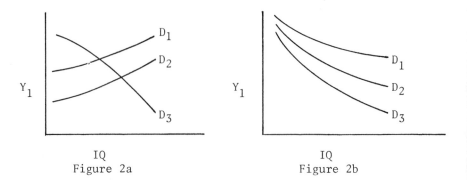

IQ
Figure 2a

IQ
Figure 2b

Figure 2. States of affairs allowing second degree interaction between Drugs and IQ (2a) and not allowing second degree interaction (2b), although
allowing a linear interaction.

second degree component, the restriction on the above Full Model

would be: $c_1 = c_2 = c_3$, resulting in a common second degree

slope of say, c_4, in the Restricted Model:

Model 12: $Y_1 = a_1 D_1 + a_2 D_2 + a_3 D_3 + b_1 F_1 + b_2 F_2 + b_3 F_3 + c_4 S_4$

$$+ E_{12}$$

where: S_4 is the squared elements of the IQ variable for
all subjects.

Model 12 would allow three separate Y-intercepts (a_1, a_2,

a_3), three separate linear slopes (b_1, b_2, and b_3), but a single

common second degree slope (c_4).

One could consider Drug dosage levels as a continuous vari-

able and still proceed to investigate a curvilinear relationship,

such as in Model 13.

Model 13: $Y_1 = a_0 U + a_4 D_4 + b_5 S_4 + b_6 (D_4 * S_4) + E_{13}$

Perhaps the interaction term is significant, and the data may even indicate that the interaction term in Model 13 is the only necessary predictor variable, resulting in a highly predictive model (R^2 close to 1.00), and simultaneously very parsimonious (only one predictor variable):

Model 14: $Y_1 = b_6(D_4 {}^*S_4) + E_{14}$

Initial reactions to the above model might be that it is quite complex, or that researchers would never have occasion to find this kind of fit to real world data. One should note that two of the most widely known "laws" of science are of just this nature ($D = 1/2 \ GT^2$ and $E = MC^2$). McNeil (1970b) gave a more complete discussion of how the former law (Newton's law) could have been developed utilizing multiple linear regression (also appearing at the end of Chapter Seven of this text).

Partial interaction

Perhaps this section should be referred to as "where the interaction is." What is being suggested here is to consider models which allow interaction to occur only in those parts of the design where interaction is suspected to be occurring. Often a researcher will suspect interaction to be occurring in only one segment of his design, and rather than testing this specific question, he will unfortunately test the overall, omnibus interaction question. Andrews, Morgan, and Sonquist (1967) present some notions which are supportive of investigating interaction in specific aspects of the design. Figure 3 depicts an extreme case

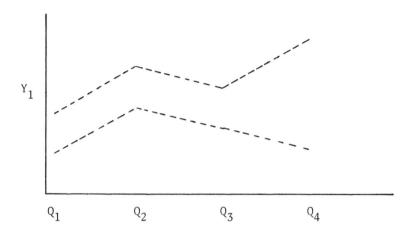

Figure 3. Interaction occurring at only one level of IQ.

wherein the distance between the levels of the one independent
variable at one point (Q_4) is quite different than at the other
points of the other independent variable. Some persons might
object to saying that this is where the interaction is, but it
is clear that what is happening at Q_4 is different than what is
happening at Q_1, Q_2, and Q_3. If the omnibus interaction question
had been calculated, the exact source of significance would not
have been ascertained, nor could a directional interpretation
have been made. But if interaction is tested by equating the
difference at Q_4 with the average of the differences at Q_1, Q_2,
and Q_3, then since only one restriction is being made, an unambiguous interpretation can be made on a significant $\underline{F}$. (Again,
unambiguous only if the <u>direction of the interaction</u> had been
specified before analysis.)

Directional interaction

The notions of directional hypothesis testing are as applicable to interaction hypotheses as to any other kind of hypothesis. In the previous example the researcher expected the difference at Q_4 to be different than at the average of Q_1, Q_2, and Q_3. More specifically, the researcher expected the superiority of Drug 1 over Drug 2 at Q_4 to be _greater_ than at the average of Q_1, Q_2, and Q_3. This specific expectation calls for a directional test of significance. For both the directional and nondirectional hypotheses the full model, restrictions, and restricted model are the same.

The Full Model for testing the interaction discussed above is:

Model 15: $Y_1 = a_1C_1 + a_2C_2 + a_3C_3 + a_4C_4 + a_5C_5 + a_6C_6 + a_7C_7$

$$+ a_8C_8 + E_{15}$$

where: Y_1 = the criterion variable;
C_1 = 1 if Drug dosage level 1 and IQ level 1,
0 otherwise;
C_2 = 1 if Drug dosage level 2 and IQ level 1,
0 otherwise;
C_3 = 1 if Drug dosage level 1 and IQ level 2,
0 otherwise;
C_4 = 1 if Drug dosage level 2 and IQ level 2,
0 otherwise;
C_5 = 1 if Drug dosage level 1 and IQ level 3,
0 otherwise;
C_6 = 1 if Drug dosage level 2 and IQ level 3,
0 otherwise;
C_7 = 1 if Drug dosage level 1 and IQ level 4,
0 otherwise; and
C_8 = 1 if Drug dosage level 2 and IQ level 4,
0 otherwise.

The restriction called for in the Research Hypothesis is:

$$(a_7-a_8) = [(a_1-a_2) + (a_3-a_4) + (a_5-a_6)] / 3$$

Imposing the above restriction on the Full Model results in an extremely complicated Restricted Model, which will not be presented. The directional research hypothesis, of course, demands that $(a_7 - a_8)$ be greater than the quantity on the right-hand side of the restriction.

If a researcher is going to treat interaction as an interesting phenomenon in its own right, then he should expect the interaction to be occurring in a certain specified way. A quick search through any content area will indicate that many researchers have expectations about how the interaction will occur. Since the probabilities recorded in most statistics books, and outputted on most computer programs for the F test are two-tailed probabilities, adjustments on the probability are called for if a directional research hypothesis is tested. McNeil and Beggs (1971) discuss these notions in more detail, and suggest that dividing the probability by two is probably still only a conservative estimate of the actual probability for the directional interaction hypothesis.

Interaction terms as covariates

Any variable may be used as a covariate, as long as that variable is not influenced by the treatments. If one considers interaction as a phenomenon in its own right, then interaction terms could be used as covariates. Some rationale should be used for including this covariate, otherwise many interactions (second degree, third degree, etc.) might be included which would drain

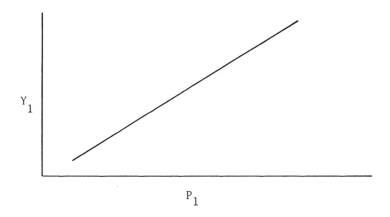

Figure 4. A single straight line, fit by Model 16:
$$Y_1 = a_0U + a_1P_1 + E_{16}$$

the degrees of freedom and spuriously generate non significance.
The regression models for the covariance analog are presented by
Kelly et al. (1969, p. 221-228). Basically, the covariates appear in both the Full and Restricted Models, with the treatment
vectors appearing in only the Full Model.

Fitting three-dimensional models

Normally when a researcher uses a continuous variable as a
predictor, he uses the originally scaled values of that predictor
in a model such as Model 16. This model is actually expressing
the linear correlation between Y_1 and P_1.

Model 16: $Y_1 = a_0U + a_1P_1 + E_{16}$

This relationship is shown in Figure 4 as a straight line.

When multiple predictors are used to predict the criterion,
most researchers still choose to use the originally scaled vari-

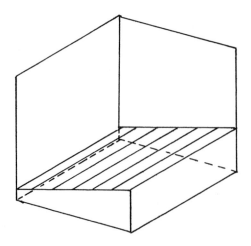

Figure 5. A plane not varying over Q_1, fit by Model 16:
$Y_1 = a_0U + a_1P_1 + E_{16}$

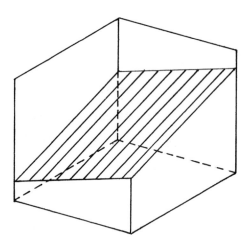

Figure 6. A plane varying over P_1 and Q_1, fit by Model 17:
$Y_1 = a_0U + a_1P_1 + a_2Q_1 + E_{17}$

ables. With more than one predictor, the graphic representation
of the relationship between the criterion and predictors may require more than two dimensions. Several such situations are discussed in this section. It must be remembered that the research
question dictates the model and graphic representation of the
model.

Model 16 above was represented in Figure 4 as a two-dimen-
sional straight line. It could also be represented in three dimensions. Suppose the researcher has knowledge of another variable, Q_1, for all subjects. If the relationship between Y_1 and P_1
is expected to be the same for all values of Q_1, then the model
to express this relationship is still Model 16; but to show the
constancy of the relationship over values of Q_1, Figure 5 would
be used. The criterion, Y_1, is on the vertical axis, and the
values for P_1 and Q_1 are along the sides of the cube. The predicted scores on Y_1 form a flat plane defined by the values of
Y_1, P_1, and Q_1.

Suppose, however, that the relationship between Y_1 and P_1 is
not expected to be constant across values of Q_1. Then Q_1 would
also have to be included in the model as a predictor. Model 17
shows this state of affairs.

Model 17: $Y_1 = a_0 U + a_1 P_1 + a_2 Q_1 + E_{17}$

Figure 6 shows that this model allows the predicted Y_1 scores to
form a flat plane which is tilted such that the criterion Y_1
value for a given P_1 value is different for each Q_1 value. (This
can also be seen by noting that at each corner of the cube the

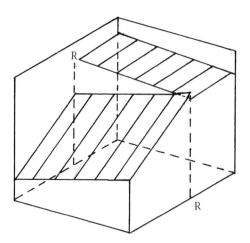

Figure 7. One plane below R on Q_1 and another plane above
R, fit by Model 18: $Y_1 = a_0U + a_1U_1 + a_2(P_1*U_1)$
$+ a_3(Q_1*U_1) + a_4U_2 + a_5(P_1*U_2) + a_6(Q_1*U_2) + E_{18}$

plane has a different value on Y_1.)

As was discussed in a previous section of this paper, the
researcher may have reason not to want to use the original values
of the predictors. He may, for example, expect a particular relationship (flat plane) to exist for values below a particular
point (R) on the Q_1 variable and expect another relationship
(flat plane) to exist for values at and above R on Q_1. Model 18
would allow this expectation.

Model 18: $Y_1 = a_0U + a_1U_1 + a_2(P_1*U_1) + a_3(Q_1*U_1) + a_4U_2 +$

$a_5(P_1*U_2) + a_6(Q_1*U_2) + E_{18}$

where: $U_1 = 1$ if the score on Q_1 is below the value of R;
        and
        $U_2 = 1$ if the score on Q_1 is at or above R.

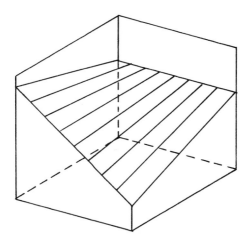

Figure 8. A twisted plane depicting "interaction" between
two continuous predictors, fit by Model 19:
$$Y_1 = a_0U + a_1P_1 + a_2Q_1 + a_3(P_1 * Q_1) + E_{19}$$

Model 18 allows for two planes, one for Q_1 values below R
and another flat plane for Q_1 values at and above R. These two
planes would not have to intersect at R as depicted in Figure 7,
although if the data were systematic, the intersection would
likely be at R.

If the researcher expects that an interaction exists between P_1 and Q_1 as they relate to Y_1, he will wish to use a model
such as Model 19.

Model 19: $Y_1 = a_0U + a_1P_1 + a_2Q_1 + a_3(P_1 * Q_1) + E_{19}$

This model is not represented by a flat plane or combination of
flat planes, but instead it forms a twisted plane of the type
pictured in Figure 8. Each edge of the plane is straight and each

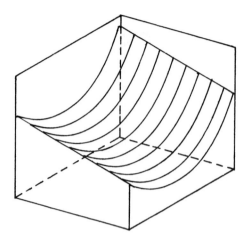

Figure 9. A plane curved on one axis, fit by Model 20:
$$Y_1 = a_0U + a_1P_1 + a_2Q_1 + a_3Q_1^2 + E_{20}$$

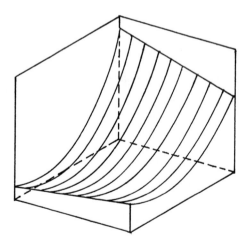

Figure 10. A twisted plane curved on one axis, fit by
Model 21: $Y_1 = a_0U + a_1P_1 + a_2Q_1 + a_3(P_1{*}Q_1)$
$+ a_4Q_1^2 + E_{21}$

line drawn across the values of Q_1 is straight, yet the plane is
twisted to allow it to intersect the four corners of the cube at
different values than could be possible if the plane were flat.

The last two models and figures to be discussed in this
section deal with those instances in which the researcher expects
a second-degree relationship between one of his predictors and
the criterion. Model 20 allows for a first- and a second-degree
relationship between Q_1 and Y_1 while the relationship between P_1
and Y_1 is linear.

Model 20: $Y_1 = a_0U + a_1P_1 + a_2Q_1 + a_3Q_1^2 + E_{20}$

This relationship is pictured in Figure 9; the predicted Y_1
scores form a curved plane, with curved edges along the Q_1 sides
and straight edges along the P_1 sides.

Model 21 adds to the above model an interaction between P_1
and Q_1.

Model 21: $Y_1 = a_0U + a_1P_1 + a_2Q_1 + a_3(P_1*Q_1) + a_4Q_1^2 + E_{21}$

The graphic representation of Model 21 is shown in Figure 10. The
curved plane formed by the predicted criterion scores has curved
edges along the Q_1 sides and straight lines with differing slopes
along the P_1 sides. Actual data for a model like Model 21 is reported in the literature by Walberg (1971) along with a detailed
figure of the data--although the figure is incorrect since the
figure does not depict one set of opposite sides being curved and
the other set of opposite sides being straight.

The above models and figures are only examples of the infinite types of relationships the researcher may expect to find

and are presented as a means of helping the researcher to visu-
alize those relationships. It must again be emphasized, though,
that being able to draw a figure of a model is not a necessary
step. If a model yields a high degree of predictability, that is
sufficient. Indeed, figures in more than three dimensions are a
little difficult to communicate, let alone comprehend.

Interpretation of complex interactions vs. predictability

Whenever one finds a significant interaction, one should try
to graph the interaction and try to interpret it. If one has hypothesized a directional interaction before the fact, then the
interpretation of the significant interaction should not be difficult. If one did not suspect interaction, then a replication
study should substantiate that interaction before one starts to
have "faith" that the interaction exists. (Actually, this goes
not only for interaction; any unhypothesized result needs replicating before being accepted.) One may continually find an interaction term which accounts for a large proportion of the criterion variance. That is, the goal of predictability has been satisfied in achieving a large R^2. But the why of this particular interaction may not be explainable in the point in time when it
was found. To throw away this high degree of predictability because of a failure to comprehend seems a little bit egotistical.
On the other hand, obtaining premature closure on interpretations
of data may be one of the most drastic mistakes researchers make.
One could argue that insightful interpretations are made as a

function of the data telling us something new, rather than as a function of our preconceived notions twisting the data so that it makes "sense" to us.

References

Andrews, F.; Morgan, J.; and Sonquist, J. Multiple Classification Analysis. Institute for Social Research, The University of Michigan, Ann Arbor, Michigan, 1967.

Bottenberg, R.A. and Ward, J.H. Applied Multiple Linear Regression. Aerospace Medical Division, Lackland Air Force Base, Texas, AD 413128, 1963.

Glass, G.V. and Stanley, J.C. Statistical Methods in Education and Psychology. Englewood Cliffs, N.J.: Prentice-Hall, 1970.

Jennings, E.E. Fixed effects analysis of variance by regression analysis. Multivariate Behavioral Research, 1967, 2, 95-108.

Kelly, F.J.; Beggs, D.L.; McNeil, K.A.; Eichelberger, T.; and Lyon, J. Research Design in the Behavioral Sciences: Multiple Regression Approach. Carbondale, Illinois: Southern Illinois University Press, 1969.

McNeil, K.A. Meeting the goals of research with multiple regression analysis. Multivariate Behavioral Research, 1970, 5, 375-386.

McNeil, K.A. The negative aspects of the eta coefficients as an index of curvilinearity. Multiple Linear Regression Viewpoints, 1970, 1, 7-17.

McNeil, K.A. Some notions about the use of statistical hypothesis testing. Paper presented at Southwest Psychological Association, May, 1971.

McNeil, K.A. and Beggs, D.L. Directional hypotheses with the multiple linear regression approach. Paper presented at American Educational Research Association, February, 1971.

Saunders, D.R. Moderator variables in prediction. Educational and Psychological Measurement, 1956, 16, 209-222.

Walberg, H.J. Generalized regression models in educational research. American Educational Research Journal, 1971, 8, 71-91.

Interacting variables--one way to view all variables

A strong case is made in the above paper for the consideration of interaction variables. All variables can be considered as
interacting variables if one remembers that any number to the
zero power is equal to one. The previous paper made the case that
any variable which is a multiplication of two or more other variables, say $X_1 X_2$, is an interaction variable. By convention, the
powered superscript is dropped when it is equal to one; there-
fore $X_1 X_2$ could also be represented as $X_1^1 X_2^1$. Likewise, $X_1^2 X_2^1$ is an
interaction variable, and $X_1^0 X_2^1$ can be considered as an interaction variable. But any number to the zero power is equal to one,
therefore $X_1^0 X_2^1$ is equal to X_2. Thus X_2 can be conceptualized as
the interaction of X_1 to the zero power and X_2 to the first
power. Similarly, U can be conceptualized as X_2^0, $X_1^0 X_2^0$, or $X_1^0 X_2^0 X_3^0$,
etc.

Mendenhall (1968, p. 94) presents the regression model allowing for interaction in a two by three analysis of variance
design as:

Model 6.9: $Y_1 = a_0 U + a_1 X_1 + a_2 X_2 + a_3 X_2^2 + a_4 X_1 X_2 + a_5 X_1 X_2^2 + E_9$

> where: X_1 = two distinct scores representing the two levels
> of the one independent variable, and
> X_2 = three distinct scores representing the three
> levels of the second independent variable.

Model 6.9 can be conceptualized as interaction variables as
in Model 6.10:

Model 6.10: $Y_1 = a_0 (X_1^0 X_2^0) + a_1 (X_1^1 X_2^0) + a_2 (X_1^0 X_2^1) + a_3 (X_1^0 X_2^2) +$
$a_4 (X_1^1 X_2^1) + a_5 (X_1^1 X_2^2) + E_{10}$

Each term in the above model is one form of interaction between one independent variable and the other independent variable.

One could extend this notion further by conceptualizing polynomial terms as interactions. For instance, X_2^2 is equivalent to $X_2^1 X_2^1$. That is to say, the second degree term is an interaction between the two linear terms. (Jack Byrne, in a personal communication, first brought this to our attention.) We don't view this as a necessary way to conceptualize all variables, but there might be some value in it for some readers.

Differences between groups, statistically controlling for possible confounding effects (Analysis of Covariance)

Analysis of covariance was discussed in detail in Chapter Five. It will now be treated here, emphasizing dichotomous variables.

A typical covariance kind of research hypothesis would be: "The experimental method is higher on the criterion Y_1 than the control method, after the effects of the covariable X_1 have been taken into consideration." The Full Model to test this Research Hypothesis would be:

Model 6.11: $Y_1 = a_0U + a_1G_1 + a_2G_2 + a_3X_1 + E_{11}$

and the restriction $a_1 = a_2$ would generate the following Restricted Model:

Model 6.12: $Y_1 = a_0U + a_3X_1 + E_{12}$

where: Y_1 = the criterion;
         U = the unit vector containing a one for each subject;

G_1 = 1 if criterion from experimental group, zero otherwise;

G_2 = 1 if criterion from control group, zero otherwise; and

X_1 = score on the suspected confounding variable.

If the $\underline{F}$ test comparison of these two models yields a directional probability level less than the value of alpha, and if the weight for the experimental group (a_1 in Model 6.11) is larger than the weight for the control group (a_2 in Model 6.11), then it can be concluded that the experimental method is higher on the criterion than the control method, over and above any effects accountable to initial differences on X_1.

It should be noted that a non-directional research hypothesis, although less desirable than a directional one, is usually discussed in the statistical literature, and consequently usually used by researchers. Suppose the following non-directional research hypothesis had been made: "The experimental method is different on the criterion than the control method, after the effects of the covariable X_1 have been taken into consideration." The Full and Restricted Models would be the same as for the directional research hypothesis. If the $\underline{F}$ test comparison of Models 6.11 and 6.12 yields a non-directional probability level less than alpha, then it can be concluded that the two methods are different on the criterion, over and above any effects accountable to initial differences on the covariable.

Chapter Five discussed in detail the covariance models when one covariable was being considered. It should be noted that the multiple regression procedure readily allows one to take into consideration more than one covariate. All of the ways in which

the groups are thought to be different at the beginning of the

experiment, whether on continuous or dichotomous variables, can

be included as predictor variables in both the full and restrict-

ed models. The only additional information appearing in the full

model but not in the restricted model is group membership infor-

mation. Therefore, the increase in R^2 from the restricted model

to the full model has to be due to the differences between

groups, over and above how their differences on the covariates

influenced the criterion scores. One could consider more than two

groups by simply adding in other group membership vectors into

the full model.

One could also hypothesize an interaction (for instance,

between teaching method and level of induced anxiety) on math

achievement, statistically controlling for the effects of a num-

ber of covariables (for instance, age, IQ, number of previous

courses, and mathematics aptitude). Suppose there are two teach-

ing methods to be used in this study, and each is represented in

the full model by a dichotomous vector (G_1 and G_2). Suppose, in

addition, that level of induced anxiety is to be measured in

terms of an arbitrary division between high and low anxiety; so

induced anxiety is represented by dichotomous vectors A_1 and A_2.

The interaction between teaching method and induced anxiety would

yield four dichotomous vectors. The covariables in this case are

measured by continuous vectors. The Research Hypothesis is:

"There is an interaction between Teaching Method and level of induced anxiety on the criterion of Math Achievement, statisti-

cally controlling for the effects of Age, IQ, Number of previous

courses, and Mathematics aptitude." The Full Model would be:

Model 6.13: $Y_1 = a_0U + a_1(G_1*A_1) + a_2(G_1*A_2) + a_3(G_2*A_1) +$

$a_4(G_2*A_2) + a_5X_1 + a_6X_2 + a_7X_3 + a_8X_4 + E_{13}$

where: Y_1 = Mathematics Achievement;
G_1 = 1 if criterion from subject in Teaching
method one, zero otherwise;
G_2 = 1 if criterion from subject in Teaching
method two, zero otherwise;
A_1 = 1 if High level of induced anxiety, zero
otherwise;
A_2 = 1 if Low level of induced anxiety, zero
otherwise;
X_1 = Age;
X_2 = IQ;
X_3 = Number of previous courses; and
X_4 = Mathematics aptitude.

If there is no interaction (statistical hypothesis) between

Teaching method and Level of induced anxiety over the above the

effects of the covariables, then the following restriction must

be true: $(a_1-a_2) = (a_3-a_4)$. Forcing this restriction onto the

Full Model results in a Restricted Model containing all covari-

ables as well as the "main effect" of Teaching method and Level

of induced anxiety:

Model 6.14: $Y_1 = a_0U + b_1G_1 + b_2G_2 + b_3A_1 + b_4A_2 + b_5X_1 + b_6X_2$

$+ b_7X_3 + b_8X_4 + E_{14}$

Determining the number of linearly independent vectors in

models such as Models 6.13 and 6.14 requires some detailed at-

tention to the nature of the vectors. In Model 6.13, each of the

covariate vectors $(X_1, X_2, X_3,$ and $X_4)$ are linearly independent;

there are four interaction vectors--one of which is linearly dependent upon the other three plus the unit vector. So for Model 6.13, one would count as good pieces of information: the four covariates, three of the four interaction vectors, and the unit vector ($m_1 = 8$). In Model 6.14, the four covariables are linearly independent, one of the two Teaching method vectors and one of the two Levels of induced anxiety vectors are linearly independent, as well as the unit vector. There are eight pieces of good information in the Full Model, and only seven in the Restricted Model; hence $df_n = (8-7) = 1$, and $df_d = (N-8)$.

The test for interaction for what some traditional analysis of variance people would call a two-factor analysis of covariance with four covariates has been briefly shown here. One would have to look long and hard in traditional analysis of covariance literature to find another discussion, let alone a computational formula, for such a research hypothesis. Yet it is possible that such a research hypothesis may be quite meaningful to some researchers.

Additional covariables do drain the denominator degrees of freedom. For example, if three more covariables had been considered in the above interaction question, the numerator degrees of freedom would still have been 1, but the denominator degrees of freedom would have been N-11 instead of N-8. If one has a lot of subjects, the inclusion of possible confounding variables is a recommended procedure. Whether a particular variable is a confounding variable is a question which can be answered empirically rather than something that can necessarily be ascertained theoretically. And as discussed in a previous section, one should not ignore interaction variables as possible confounding variables. An interaction variable can be thought of as just another vector. As long as the variables being interacted have been measured before the application of the treatments and hence are not influenced by the treatments, the interaction vector can be legitimately included as a covariate.

The notions in this section lead to Generalized Research Hypothesis 6.4. GRH 6.4 could be directional if only one restriction were made (e.g., Group 1 hypothesized to be better than Group 2, or some kind of comparison such as the average effect of Groups 1, 2, and 3 hypothesized to be better than the average effect of Groups 4, 5, 6, 7, and 8).

Dichotomous Criterion Variables

None of the models discussed in this text are restricted to using continuous variables as the criterion variable. The criterion variable may be identifying membership in two groups. The multiple linear regression procedure is limited to a single criterion variable, and hence a criterion specifying two groups can be analyzed. The criterion would consist of two different numbers, each identifying membership in one of the two groups. (The most logical choice would be a dichotomous vector of ones and zeroes, but any two numbers would work.)

If more than two groups were under consideration, one could posit the ordering of those k groups in a single criterion vec-

Generalized Research Hypothesis 6.4

Directional Research Hypothesis: (See text.)

Non-directional Research Hypothesis: k groups are not all the same on the criterion Y_1, after the effects of m possibly confounding variables are taken into consideration.

Statistical Hypothesis: k groups are all the same on the criterion Y_1, after the effects of m possibly confounding variables are taken into consideration.

Full Model: $Y_1 = a_0U + a_1X_1 + a_2X_2 \ldots + a_kX_k + b_1Z_1 + b_2Z_2$

$\ldots + b_mZ_m + E_1$

Restrictions: $a_1 = a_2 \ldots = a_k$

Restricted Model: $Y_1 = a_0U + b_1Z_1 + b_2Z_2 \ldots + b_mZ_m + E_2$

where:

Y_1 = criterion;
X_1 = 1 if criterion from subject in Group 1, 0 otherwise;
X_2 = 1 if criterion from subject in Group 2, 0 otherwise;
.
.
.
X_k = 1 if criterion from subject in Group k, 0 otherwise;
Z_1 = score on the confounding variable Z_1;
Z_2 = score on the confounding variable Z_2;
.
.
.
Z_m = score on the confounding variable Z_m.

Degrees of freedom numerator = $[(k+m)-(m+1)] = (k-1)$

Degrees of freedom denominator = $[N-(k+m)] = N-k-m$

where: N = number of subjects
         k = number of groups
         m = number of (continuous) covariables

Note: Dichotomous variables can be used as covariates, and their use will affect calculation of degrees of freedom. The concept of linearly independent vectors still holds, though.

tor. Most often, researchers are not able to be that specific. In the case that the ordering is not known, the k groups would have to be specified with k group membership vectors. Since the MLR approach can analyze only one criterion at a time, the technique of discriminant analysis would have to be utilized for analyzing the multiple criterion vectors (see Veldman (1967)). The reader should be aware, though, that the tests of significance that presently exist for discriminant analysis are treated within a non-directional framework.

Because of numerous applications of certain predictor con-figurations, some models have achieved the notoriety of special names. Those special cases will now be discussed, not in hopes of reinforcing those special names, but instead, of establishing in a single source the underlying similarity of all the least squares techniques.

A single continuous predictor predicting a dichotomous criterion

A specific Research Hypothesis might be: "IQ is predictive of whether the student will Graduate or not." Some researchers would turn to the "t test" for the difference between two means (Graduating versus Non-graduating) to test the Research Hypoth-esis. Other researchers would compute a "point-biserial correla-tion coefficient" to test the Research Hypothesis. Most statis-tics texts present the "t test" and the "point-biserial correla-tion coefficient" in two very different sections. Some statistics authors hint at the similarity, but the two techniques are ex-

---

Generalized Research Hypothesis 6.5

Directional Research Hypothesis: The continuous variable X_1 is positively (or negatively) correlated with the dichotomous variable Y_1.

Non-directional Research Hypothesis: The continuous variable X_1 is correlated with the dichotomous variable Y_1.

Statistical Hypothesis: The continuous variable X_1 is not correlated with the dichotomous variable Y_1.

Full Model: $Y_1 = a_0U + a_1X_1 + E_1$

Restrictions: $a_1 = 0$

Restricted Model: $Y_1 = a_0U + E_2$

  where:

  Y_1 = dichotomous criterion; and
  X_1 = continuous predictor.

Degrees of freedom numerator = (m_1-m_2) = $(2-1)$ = 1

Degrees of freedom denominator = $(N-m_1)$ = N-2

  where:

  N = number of subjects

---

actly equivalent in terms of the resultant probability values (Kelly et al., 1969).

The regression formulation is presented in its generalized form as Generalized Research Hypothesis 6.5. The careful reader will note the similarity in this hypothesis and the one in GRH 6.1. The results will be the same. The only difference is in the

choice of either the continuous variable or the dichotomous variable as the criterion. And that choice is made by the researcher,
not on statistical grounds, but depending upon the variable that
the researcher wants to consider as his criterion variable. Often
the time at which the variables are observed will play a crucial
part in the decision, as variables obtained later in time are
usually thought to be results of previously measured causes. On
other occasions, the researcher will have to rely upon his nomological net to guide his decision. Most correlational studies do
not make the cause-and-effect distinction. The correlational model does not preclude the cause-and-effect inference, but neither
does it mandate such an inference. The cause-and-effect relationship must be theoretically appealing and it must undergo empirical verification--actual manipulation of the causative factor to
ascertain the change in the criterion. Empirical manipulation,
not statistical significance, is the ultimate test of a cause-
and-effect relationship. Statistical significance is a necessary,
but not sufficient condition for causality.

A single dichotomous predictor predicting a dichotomous criterion

Once the viability of predicting a dichotomous criterion
from a continuous predictor has been established as in the previous section, it is no great intellectual leap to develop regression models which predict a dichotomous criterion from a
dichotomous predictor. Persons influenced by correlational thinking would compute a "phi coefficient" on this kind of hypothesis,

---

Generalized Research Hypothesis 6.6

Non-directional Research Hypothesis: The dichotomous variable D_2 is correlated with the dichotomous variable D_1.

Directional Research Hypothesis: The dichotomous variable D_2 is positively (or negatively) correlated with the dichotomous variable D_1.

Statistical Hypothesis: The dichotomous variable D_2 is uncorrelated with the dichotomous variable D_1.

Full Model: $D_1 = a_0 U + a_1 D_2 + E_1$

Restrictions: $a_1 = 0$

Restricted Model: $D_1 = a_0 U + E_2$

where:

D_1 = dichotomous criterion; and
D_2 = a dichotomous variable.

Degrees of freedom numerator = $(m_1 - m_2) = (2-1) = 1$

Degrees of freedom denominator = $(N - m_1) = N-2$

where:

N = number of subjects

---

while analysis of variance-trained persons would compute a "chi square." Neither of these two special terms are needed, as they are both computational simplifications of the general least squares procedure. A research hypothesis of the nature of GRH 6.6 will yield a probability value exactly the same as the test for the phi coefficient, and more exact than that for the chi square analysis (McNeil, 1974).

The R^2 of the Full Model in GRH 6.6 is equal to (phi)^2 and is also equal to X^2/N. The multiple linear regression approach to hypothesis testing is a better method for at least four reasons:

(1) The multiple linear regression approach forces the researcher to state the research hypothesis. Unfortunately, this is not always done in chi square analyses and in phi coefficient analyses.

(2) The multiple linear regression approach is easily generalized to all other least squares hypotheses. Separate computing formulas and different rules for calculating degrees of freedom are not necessary.

(3) The stating and testing of directional hypotheses is encouraged by the multiple linear regression approach, whereas directional chi square analyses and directional phi coefficient analyses are at best only briefly mentioned in statistics books.

(4) The multiple linear regression approach considers the number of subjects and the number of categories in the predictor variable in the calculation of the denominator degrees of freedom. Since the chi square test of significance assumes a large number of subjects, the probability statement is more exact when it is calculated from the F test resulting from the multiple linear regression approach.

Many predictors predicting a dichotomous criterion

One could, of course, use a number of dichotomous predictor

variables to predict a single dichotomous criterion. Or one could use a number of dichotomous predictors and a number of continuous predictors to predict a single dichotomous criterion. If the dichotomous criterion represents two kinds of subjects, say male and female, then one could posit the following Research Hypothesis: "Males can be discriminated from Females on the basis of Handedness, Hair color (Blond vs. Non-blond), IQ, Anxiety, and Motivation." Notice that the first two variables (Handedness and Hair color) could be considered dichotomous, and the last three (IQ, Anxiety, and Motivation) could be considered continuous. One could also think of this analysis as comparing the profiles of Males to the profiles of Females. Again, if more than two groups are going to be compared or discriminated, one must turn to discriminant analysis, as discussed and computerized by Veldman (1967).

There is no requirement that the number of ones be the same as the number of zeros in the criterion. Furthermore, there is no requirement to have the same proportion of ones in the dichotomous predictors as in the criterion.

Dichotomous vs. continuous variables

There is a lot of literature expounding the limits of predicting a dichotomous criterion, especially when the predictor variable approximates a normal distribution. There is indeed a limit to the predictability when one uses a continuous predictor to predict a dichotomous criterion. Given that a dichotomous cri-

terion is to be predicted, it might make more sense to search
among the dichotomous predictors, or to artificially dichotomize
some continuous predictor, rather than limiting predictability by
using a continuous predictor. With reference to the correlations
in Table 6.2, one of the "real dichotomies," Sex, accounts for
more variance in the criterion than does the continuous IQ vector
[i.e., $r_{(D_1 \text{ vs Sex})} = .75$; $r_{(D_1 \text{ vs IQ})} = .65$]. But the artificially dichotomized variable of IQ (X_4) correlates perfectly with
the dichotomous criterion [$r_{(D_1 \text{ vs IQ-dichotomized})} = 1.00$]. This
is not to lead the reader to expect that in the prediction of a
dichotomous criterion he can always artificially dichotomize a
variable and end up with perfect predictability. Furthermore,
since the cutting point for the artificial dichotomization is entirely arbitrary, and dependent upon the sample data, the need
for replication should be quite obvious.

Whether a dichotomous variable or continuous variable
should be used will always be a function of the data. The statistician cannot have insight into content-specific data to determine the kinds of variables to be used. The decision is based
upon the end product--the highest R^2. In Figure 6.2 are two sets
of data, both admittedly extremely systematic to emphasize the
point. The dichotomous prediction model yields a perfect fit for
the data at the left of Figure 6.2, whereas the continuous prediction model yields a better fit for the data at the right of
Figure 6.2

There are three reasons why continuous predictor variables

are usually preferred over dichotomous predictors. Most of the variables in the real world are continuous, which means that criterion variables will usually be continuous. And since the real world usually varies systematically (right side of Figure 6.2) rather than by leaps and bounds (left side of Figure 6.2), the continuous predictors will usually yield a higher R^2.

Secondly, each additional categorization requires an additional dichotomous vector in the model. When degrees of freedom denominator are calculated, one less degree of freedom exists for each categorization. If a lot of subjects are available, this does not become much of a problem. But if an extremely large number of categories are identified (the limit being a category for each score on the X_1 variable--the "eta coefficient" model), the drain on the degrees of freedom may be excessive. Furthermore, the replicability of the weighting coefficients for each category becomes more unlikely with each additional category.

Thirdly, if a researcher has a continuous variable, he can always dichotomize it if he finds that prediction with the continuous variable does not yield a high enough R^2. Recapturing the continuous scores from dichotomous vectors is not always an easy process, and indeed sometimes is impossible. If a variable has originally been collected as continuous, then the researcher probably thought it to be systematically related to the criterion. It remains the task of the researcher to discover that systematic relationship. More discussion on finding systematic relationships will be presented in Chapter Seven.

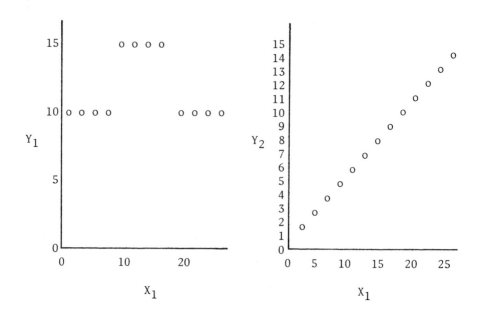

$$R^2 = 1.00 \qquad Y_1 = a_1D_1 + a_2D_2 + a_3D_3 + E_1$$

$$R^2 = .89 \qquad Y_2 = a_1D_1 + a_2D_2 + a_3D_3 + E_2$$

$$R^2 = .00 \qquad Y_1 = a_0U + a_1X_1 + E_3$$

$$R^2 = 1.00 \qquad Y_2 = a_0U + a_1X_1 + E_4$$

where: Y_1 = criterion in left-hand figure;
Y_2 = criterion in right-hand figure;
X_1 = a continuous variable;
D_1 = 1 if X_1 is less than 10, zero otherwise;
D_2 = 1 if X_1 is between 10 and 20, zero otherwise; and
D_3 = 1 if X_1 is greater than 20, zero otherwise.

Figure 6.2. Two sets of data, the left-hand side requiring dichotomous predictors for perfect prediction, the right-hand side requiring a continuous predictor for perfect prediction.

Table 6.2 Raw data and correlations indicating the value of
artificially dichotomizing data.

Criterion (D_1)	IQ (X_1)	Male (X_2)	Handedness (X_3)	IQ (X_4) 1 if > 90 0 if ≤ 90
1	120	1	1	1
1	110	1	1	1
1	91	1	1	1
1	91	0	0	1
0	90	0	1	0
0	85	0	0	0
0	83	0	0	0

	D_1	X_1	X_2	X_3	X_4
D_1	1.00	.65	.75	.41	1.00
X_1		1.00	.76	.63	.65
X_2			1.00	.75	.75
X_3				1.00	.41
X_4					1.00

7
Non-linear Relations

Some researchers assume linear regression analysis deals only with straight lines. This chapter illustrates some of the non-rectilinear (also referred to as non-linear or curvilinear) forms that one may encounter in his research and analyze with multiple linear regression. Observed non-linear sample data may be due to at least two conditions. (1) The theoretic expectation is reasonable regarding a non-linear function existing among the predictor(s) and the criterion (e.g., the curvilinear relationship in Newton's Law, $d=1/2gt^2$). (2) The scaling of the X and Y variables is arbitrary, so departures from rectilinearity may be a scaling artifact. The two major sections of this chapter deal with these two conditions.

Before entering into these two sections, a discussion regarding homoscedasticity and heteroscedasticity is provided. Some readers may wish to skim this section and get on to curviness, but one should at least skim the material because an understanding of these concepts may be needed to communicate with statisticians. This discussion treats the violations of assumptions as the basis for good information for use in prediction rather than as conditions which preclude statistical analysis.

254

Homoscedasticity and Heteroscedasticity

In Chapter Two the assumption of homogeneity of variance was discussed in relation to analysis of variance. Given two samples which receive different treatments, least squares procedures assume that each sample is from a common population and hence these samples come from "populations" with equal variance. Violation of this assumption in most cases does _not_ upset the inferences made when using the F distribution. When deriving a line of best fit, homogeneity of variance is reflected by equal variance about the line of best fit for each scale point on the X axis (see Figure 2.3a). This equal variance about the line of best fit is what is called "homoscedasticity." Figure 7.1 illustrates a case where homogeneity of variance is observed in the sample, and is thus a reasonable expectation for the population.

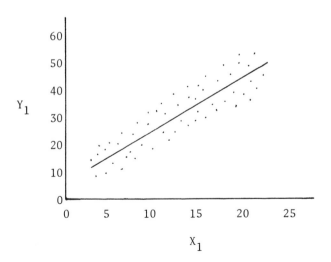

Figure 7.1. A representation of a line where the Y_1 observations are homogeneously distributed about the line for all values of X_1.

Note in Figure 7.1 that at scale point 10 on the X axis
the observed Y_1 scores are distributed in about the same way as
they are for observations at scale point 20 on the X axis. Given
such a data plot, one can assume that homoscedasticity exists.

When scattergrams (Figure 7.1 is a scattergram) depart from
the ideal and distributions are not symmetric about the hypothesized line of best fit, most statisticians will admonish the researcher to be cautious in interpretation of that line because
the observed heteroscedasticity suggests the line has different
errors across scale values on the X axis. (The plotting subroutine provided in the computer program in Appendix A can easily
provide the scattergram for inspection to ascertain whether the
data is heteroscedastic.) Figure 7.2 illustrates three cases
where heteroscedasticity exists.

If one views the scattergrams in Figure 7.2 from a straight-
line point of view, he may be upset because an assumption is violated, and the R^2 is small due to heteroscedasticity. From the
position of a researcher, systematic departures around the line
of best fit as illustrated ought to be seen as a starting point
for inquiry--it is very likely that in all three cases the
straight-line model constructed to fit the data is not appropriate. Given the systematic departures from homoscedasticity in
Figure 7.2, one might suspect that a theoretically unexpected
interaction (see Chapter Four) between the X_1 variable and some
other unknown variable(s) yields the odd distributions.

For the sake of simplicity, suppose all three cases illus-

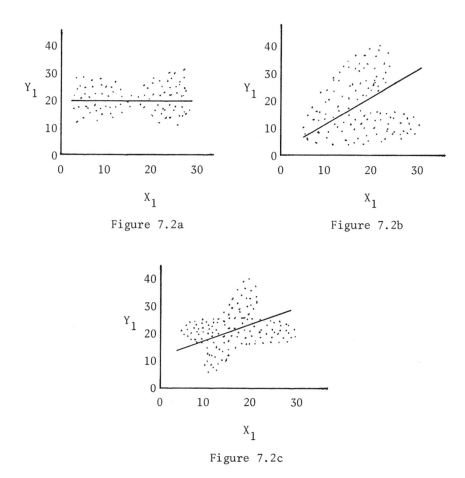

Figure 7.2a

Figure 7.2b

Figure 7.2c

Figure 7.2. Three hypothetical cases of heteroscedasticity:
a, where R^2 = 0; and b and c, where R^2>0 but
with different forms.

trated are due to what has been called a treatment-by-aptitude

interaction, and that the researcher was unaware of the treat-

ment. Consider Figure 7.2a. If one was interested in the rela-

tionship of ability (X_1) to criterion performance (Y_1), one may

ask, "Why is there so much variability at the extremes of the

ability continuum?" In the hypothetical case under consideration,
the sample may be drawn from two different classes of students
with two teachers. Can the results be due to teacher differences?
For example, perhaps Teacher A is great for students who are below average, and Teacher B is great for students who are above
average. If one ignored teacher effects, the one line in Figure
7.2a would be the outcome. If one expanded the simple linear
equation $(Y_1 = a_0U + b_1X_1 + E_1)$ to include two lines, one for
students of Teacher A and one for students of Teacher B, Figure
7.3 might be the outcome state.

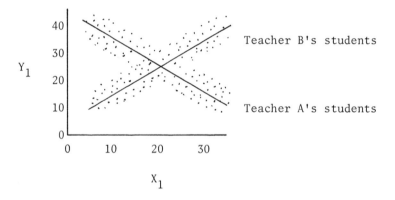

Figure 7.3. Data from Figure 7.2a where the observed variability in extremes is explained by teacher
effects. The ellipses about the two lines reflect homoscedastic data points.

As depicted in Figure 7.3, the heteroscedasticity is re-
moved because the systematic teacher effects are accounted for,
and the R^2 will be dramatically increased from zero to an R^2
substantially greater than zero.

The apparent heteroscedasticity in Figure 7.2b might be due
to an ordinal interaction (two or more lines which are not parallel, but which do not cross in the range of interest). Figure
7.2c may be due to one treatment which yields strong predictability of Y_1 from X_1 and another treatment which yields little relationship between X_1 and Y_1. Figure 7.4 illustrates these two
conditions which might account for the observed heteroscedasticity in Figures 7.2b and 7.2c.

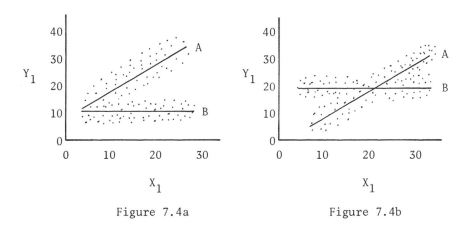

Figure 7.4a   Figure 7.4b

Figure 7.4a. Representation of Figure 7.2b where the vari-
            ability at the higher end of the X axis is re-
            vealed to be due to a systematic effect where
            Treatment A yields high scores at the upper end
            of X_1, whereas Treatment B yields very little
            effect on criterion performance across the
            total range of X_1.

Figure 7.4b. An interaction where Treatment B has no differ-
            ential effect across X_1, but Treatment A yields
            a very strong relationship between X_1 and Y_1.

So far this discussion of heteroscedasticity has dealt with
straight lines of best fit. Consider the conditions observed in

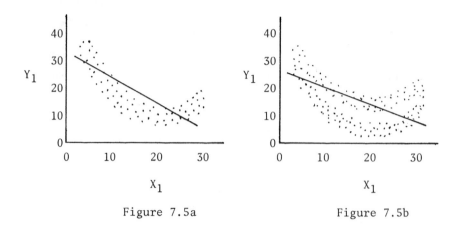

Figure 7.5a   Figure 7.5b

Figure 7.5. Scattergrams a and b illustrate hetero-
            scedasticity which suggests that a straight
            line is not the best functional relationship
            between X_1 and Y_1.

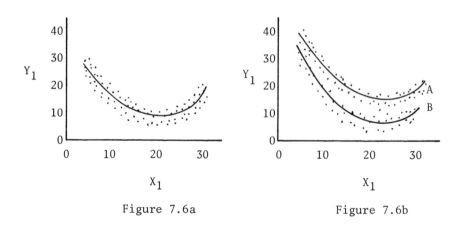

Figure 7.6a   Figure 7.6b

Figure 7.6a. The removal of unequal variability about the
             line of best fit given in Figure 7.5a by
             fitting a curved line to the data.

Figure 7.6b. The removal of unequal variability by fitting
             two curved lines.

Figures 7.5 and 7.6. The "a" portions of both Figures 7.5a and 7.5b indicate that unequal variability across X_1 might be due to a non-rectilinear relationship between X_1 and Y_1. The "b" portions of the Figures 7.6a and 7.6b illustrate a situation where Treatments A and B have two different non-rectilinear relationships between X_1 and Y_1.

Whether the curves represented in Figure 7.6 are due to a theoretic expectation or are due to poor scaling is undetermined in this discussion. The rest of this chapter deals with these two notions.

In summary, homoscedasticity is statistically desirable because the R^2 is enhanced. Furthermore, the homoscedasticity assumption implicitly assumes that the proper form of regression equation has been fitted (Winer, 1962, p. 586). When the equal variability about the line of best fit is lacking, searching for unknown variables which may account for the variability and including those variables in the analysis seems to be the best research strategy. Ideally, if one has all the relevant predictor variables, there will be no variability about the line of best fit, an R^2 of 1.0 will be observed, and this whole section would be superfluous.

Fitting Expected Non-linear Functional Relationships

The second degree relationship

In many psychomotor skill learning conditions, the theory

postulates a positively accelerating curvilinear relationship.
For example, Hours of practice and Key punch card production
might take a form such that little gain in Key punch card production is observed for the first few Hours of practice, and then a
spurt is observed. Figure 7.7 represents this expectation.

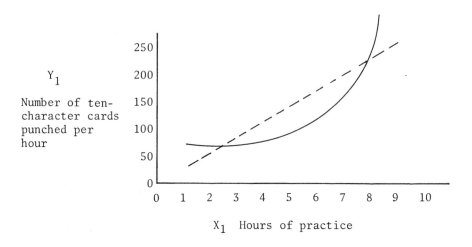

Figure 7.7. The line of best fit which represents a curved
          (solid line) and a rectilinear (dashed line)
          fit to the data.

In Figure 7.7 the solid line represents the curvilinear fit.
As shown, the straight line may do a fairly decent job of representing the curved data; but if the curved line is the best fit,
then at the extremes (1-2 and 9-10 Hours of practice) errors of
underprediction will be observed (the solid line is above the
dashed), and the middle Hours of practice will yield errors of
overprediction (the dashed line is above the line of best fit).

A linear regression equation which includes a second order

polynomial of the Hours of practice vector (X_1^2) will fit the line

in Figure 7.7 fairly well. The linear model is:

Model 7.1: $Y_1 = a_0U + b_1X_1 + s_2X_1^2 + E_1$

    where: Y_1 = Number of ten-character cards punched in an
                  hour;

              X_1 = Number of hours of practice; and

              X_1^2 = the square of the elements in vector X_1.

        [Note: In this and succeeding models, the unit
                vector (U), the weighting coefficients
                (here, a_0, b_1, and s_2) and the error vector
                (E_1) will not generally be defined.]

The vectors would look like those in Table 7.1.

Table 7.1 Illustration of the vectors representing the
              example curvilinear model with possible observed
              scores.

Person	Y_1	$=$		a_0U	$+$	b_1X_1	$+$	$s_2X_1^2$	$+$	E_1
1	50			1		1		1		?
2	51			1		2		4		?
3	75	$= a_0$		1	$+ b_1$	5	$+ s_2$	25	$+$	?
4	170			1		9		81		?
.	.			.		.		.		.
.	.			.		.		.		.
60	250			1		10		100		?

Note that in Table 7.1 the first score represented in Y_1 is

50 cards punched in an hour with one hour practice $(X_1 = 1$ and

$X_1^2 = 1)$. The 60th score in Y_1 is 250 and $X_1 = 10$; thus, 250 cards

were punched in an hour for the person with 10 hours practice

$(X_1^2 = 10*10 = 100)$.

Suppose a researcher had available 60 naive students in a card-punching class and he wanted to determine if his expectations regarding the surge in performance is a likely non-chance event. In a real situation, he may monitor each student's performance after each hour of practice and then use a repeated measures test for curvilinearity. In order to remain conceptually simple, the present example will not include the repeated measures design (Chapter Eight discusses the repeated measures design).

To test the expected curvilinear relationship, the researcher may randomly assign six students to each of the ten practice conditions (6 for one hour, 6 for two hours, etc.). The researcher would then give the practice and record the number of cards punched for each student. The example <u>Research Hypothesis</u> may be:

> For the specified population, there is a positive second degree curvilinear relationship between Hours of practice and Number of ten-character cards punched, over and above a positive rectilinear relationship.

The example <u>Statistical Hypothesis</u> is:

> For the specified population, there is <u>no</u> positive second degree curvilinear relationship between Hours of practice and Number of ten-character cards punched, over and above a positive rectilinear relationship.

Model 7.1 above $(Y_1 = a_0 U + b_1 X_1 + s_2 X_1^2 + E_1)$ is the appropriate Full Model and the vectors are as described previously. The Research Hypothesis states $s_2 > 0$ and the Statistical Hypothesis sets $s_2 = 0$; therefore, the Restricted Model is:

Model 7.2: $Y_1 = a_0 U + b_1 X_1 + E_2$

There are three linearly independent vectors in the Full

Model and two in the Restricted Model. Suppose $R_f^2 = .75$ and $R_r^2 = .60$; the same general F ratio would be used to test the research hypothesis:

$$F_{(m_1-m_2, N-m_1)} = \frac{(R_f^2 - R_r^2)/(m_1-m_2)}{(1 - R_f^2)/(N-m_1)}$$

$$F_{(1,57)} = \frac{(.75 - .60)/(3-2)}{(1 - .75)/(60-3)} = \frac{.15/1}{.25/57} = \frac{.15}{.0004} = \frac{\hat{V}_s}{\hat{V}_w} \simeq 34.1$$

An F of 7.12 with 1 and 57 degrees of freedom, when the weights are in the hypothesized direction, is observed less than 5 times in a thousand ($p<.005$). If the researcher had set his alpha at .01 and the weights b_1 and s_2 are positive, then the Statistical Hypothesis must be rejected and the Research Hypothesis accepted. The weights need to be positive because the researcher expects a positive second degree curvilinear relationship, over and above a positive linear relationship. Indeed, if s_2 was negative, the curve would go up and then descend, an inverted U which would be upsetting to this researcher. Continued practice would yield less production beyond a certain point. Such a finding would go contrary to all expectations regarding skill learning, and if observed should call for an immediate replication so a strong empirical base can be referred to.

The reader may note that 10 groups were formed (one for each of the whole hours of practice), yet these 10 groups were cast into one continuous vector with values 1 through 10. One would expect that 5 1/2 hours of practice will yield production somewhere between the production observed by the 5 and 6 hour prac-

tice groups, and so on for all half-hour periods; therefore, it
is reasonable to consider practice to be represented on a continuum.

Applied Research Hypothesis 7.1

Directional Research Hypothesis: For a given population, there
is a positive second degree
effect of X_3 on X_2, over and
above the linear effect of X_3.

Statistical Hypothesis: For a given population, there is not a
positive second degree effect of $\overline{X_3}$ on
X_2, over and above the linear effect
of X_3.

Full Model: $X_2 = a_0U + a_3X_3 + a_{16}X_{16} + E_1$

where: $X_{16} = X_3{}^*X_3$

Restriction: $a_{16} = 0$

Restricted Model $X_2 = a_0U + a_3X_3 + E_2$

alpha = .01

$R_f^2 = .98$ $R_r^2 = .92$

$\underline{F}_{(1,57)} = 159$ Directional p < .0000001

Interpretation: Since the weighting coefficient for the second-
degree component (X_{16}) is positive and p <
alpha, the Research Hypothesis is tenable: For
a given population, there is a positive second-
degree effect of X_3 on X_2, over and above the
linear effect of X_3.

The sign of the linear component does not have to be hypothesized, as Applied Research Hypothesis 7.1 illustrates. The second
degree component here only accounts for an additional 6% of the

criterion variance, but that 6% is highly significant because
only 8% was left unaccounted for by the rectilinear fit.

Inverted U-shaped curve--the expected relationship

In a number of areas of research, the relationship between
X and Y is expected to be a specific curvilinear relationship.
For example, arousal (central neural excitement) has been found
to yield an inverted U-shaped relationship to cognitive performance on a moderately difficult task. That is, those who have
either low or high arousal levels score poorly, and those who
are moderately aroused produce the highest responses. Figure 7.8
represents the inverted U relationship.

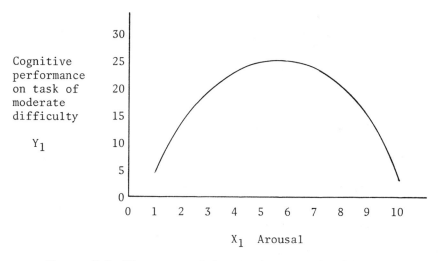

Figure 7.8. The expected inverted U relationship between
Arousal and Cognitive performance.

As depicted in Figure 7.8, individuals who score 1 or 10 on

Arousal (X_1) score 5 on the cognitive performance task. Subjects with moderate Arousal levels (5 and 6) score above 20 on the Cognitive performance task. The theory claims that low aroused and high aroused subjects do poorly on the cognitive performance task for different reasons. The person who is barely aroused has insufficient central excitation for focusing on the task. On the other hand, the highly aroused individual has such a high level of central excitation that trivial stimuli distract the person from the primary cognitive task. Moderate levels provide suffi- cient central excitation to focus on the task, but not so much that it yields distraction.

Given such theoretic expectations, in an experimental situ- ation the researcher might hypothesize:

There is an inverted U relationship between measured Arousal level and Performance on a cognitive task of moderate diffi- culty for a specific population of humans.

The researcher might then select a sample of subjects which represents a specific population (e.g., 16 year old males in high school), and provide the moderately difficult task while monitoring the individuals' Arousal level. The research expecta- tion might provide the following Research Hypothesis:

Among 16 year old high school boys, there is an inverted U relationship between measured Arousal level and Perform- ance on a cognitive task of moderate difficulty.

The inverted U relationship is a very specific expectation. In order to obtain the inverted U, in the upper right hand quad- rant as in Figure 7.9a, the weight associated with the Arousal vector must be positive and the weight associated with the vector containing squared Arousal scores must be negative. The Y-inter-

cept, a_0, may be positive, negative, or zero. A plot of the data

must be made to determine if the change in direction is within

the range of observed X_1 scores.

Table 7.2 provides an example to show the weights for an

inverted U relationship with a perfect fit of the data.

Table 7.2 Vector illustration of an example perfect inverted U relationship with a_0 = 0, b_1 = +5, and
s_1 = -1.

Subject	Y_1	=	$a_0 U$	+	$b_1 X_1$	+	$s_1 X_1^2$	+	E_1
1	0		1		0		0		0
2	4		1		1		1		0
3	6		1		2		4		0
4	6	=0	1	+5	3	+(-1)	9	+	0
5	4		1		4		16		0
6	0		1		5		25		0

If the three elements for subject 1 in vectors U, X_1, and X_1^2 are

multiplied by the appropriate weights, the predicted score is 0

and the observed score is 0. If the same procedure is done for

all six subjects, the predicted score goes up and then down as

X_1 increases. In the example case in Table 7.2, inspection of the

values for Y_1 confirms that the change in direction is within the

range of observed X_1 scores.

The general linear model allowing for a second degree rela-

tionship is:

Model 7.3: $Y_1 = a_0 U + b_1 X_1 + s_1 X_1^2 + E_3$

Figure 7.9 shows a number of second degree relationships which may be fit with the "second degree model."

b_1 = positive
s_1 = negative

b_1 = negative
s_1 = positive

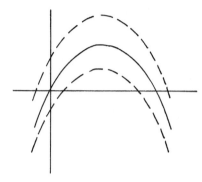

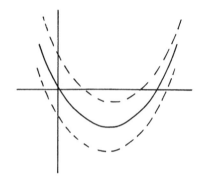

Figure 7.9a

Figure 7.9b

b_1 = negative
s_1 = negative

b_1 = positive
s_1 = positive

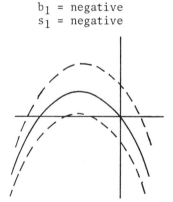

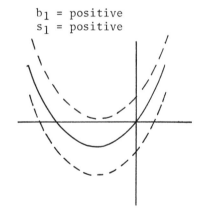

Figure 7.9c

Figure 7.9d

Figure 7.9. Changes in the second degree curve as weights
for a_0, b_1, and s_1 change sign (the dark line
represents $a_0=0$, while the upper dashed line
represents $a_0>0$ and the lower dashed line represents $a_0<0$).

A few general observations can be made from the curves of Figure 7.9.

1. When s_1 is negative, the open end of the curve is down, and when s_1 is positive, the open end is up.

2. When s_1 and b_1 are of the same sign (both positive or both negative) the maximum (or minimum) is at a negative value on the X axis. When s_1 and b_1 are of opposite signs (one negative and one positive) the maximum (or minimum) is at a positive value on the Y axis.

Figure 7.9a represents an inverted U outcome state. If the values for X range from 0 to 7, then the inverted U relationship is obtained. However, if the X values range from 0 to 4, what can one say? Figure 7.10 is a representation of Figure 7.9a over the range of scores from 0 to 4.

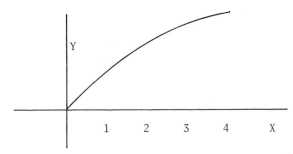

Figure 7.10. Data from Figure 7.9a with the range of ob-
          served scores limited from 0 to 4; the second
          degree curve with weights $a_0 = 0$, $b_1 = $ a posi-
          tive value, and $s_1 = $ a negative value.

In Figure 7.10 a positive b_1 and a negative s_1 does <u>not</u>

yield an inverted U. If one observed such a plot in the arousal study, one might conclude that the population the sample represents does <u>not</u> have individuals with high levels of Arousal. Indeed, if a larger population is defined that included Arousal levels up to scale value 7 as in Figure 7.9a, one might still expect to find the inverted U relationship to hold.

All curves in Figure 7.9 show all four quadrants. In most research in the behavioral sciences, X values are positive and Y values are also positive; therefore, most research would be concerned solely with the upper right-hand quadrant.

Testing the inverted U

To test the Research Hypothesis: "Among 16 year old high school males, there is an inverted U relationship between measured Arousal level and Performance on a cognitive task of moderate difficulty," three conditions must be met to yield an affirmative answer.

The linear equation which allows for an inverted U relationship is:

Model 7.4: $Y_1 = a_0 U + b_1 X_1 + s_1 X_1^2 + E_4$

$$\text{where:} \quad Y_1 = \text{Performance on the cognitive task of diffi-culty;}$$

$$X_1 = \text{the continuous predictor variable (Arousal level); and}$$

$$X_1^2 = \text{a vector containing the square of the elements in vector } X_1.$$

Condition 1: The weight (b_1) associated with the linear component (X_1) must be positive.

Condition 2: The weight (s_1) associated with the second degree polynomial (X_1^2) must be negative.

Condition 3: If conditions 1 and 2 are met, then a plot of the line of best fit must show that those individuals with the lowest and highest scores on the X_1 variable have relatively the same (minimum) score on Y_1, and the maximum score on Y_1 is approximately midway between the lowest and highest scores on X_1.

To meet conditions 1 and 2, two research hypotheses and two statistical hypotheses must be cast:

Directional Research Hypothesis for testing condition I:

Among 16 year old high school males, there is a positive linear relationship between measured Arousal level and Performance on a cognitive task of moderate difficulty, over and above the second degree polynomial of Arousal and the regression constant.

Given the Full Model (Model 7.4), the Research Hypothesis implies that $b_1 > 0$.

Statistical Hypothesis:

Among 16 year old high school males, there is no linear relationship between measured Arousal and Performance on a cognitive task of moderate difficulty, over and above the second degree polynomial of Arousal and the regression constant.

The Statistical Hypothesis implies that $b_1 = 0$, resulting in the following Restricted Model:

Model 7.5: $Y_1 = a_0U + s_1X_1^2 + E_5$

If the weight b_1 is positive and the directional probability
is less than the stated alpha, the first condition is met. If
not, then the inverted U shaped expectation is untenable.

Given an affirmative answer to condition 1, then the second
condition must be verified:

Directional Research Hypothesis for testing condition 2:

Among 16 year old high school males, there is a negative
relationship between the second degree polynomial of Arousal
and Performance, over and above the positive linear component of Arousal and the regression constant. (This hypothesis states $s_1 < 0.$)

Statistical Hypothesis:

There is no relationship between the second degree polynomial of Arousal and Performance, over and above the positive linear component of Arousal and the regression constant.

The Full Model is the same as Model 7.4, and the restriction
implied by the Statistical Hypothesis is $s_1 = 0$, resulting in the
following Restricted Model:

Model 7.6: $Y_1 = a_0U + b_1X_1 + E_6$

If the weight s_1 is negative and the directional probability
is less than the stated alpha, the second condition is met.

A plot of the line of best fit could yield a number of relationships, but all will be some form of the three lines depicted in Figure 7.11.

Figure 7.11a shows a case where the inverted U relationship exists, but the upper end of the Arousal continuum is not
present in the population. The dashed line as extrapolated, does
yield the inverted U. Given such a finding, the researcher may
wish to seek a population which includes the higher end of the

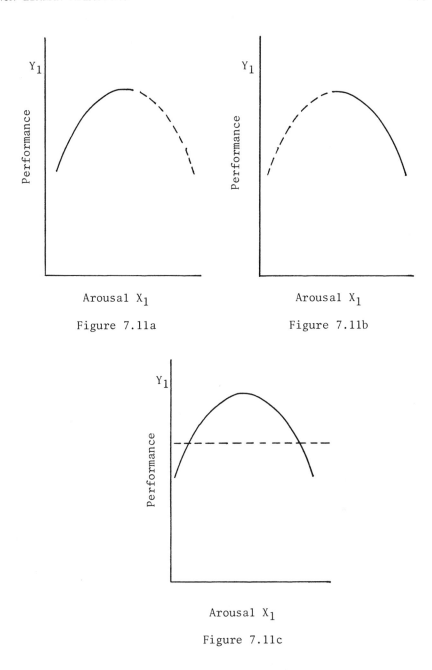

Figure 7.11a

Figure 7.11b

Figure 7.11c

Figure 7.11. Three possible outcome states from the second
degree model: $Y_1 = a_0U + b_1X_1 + s_1X_1^2 + E_1$,
when b_1 is positive and s_1 is negative.

Arousal scale; or he may conclude that for the specified population an inverted J relationship exists. (No very highly aroused
subjects are observed in the population of 16 year old high
school males.)

Figure 7.11b is the same as Figure 7.11a, except that no
lowly aroused 16 year old high school males are observed.

Figure 7.11c represents the idealized inverted U relationship. Given 7.11c, all three conditions are met and one may conclude, "Among 16 year old high school males, an inverted U relationship exists between measured Arousal level and Performance
on a moderately difficult cognitive task."

Note of Caution! (1) The testing for the
signs of the regression weights must be done
within the context of making one restriction
at a time upon the same full model because
the sign of each weighting coefficient was
posited within the context of the other
variables in the Full Model. An inverted U
requires both a linear component and a second degree component. (Whether a non-zero
Y-intercept is necessary is a function of
the data and was of no interest to this particular application.)

In the case just presented, if the
sign for b_1 was tested using a Full Model

that did not contain X_1^2 ($Y_1 = a_0U + b_1X_1 +$ E_2), then setting $b_1 = 0$ would yield a non-significant F value because there is no linear trend independent of the second degree polynomial. See the dashed line in Figure 7.11c.

Similarly, the second degree polynomial must be tested as a restriction on the Full Model, because a restriction of $s_1 = 0$ on a model not containing X_1 ($Y_1 = a_0U + s_1X_1^2 + E_3$) might not be significant if a perfect inverted U relationship exists. Figure 7.12 shows two possible lines using $Y_1 = a_0U + s_1X_1^2 + E_3$. When b_1 is not included in the Full Model, the curve is symmetrical about the Y axis and thus will not make a good fit when the maximum point departs markedly from the Y axis (e.g., the data in Figure 7.11c).

(2) In the case where more than one continuous predictor variable is used in the Full Model, one should not interpret the magnitude of the regression weight because small variations from sample to sample may change these numerical values greatly, even though the signs should be stable. (The

weights could be used in a predictive situa-

tion as the predicted criterion would be ap-

proximately the same from one set of weights

to another.) This is due to the fact that

continuous vectors tend to be correlated,

and small sample variations may yield large

changes in the weighting coefficients. The

magnitude of the R^2 will remain relatively

constant, however. McNeil and Spaner (1971)

present a more generally applicable argument

for including correlated predictor variables

in regression models.

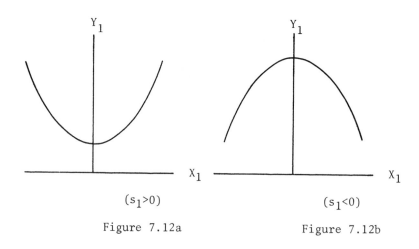

$(s_1 > 0)$ $(s_1 < 0)$

Figure 7.12a   Figure 7.12b

Figure 7.12. Curves reflected by the model: $Y_1 = a_0 U +$
$s_1 X_1^2 + E_3$, indicating that the curve must be
symmetric about the Y axis if the linear
component is not in the model.

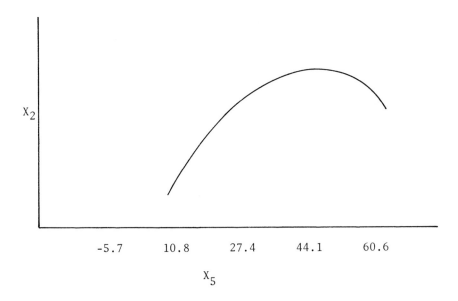

Figure 7.13. The obtained (J-shaped) relationship between
X_5 and X_2.

Third degree polynomial line, fitting

Curved lines of best fit may take many forms, and the second degree polynomial just presented is only one. In many psychomotor studies, an S-shaped curve is expected. For example,
consider the problem in this chapter regarding the relationship
between key punch card production and hours of practice. After
several hours of practice, a surge in production was observed.
Given further practice, improvement in production will sooner or
later level off due to the limits of the method, key puncher, and
key punch machine. Figure 7.14 illustrates a possible outcome
state.

Applied Research Hypothesis 7.2

General Hypothesis: For a given population, there is an inverted U relationship between X_5 and the
criterion X_2.

Condition 1 Research Hypothesis: There is a positive linear relationship between X_5 and X_2,
over and above the predictability of the second degree
polynomial of X_5 and the
regression constant.

Condition 1 Statistical Hypothesis: For a given population,
there is no linear rela-
tionship between X_5 and
X_2, over and above the
predictability of the second degree polynomial of X_5
and the regression constant.

Full Model for Condition 1: $X_2 = a_0U + a_5X_5 + a_{15}X_{15} + E_1$

where:

$$X_{15} = X_5 * X_5$$

Restriction: $a_5 = 0$

Restricted Model for Condition 1: $X_2 = a_0U + a_{15}X_{15} + E_2$

alpha = .01

$R_f^2 = .95$ $R_r^2 = .38$

$F_{(1,57)} = 755$ directional p < .000001

Interpretation: Reject Condition 1 Statistical Hypothesis and
accept Condition 1 Research Hypothesis.

Condition 2 Research Hypothesis: For a given population, there
is a negative relationship be-
tween the second degree polynomial of X_5 and X_2, over and
above the predictability of X_5
and the regression constant.

Condition 2 Statistical Hypothesis: For a given population,
there is no relationship

---

between the second degree
polynomial of X_5 and X_2,
over and above the predictability of X_5 and the regression constant.

Full Model for Condition 2: $X_2 = a_0U + a_5X_5 + a_{15}X_{15} + E_1$

Restriction: $a_{15} = 0$

Restricted Model for Condition 2: $X_2 = a_0U + a_5X_5 + E_3$

alpha = .01

$R_f^2 = .95$ $R_r^2 = .55$

$F_{(1,57)} = 535$ directional p < .00001

Interpretation: Reject Condition 2 Statistical Hypothesis and
accept Condition 2 Research Hypothesis. Based
upon the plot of X_5 against X_2, pictured in
Figure 7.13, the data reflect a J-shaped curve,
rather than a U-shaped curve. There is clearly
a maximum, but the line of best fit did not
continue downward enough to justify referring
to this curve as J-shaped. Therefore the
General Hypothesis cannot be accepted.

To fit the data for the first 10 Hours of practice, a simple
second degree polynomial will be adequate ($Y_1 = a_0U + b_1P_1 + s_1P_1^2$
$+ E_1$). Beyond 10 Hours of practice, the curve is expected to
change direction. In order to investigate such a situation, the
researcher who conducted the previous study could use that data
and could select 50 more subjects, randomly assign these to five
practice groups (Hours of practice 11, 12, 13, 14, and 15), and
test the Research Hypothesis, "For naive key punching students,
the Number of ten character cards punched per hour slowly increases, spurts, and then levels off with increased Hours of

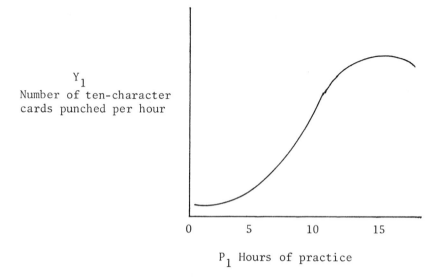

$$Y_1$$
Number of ten-character
cards punched per hour

$$0 \qquad 5 \qquad 10 \qquad 15$$

P_1 Hours of practice

Figure 7.14. An S-shaped curve resulting from the functional
relationship between Hours of practice and number of cards produced per hour.

practice."

A linear regression model which includes the regression con-

stant, Hours of practice (P_1), Hour of practice squared (P_1^2), and

Hours of practice cubed (P_1^3) will allow the line of best fit to

reflect the S-shaped curve:

Model 7.7: $Y_1 = a_0 U + b_1 P_1 + s_2 P_1^2 + t_3 P_1^3 + E_7$

where: Y_1 = Number of ten-character cards punched per hour;

P_1 = Hours practiced;

P_1^2 = squared elements of P_1; and

P_1^3 = cubed elements of P_1.

In order for the particular S-shaped curve in Figure 7.14 to

be obtained, t_3 must be negative and s_2 positive. This is so because the criterion scores are lower for large values of P_1. Hence
the highest powered term, P_1^3, must have a negative weight. In order to get the inflection point over the P_1 value of approximately
12, the second degree term must be opposite to that of the third
degree term. If the curve is horizontal at the Y axis, then the
linear weight (b_1) will be zero. If the curve is slanted downward
from left to right, b_1 will be negative, and if slanted upward, b_1
will be positive. The sign and magnitude of the linear term and
the Y-intercept are not relevant to the determination of the S-
shaped curve (see Figures 7.9 and 7.12). One may test each of the
weights by restricting out one weight at a time from the Full
Model. If all variables add to predictability and each of the
weights has the appropriate sign, then the Full Model is the preferred model to fit the data. If $t_3 = 0$, then the expected level-
ing has not yet occurred and the researcher might wish to extend
the study in order to determine where practice should stop. Chapters Eight and Ten deal more fully with such trend analysis procedures.

When dealing with higher order relationships, the signs of
the weights become rather difficult to specify without resort to
analytic geometry. If one has a mastery of analytic geometry, he
may be able to specify the expected signs of weights and go about
testing his expectations. Without such training, the researcher
can use empirical procedures to develop models and test expectations. Indeed, it can be very difficult to ascertain the curve of

best fit. Finding the best curve usually requires a data snooping

exercise. Curve fitting is complicated even more by the prospect

that powers other than whole numbers might be operating. Suppose,

for instance, that the exponent is unknown in the following equa-

tion:

Equation 7.1: $Y_1 = a_1 X_1^m$

Taking the logarithm of both sides of Equation 7.1 results in:

Equation 7.2: $\log_e Y_1 = m(\log_e X_1) + \log_e a_1$

If $\log_e X_1$ and the unit vector are used to predict $\log_e Y_1$, then

the following model results:

Model 7.8: $\log_e Y_1 = a_0 U + b_1 \log_e X_1 + E_8$

The exponent for Equation 7.1 is b_1, and a_0 is $\log_e a_1$.

Many researchers have been mistakenly told a number of mis-

conceptions regarding curve fitting. First, not all of the lower

powers need to be included. That is to say, the data may be well

fit by the second degree term only.

Second, the lower powered terms are not the "simplest" vari-

ables. Complexity should be viewed as the number of predictor

variables needed to satisfactorily account for citerion variance

(McNeil and McShane, 1974). Researchers have for too long now

first investigated linear relationships because they thought that

those were the simplest.

Third, the functional fit over the range of sample values

is applicable to that range. Generalizations beyond that range

are good guesses, but are based on the assumption that the same

functional relationship occurs as in the range originally studied.

That is an empirical question which additional data gathering can answer. The third degree function in Figure 7.14 is essentially linear from 7 to 12 Hours of practice. And it is a second degree function from 0 to 10 Hours of practice.

Curve fitting, when using more than one predictor variable, is referred to as fitting a response surface. Interest may focus on maximums, minimums, the similarity of one group's surface to another group's, or simply the nature of the surface. Mendenhall (1968) goes into more detail than this text. As indicated in Chapter Six, Walberg (1971) gives an interesting example of such response surface fitting.

Hypothesis seeking through the descriptive use of multiple regression

In the spirit of Chapter One, the task of the researcher is to build theoretic models and cast linear equations that will encompass the variables and functional relationships which account for a large proportion of observed criterion variance. Ultimately one hopes to approximate an R^2 of 1.00. In the quest to increase predictability, one may use linear equations descriptively in order to find the most parsimonious predictive model. With regard to the data presented in Figure 7.14, one could develop a regression model with 15 linearly independent vectors:

Model 7.9: $Y_1 = a_0 U + a_1 X_1 + a_2 X_2 + \ldots + a_{15} X_{15} + E_9$

where: Y_1 = the criterion;

X_1 = a one if the criterion is from a person with one Hour of practice, zero otherwise;

X_2 = a one if the criterion is from a person with two Hours of practice, zero otherwise;

.
.
.

X_{15} = a one if the criterion is from a person with fifteen Hours of practice, zero otherwise.

The above model will be the best fitting model (in terms of R^2) to any set of data points because each X axis value is allowed to have its own mean. Model 7.9 yields maximum curvilinearity and is referred to as the eta coefficient model. Even though the eta coefficient model yields maximum curvilinearity, there are several drawbacks to the model (McNeil, 1970b). First, there are a number of weighting coefficients to be calculated, and hence the likelihood of being able to replicate those weights on successive samples is low. Second, the eta coefficient model does not allow for inferences about Y_1 values for values between the specified X axis values, whereas using continuous predictor variables does allow for such generalizations.

If the S-shaped curve is the relationship the data takes, the third degree model (Model 7.7) will account for almost as much criterion variance as the eta coefficient model, but Model 7.7 will do so in a more parsimonious fashion since only four pieces of information are needed, whereas fifteen are needed in the eta coefficient model. Furthermore, the elimination of a_0 and/or b_1 may yield as predictive a model as Model 7.7 and will be a more parsimonious model containing only three pieces of information:

Model 7.10: $Y_1 = b_1P_1 + s_2P_1^2 + t_3P_1^3 + E_{10}$ (where $a_0 = 0$)

Model 7.11: $Y_1 = a_0U + s_2P_1^2 + t_3P_1^3 + E_{11}$ (where $b_1 = 0$)

That model which is found through this process to be the most predictive yet parsimonious model (as determined by the researcher) should be used as the basis for a cross-validation study (see Chapter Nine). It is the best model for the sample data. However, it may not turn out to be generalizable to the population. The lowered likelihood of the model being generalizable is due to the fact that it was not hypothesized beforehand and it was discovered by searching through the sample data.

In a case where a researcher has knowledge of sex, treatment, and some continuous covariate, but little expectation regarding how these variables interrelate to criterion performance, the following procedure may prove valuable in formulating an hypothesis that the researcher can then test in a future study. Suppose there are three treatment conditions, two sexes, and an ability measure, plus a criterion measure. One may cast bivariate plots between the ability measure and the criterion for each of the possible groups (Males in Treatment I; Males in Treatment II; Males in Treatment III; Females in Treatment I; Females in Treatment II; Females in Treatment III). The plots might look like Figures 7.15a through 7.15f.

Each of the lines of best fit are the most parsimonious when dealt with separately; however, an inspection of the figures might lead to a more parsimonious model for the entire sample. There are 13 linearly independent vectors which describe the six lines (note that for Figure 7.15e a second degree polynomial is

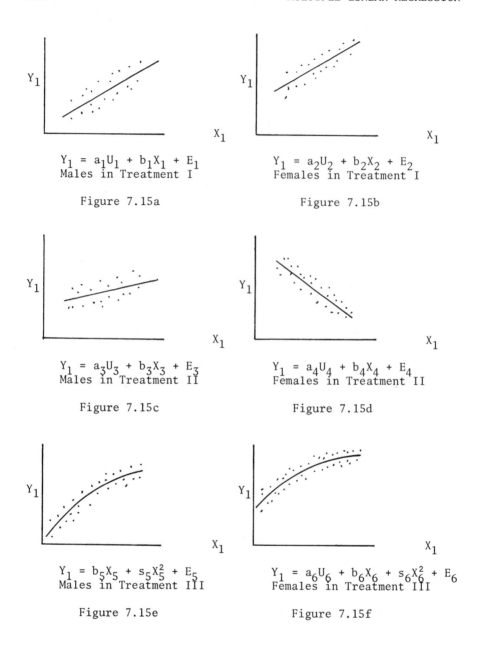

$Y_1 = a_1 U_1 + b_1 X_1 + E_1$
Males in Treatment I

Figure 7.15a

$Y_1 = a_2 U_2 + b_2 X_2 + E_2$
Females in Treatment I

Figure 7.15b

$Y_1 = a_3 U_3 + b_3 X_3 + E_3$
Males in Treatment II

Figure 7.15c

$Y_1 = a_4 U_4 + b_4 X_4 + E_4$
Females in Treatment II

Figure 7.15d

$Y_1 = b_5 X_5 + s_5 X_5^2 + E_5$
Males in Treatment III

Figure 7.15e

$Y_1 = a_6 U_6 + b_6 X_6 + s_6 X_6^2 + E_6$
Females in Treatment III

Figure 7.15f

Figure 7.15. Six groups and six lines which might reflect
the lines of best fit (see Model 7.12 for
definition of vectors).

included, but the Y-intercept is zero, thus $a_5 = 0$; and Figure

7.15f has three linearly independent vectors since $a_6 \neq 0$). The

Males and Females who had Treatment I look like they have the

same slope $(b_1 = b_2)$, but have different Y-intercepts. Further-

more, the curves for Males and for Females who had Treatment III

seem to be the same but $a_5 = 0$ for Males and $a_6 > 0$ for Females,

and thus $b_5 = b_6$ and $s_5 = s_6$ seem like reasonable restrictions.

Model 7.12 is one possible model reflecting all six lines in

one linear equation. It was derived from an inspection of the

data rather than from a prior research hypothesis. Given this

possible model (7.12), the expected equalities can be inspected

by imposing the appropriate restrictions on Model 7.12 and then

comparing the resulting R^2 values for each model.

Model 7.12: $Y_1 = a_1U_1 + b_1X_1 + a_2U_2 + b_2X_2 + a_3U_3 + b_3X_3 + a_4U_4$

$\qquad\qquad + b_4X_4 + b_5X_5 + s_5X_5^2 + a_6U_6 + b_6X_6 + s_6X_6^2 + E_{12}$

where: Y_1 = the criterion;

$\qquad U_1$ = 1 if Y_1 comes from a Male in Treatment I, zero
$\qquad\qquad$ otherwise;

$\qquad X_1$ = Ability if Y_1 comes from a Male in Treatment I,
$\qquad\qquad$ zero otherwise;

$\qquad U_2$ = 1 if Y_1 comes from a Female in Treatment 1,
$\qquad\qquad$ zero otherwise;

$\qquad X_2$ = Ability if Y_1 comes from a Female in Treatment
$\qquad\qquad$ I, zero otherwise;

$\qquad U_3$ = 1 if Y_1 comes from a Male in Treatment II, zero
$\qquad\qquad$ otherwise;

$\qquad X_3$ = Ability if Y_1 comes from a Male in Treatment II,
$\qquad\qquad$ zero otherwise;

$\qquad U_4$ = 1 if Y_1 comes from a Female in Treatment II, zero otherwise;

X_4 = Ability if Y_1 comes from a Female in Treatment II, zero otherwise;

X_5 = Ability if Y_1 comes from a Male in Treatment III, zero otherwise;

X_5^2 = Ability squared if Y_1 comes from a Male in Treatment III, zero otherwise;

U_6 = 1 if Y_1 comes from a Female in Treatment III, zero otherwise;

X_6 = Ability if Y_1 comes from a Female in Treatment III, zero otherwise; and

X_6^2 = Ability squared if Y_1 comes from a Female in Treatment III, zero otherwise.

Since empirical line fitting as seen in this context is basically an attempt to parsimoniously describe the data, no research hypothesis is necessary. One may successively cast models which reflect $b_1 = b_2$, $b_5 = b_6$, and $s_5 = s_6$ and compare those models with Model 7.12.

In the context of Model 7.12, one may investigate the possibility that $b_1 = b_2$, both equal to b_7, a common slope. If vectors X_1 and X_2 are added together to form X_7, then b_7 is a common slope for Males and Females under Treatment I. The resulting model is:

Model 7.13: $Y_1 = a_1U_1 + a_2U_2 + b_7X_7 + a_3U_3 + b_3X_3 + a_4U_4 + b_4X_4$
$$+ b_5X_5 + s_5X_5^2 + a_6U_6 + b_6X_6 + s_6X_6^2 + E_{13}$$

There are 13 linearly independent vectors in Model 7.12 and 12 in Model 7.13; thus Model 7.13 is more parsimonious. If Model 7.13 is judged by the researcher to be as predictive (has as high an R^2) as Model 7.12, then the researcher would prefer Model 7.13

over Model 7.12. If this is the state of affairs, Model 7.13 be-

comes the preferred model to be compared against further models.

The researcher may then wish to investigate the possibility

that $b_5 = b_6$, both equal to b_8, a common weight; and $s_5 = s_6$,

both equal to s_8, a common weight.

These two restrictions can be made one at a time, sequen-

tially, or both at once. If done together, a large loss in R^2

would not tell the researcher whether one or the other or both of

the restrictions caused the loss in R^2. On the other hand, if the

R^2 loss is very small from Model 7.13 (or 7.12) to Model 7.14,

then one could conclude that $b_5 = b_6$ and $s_5 = s_6$.

To gamble a bit, since the researcher has a hunch that both

restrictions are reasonable, one may apply both restrictions at

once. By adding vectors X_5 and X_6 to get vector X_8 and adding X_5^2

and X_6^2 to get X_8^2, the weights b_8 and s_8 will give the common lin-

ear and second degree slopes for Males and Females given Treat-

ment III. The resulting model for comparison with Model 7.13 is:

Model 7.14: $Y_1 = a_1U_1 + a_2U_2 + b_7X_7 + a_3U_3 + b_3X_3 + a_4U_4 + b_4X_4$

$$+ b_8X_8 + s_8X_8^2 + a_6U_6 + E_{14}$$

There are 12 linearly independent vectors in Model 7.13 and

10 in Model 7.14. If the R^2 value for Model 7.14 is the same as

or close to (as judged by the researcher) the R^2 value for Model

7.13, then the researcher would conclude that Model 7.14 was a

better description of the data than Model 7.13 since Model 7.14

is as predictive but more parsimonious than Model 7.13. The re-

searcher could then formulate an hypothesis based on Model 7.14;

and he could test this hypothesis on new, "unsnooped" data.

When using the empirical line fitting procedure just presented, one should be aware that he is data snooping and hypothesis seeking. Under such circumstances, the obtained R^2 must be seen as a descriptive statistic, which can be used for <u>future</u> hypothesis testing. One should <u>not</u> use the data inferentially. A replication with new subjects is necessary before generalizing to the population. A more extensive discussion of this concept appears in Chapter Nine.

Rescaling Observed Non-linear Relationships to the Criterion

The preceding sections of this chapter have discussed transforming (squaring, cubing) variables in order to reflect the hypothesized non-linear functional fit between the predictor set and the criterion. In those instances, it was expected that the <u>actual constructs</u> (those underlying phenomena that the predictors and criteria were measuring, however imperfectly) were related to one another in a non-linear manner. When a non-linear relationship between variables is observed, however, it may not be due to the constructs being related in that way; it could be due instead to a variable being a poor measure (or map) of the construct it represents.

The concern of these last sections of Chapter Seven will be with transforming a variable such that the numbers more accurately reflect the construct under consideration. Since constructs, by definition, cannot be observed, researchers must realize that the measurements they use to indicate those constructs are arbitrary numbering systems that someone has decided are "good."

The "goodness" of the arbitrary scoring could be ascertained by observing the overlap in variance between the numbers and the construct. Figure 7.16 represents what researchers usually assume as the state of affairs, whereas Figure 7.17 represents what may be a more likely state of affairs. The process of transforming variables is an attempt to reach the state of affairs in Figure 7.16.

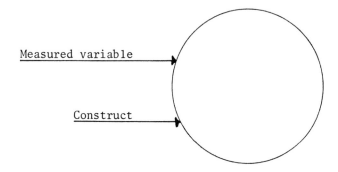

Figure 7.16. The assumed overlap between the variable and the construct that variable measures--100% overlap.

There is no conventional or obvious way to transform a variable such that the numbers will better match the construct. A thorough knowledge of the content field will certainly help. A

well-thought-out nomological net would also be of some assistance.

But since the transformations must ultimately be tested by suc-

cessive empirical verifications, perhaps the most efficient meth-

od would be through data snooping. Data snooping was briefly dis-

cussed earlier and will be more fully developed in Chapter Nine.

First, several examples will be given to illustrate the need for

transformation of variables--a process which will hereafter be

referred to as rescaling.

Rescaling example one--difficulty of items

The numbers assigned in order to measure a construct should:

(1) reflect that the scale points are in order (e.g., a score of

5 reflects more of the construct than does 4 and less than does

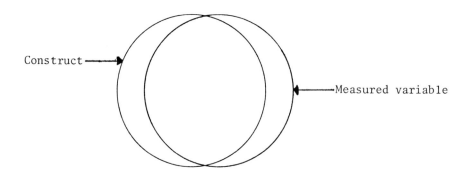

Figure 7.17. A more likely correspondence between the measured variable and the construct than Figure
7.16.

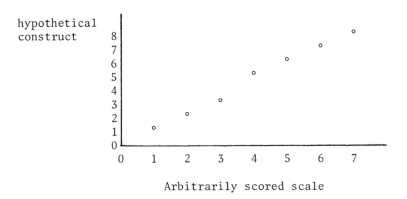

Figure 7.18. An Arbitrarily scored scale which does not
accurately reflect construct distances between
scale points.

6); and (2) reflect the distance between scale points (e.g., the

construct distance between scores of 4 and 6 is the same as the

construct distance between 8 and 10).

The Arbitrarily scored soale in Figure 7.18 has seven items,

but it covers 8 units on the construct. (Note that the convention

in this text of capitalizing actual measures and not capitalizing

constructs is retained here. It must be emphasized that attain-

ment of actual construct scores--the criterion in this part of

Chapter Seven--would most likely not be possible.) If a unit

weight is assigned to each item on the scale, it will not ac-

curately map the construct. Persons who get scores of 4, 5, 6, or

7 on the Arbitrarily scored scale really have 5, 6, 7, or 8 units

of the hypothetical construct. In order for the Arbitrarily

scored scale to map the construct, one unit would have to be add-

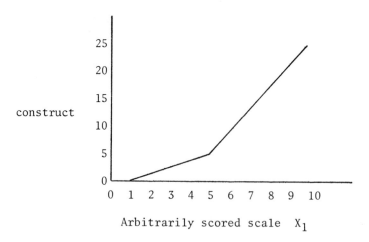

Arbitrarily scored scale X_1

Figure 7.19. Relationship between an Arbitrarily scored
            scale and the construct when only a rela-
            tively few difficult items are included.

ed to the Arbitrarily scored scale if the Arbitrarily scored

scale score is greater than three. Another possibility would be

to add another item to the seven-item test.

Rescaling example two--difficulty of items

Suppose that the scale developer had done a relatively good

job in writing items and assigning numbers to them, except with

the very difficult items. Only a few difficult items were in-

cluded, and they were really difficult. Figure 7.19 might be the

result of this scale. Items 1 through 5 each represent one unit

on the construct, but items 6 through 10 cover a distance from

above 5 to 25 on the construct. Assigning a unit weight to each

item will not map the construct accurately.

If the Arbitrarily scored scale is five or below, it does

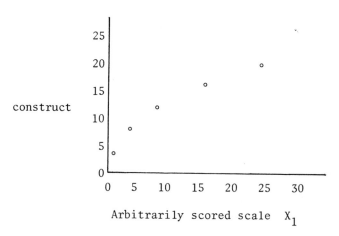

Figure 7.20. Data depicting a "ceiling effect."

not have to be transformed. If the Arbitrarily scored scale is
greater than five, then the following data transformation would
map the Arbitrarily scored scale onto the construct:

IF(X(1).GT.5.0) X(1) = (4.0 * X(1)) - 15.0

The values of 4 and -15 were obtained by ascertaining the
line of best fit over the range from 5 to 10. This line has a
slope of 4 and if extended to the Y axis, would have a Y-intercept of -15.

The reader should note in the above two examples that the
rescaling suggested would simply result in all data points falling on a straight line. The rescaled Arbitrarily measured scale
rectilinearly maps the criterion. The resultant scoring system is
also arbitrary but has an added feature--a rectilinear fit with
the construct. That perfect mapping with the construct is the
goal of this discussion.

Rescaling example three--ceiling effect

The term "ceiling effect" implies that there is not enough discrimination at the high end of the Arbitrarily scored scale, as in Figure 7.20. One could overcome this problem by including additional difficult items, or by rescaling the existing Arbitrarily scored scale.

The construct can be better mapped by taking the square root of X_1 and then multiplying by 4. The multiplication by four is not a necessary step, since the first step (that of taking the square root) results in the scores falling on a straight line. The multiplication by four simply results in the rescaled scores matching the (arbitrarily scored) construct.

Rescaling example four--guessing effect

If a 100 item, 4 choice, multiple choice test is developed, some 25% of the items can be correctly answered solely by guessing. Any score of 25 or below reflects a construct score of 0. The remaining scores should also be adjusted for the guessing effect. The required adjustments to make the data in Figure 7.21 reflect the construct are:

    IF(X(1).LE.25.0) X(1) = 0

    IF(X(1).GT.25.0) X(1) = (4./3.)*(X(1)-25.0)

These data transformations set the rescaled score equal to zero if the original score was less than or equal to 25. If the original score was greater than 25, the rescaled score is adjusted for guessing.

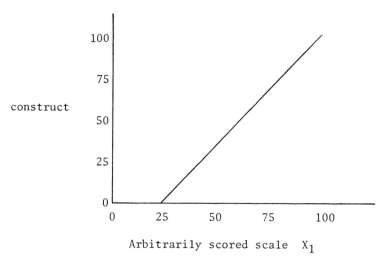

Figure 7.21. The possible relationship between a 100-item
multiple choice test and the criterion.

Rescaling example five--conceptual rescaling

J. McNeil (1974) reported a study involved with the rescaling of Arbitrarily scored scales. No criterion (or construct) was
available to verify the rescaling, although some conceptual notions were available which allowed for some attempts at rescaling.
One of the items for which such rescaling was done was a measure
of infant general muscle tonus. On the original, descriptive
scale, a "1" was assigned if the infant was flaccid and limp; the
numbers progressed through more and more muscle tone up to "5"
for "responsive with good tone...approximately 75% of the time,"
and then on up to "9" for extremely hypertonic. These numbers may
map the construct of degree of muscle tone, but the scale was
being used as a measure of the construct, "level of development."

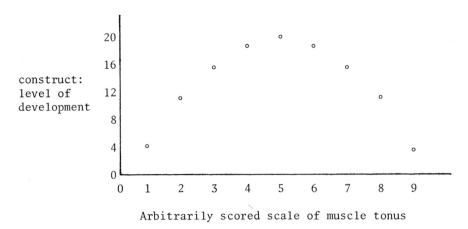

Arbitrarily scored scale of muscle tonus

Figure 7.22. An Arbitrarily scored scale which needs rescaling based on conceptual notions about the
construct.

For mapping this construct, the logical best score would seem to
be a "5". Moderate muscle tonus, rather than extremely limp or
extremely rigid tonus, would seem to reflect better infant development, as shown in Figure 7.22.

The data transformation needed to rescale the muscle tonus
scale to map the criterion of development would be:

$$X(1) = (10.0*X(1)) - (X(1)*X(1)) - 5.0$$

Note that this is equivalent to the vector expression:
$X_1 = -5U + 10X_1 - X_1^2$, which was shown earlier in this chapter
to be the model for an inverted U relationship, with $a_0 = -5$,
$b_1 = 10$, and $s_2 = -1$.

The Spearman Rank Correlation

Researchers are often encouraged to rank their data when

they have no faith in the interval property of their data and

then to use the Spearman Rank Correlation if an index of rela-

tionship is desired. The lowest score is assigned a value of "1",

the next lowest a value of "2", and so on. If one wanted to find

the Rank Correlation between, say, X_1 and X_2, one could ascertain

the rank of each score on each variable, and then obtain the R^2

for the following model:

Model 7.15: $R_1 = a_0U + a_1R_2 + E_{15}$

      where: R_1 = the ranked X_1 values, and

              R_2 = the ranked X_2 values.

The square root of R^2 would be the desired numerical value

for the Spearman Rank Correlation. The important point to realize

is that the assigning of "1", "2", etc. is one arbitrary way of

assigning scores.

Square root transformation

Researchers are sometimes admonished, when they discover

unequal variances in several groups, to "make a square root

transformation." Depending upon the nature of the data, this pro-

cedure may serve the purpose of equalizing the (sample) vari-

ances. Several points should be made. First, the statistical

assumption is about the population variances, not about the sam-

ple variances. Second, blindly applying a square root transforma-

tion will not always produce the desired end result. The kind of

transformation required on the criterion to obtain homogeneous

variances is a function of the data, and will not always be a
square root function. Indeed, one interpretation of this chapter
would be that the unequal variances might be due not to scaling
considerations but to inherent interactions.

Functional fit or rescaling?

When transformations on the arbitrarily scored variables
are made, either the researcher is rescaling his variables, or he
is allowing for a given functional fit. Which of the two he is
doing is not important from a predictive point of view--the crucial concern is whether the R^2 has been increased by the transformation. A simple example may be of value here. Suppose that
Model 7.16 yields an R^2 of .60, but that Model 7.17 yields an R^2
of 1.00.

Model 7.16: $Y_1 = a_0 U + a_1 X_1 + E_{16}$

Model 7.17: $Y_1 = a_0 U + a_2 X_1^2 + E_{17}$

These two models cannot be compared by the general F test
because one is not a restriction of the other; indeed, both have
two predictor pieces of information. These two models can be compared, however, on the basis of degree of predictability. Based
upon the goal of predictability, Model 7.17 would be the preferred model because it yields a higher R^2 value. If the researcher felt that X_1 was a good rectilinear mapping of the construct, then Model 7.17 would reflect a functional fit as discussed in the early part of this chapter. If the squaring of X_1
was a result of acknowledging the arbitrariness of the original

measure, then Model 7.17 would reflect the rescaling notions also

discussed in this chapter.

The criterion can be transformed rather than the predictor,

although Model 7.18 would yield a different R^2 than would Model

7.17.

Model 7.18: $\sqrt{Y_1} = a_0U + a_1X_1 + E_{18}$

Whether the transformation represents a functional fit or

rescaling is important from an "understanding" point of view.

Suppose that a given criterion is predicted quite well by a given

predictor as in Figure 7.23. The researcher's theory says that

the same functional fit ought to be the case for higher values of

the predictor. As can be seen in Figure 7.24, the linear fit ob-

served over the lower values of the predictor does not generalize

to higher values. Indeed, the relationship over the entire range

of values in Figure 7.24 is better depicted by a second degree

curve. If other values of the predictor variable were investi-

gated, perhaps the functional fit would be other than a second

degree one. The functional fit which is claimed must be empiri-

cally verified on new values of the predictor variable, just as

it must be verified on the population.

The important concern of this chapter is to emphasize the

notion that numbers put on measurements are arbitrary and that

initial attempts at numbering most likely do not map the con-

struct. Given that the researcher has rescaled his variable and

feels secure that the rescaled variable maps the construct, the

concern then is to reflect (with regression models) the hypothe-

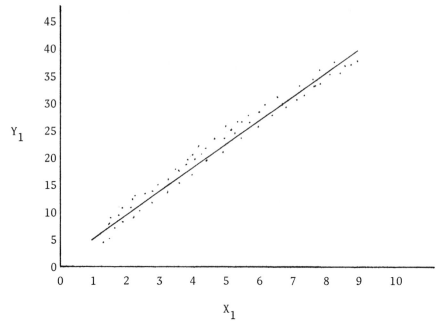

Figure 7.23. An observed rectilinear fit over the range of
X_1 values 0-10.

sized functional fit. Using the X_1 variable allows only for a

linear fit, whereas other fits may be inherent in the data.

Given that the scaling is arbitrary, there is no more reason to

expect a linear fit than a non-linear fit. X_1 is actually X_1^1,

and should indicate that there are an infinite number of other

exponents which can be investigated (e.g., X_1^2, X_1^3, $X_1^{1.5}$, X_1^3,

$X_1^{.313}$, etc.). If the linear component is used, then some justi-

fication should be provided--just as justification should be

provided for any other exponent.

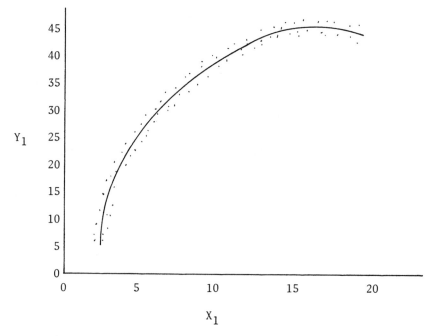

Figure 7.24. An observed non-linear fit over the range of X_1 values 0-20.

Reliability and validity

Reliability is the extent to which a variable yields the same value on repeated measures. Validity is the extent to which a variable measures what it is supposed to measure. The regression approach can be used to ascertain the magnitude of both of these concepts.

Test-retest reliability is simply the correlation between two administrations of the same test. Applied Research Hypothesis 7.3 provides practice with this kind of question. Note that

---

Applied Research Hypothesis 7.3

Directional Research Hypothesis: For a given population, the
                                 test-retest reliability of X_2
                                 (the correlation between X_2
                                 and X_3) is greater than .90.

Statistical Hypothesis: For a given population, the test-retest
                        reliability of X_2 is equal to .90.

Full Model: $X_2 = a_0 U + a_3 X_3 + E_1$

Restricted Model: $R_r^2 = (.90)^2 = .81$

alpha = .02

$R_f^2 = .92$ $R_r^2 = .81$

$F_{(1,58)} \simeq 80$ required $F_{(1,40)} = 7.31$

Interpretation: Since the F must be calculated by hand, the
                approximate F of 80 must be compared to the
                tabled F value of 7.31. Since the calculated
                F is larger than the tabled F, and since the
                correlation between X_2 and X_3 is positive, the
                Research Hypothesis is tenable. For a given
                population, the test-retest reliability of X_2
                is greater than .90.

Note: Most F tables have values for selected degrees of free-
      dom. If the desired denominator degrees of freedom does
      not exist, then one should move in the conservative
      direction and refer to an F with fewer degrees of free-
      dom.

---

the Restricted Model is forced to have an R^2 of .81, since that
is the R^2 specified by the Statistical Hypothesis. The F would
most likely have to be computed by hand, as most computer programs do not allow for the insertion of a specified R^2 value.

Reliability is of minor concern, though, as validity is the

ultimate test of any variable. The notion of rescaling is to get
the measure to be a valid indicator of the construct. The notion
of functional curve fitting is to get the predictor set to be a
valid indicator of the criterion. Most measurement texts correctly indicate that a variable can be highly reliable, yet have
no validity for a particular construct. Shoe size can be reli-
ably measured, yet it is not considered by many to be a valid
measure of IQ.

Furthermore, many constructs meaningfully fluctuate over a
short period of time. Breathing rate changes from one hour to the
next; anxiety fluctuates from one task to another; and reading
interest fluctuates from one comic strip to another. The concern
in each case should be, "Does the measure accurately reflect the
criterion?"

It is the position of this text that too much attention has
been placed on the reliability of measures, and not enough on the
validity. Since some form of reliability is a necessary but not
sufficient condition for validity, and since validity is the ultimate goal, more attention should be placed on validity. One
could go so far as to say that this text takes the position that
reliability is not a valid construct.

The criterion as an approximation of the construct

In the past, cause-and-effect has been interpreted on data
which, although highly significant, yields small R^2 values. One
of the positions of this text is that researchers can more fruit-

fully spend their time by finding variables which will increase
the R^2, rather than trying to understand how that small proportion of criterion variance has been predicted. Related to this
problem, though, is the adequacy of the overlap of the criterion
measure being used with the construct one has chosen to consider.
Too often the criterion is equated with the construct, when in
reality it must be realized that the criterion is only an approximation of the construct. Unfortunately, there is no way at the
present to know what overlap exists between the construct and the
criterion variable. Figure 7.25b is the state of affairs assumed
by most researchers, whereas Figure 7.25a is more likely the
state of affairs--that is, there is only a partial overlap between the criterion and the construct. One may do well to continue trying to account for the criterion variance beyond that
accounted for by predictors 1, 2, and 3 in Figure 7.25a because
of the remaining overlap between construct and criterion that has
yet to be accounted for.

Figures 7.25c and 7.25d may also be the existing state of
affairs. In Figure 7.25c, the common overlap between the criterion and the construct has yet to be accounted for. That part
of the criterion which is not predictive of the construct has
(unfortunately) been well accounted for. Any interpretation regarding the causers of the construct in this situation will be
grossly mistaken. If the predictor variables should really map
the construct, according to one's theory, then the predictor
variables should probably be rescaled.

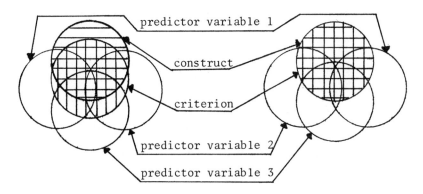

criterion approximates
construct

criterion and construct
isomorphic

Figure 7.25a   Figure 7.25b

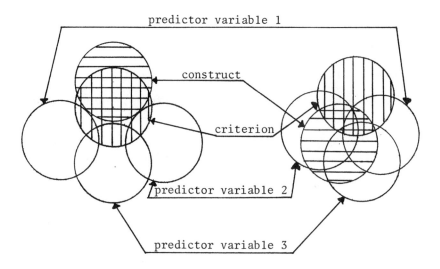

predicted variance not
overlapping with
construct variance

all of the common variance
between criterion and construct is accounted for

Figure 7.25c   Figure 7.25d

Figure 7.25. Some possible consternating relationships
            between criterion, construct, and predictor
            variables.

Figure 7.25d presents an entirely different picture. All of the common variance between the criterion and the construct has been accounted for. To attempt to build up the predictability of the criterion would in this case be wasted effort, since the criterion is mapping no more of the construct. Interpretations of causality in this case may be of some value, but one must note that complete accounting of the construct (as indexed by R^2) will not be possible. The situation in Figure 7.25d may call for the rescaling of the criterion measure. If various rescaling measures do not map the construct better, then the researcher should either turn to another criterion measure or realize that the construct he is concerned about is multidimensional. In this latter case, multiple criterion measures would be necessary, an extension of multiple regression notions. Many constructs in the behavioral sciences may well be multidimensional, but unfortunately the accounting for multiple criterion variables has not been fully developed by statisticians. Many multivariate procedures do exist, but are extensions beyond the scope of this text. The interested reader may refer to Kerlinger and Pedhazur (1973), Cooley and Lohnes (1962), Cattell (1966), and Veldman (1967).

8

Detection of Change

Someone engaged in research usually wants to do more than describe a situation at a particular point in time. He wants to detect when changes occur; and he wants to determine what caused those changes so that in the future those causal factors (or causers) can be purposefully manipulated to bring about desirable changes. The agricultural scientist may find that soil composition is a predictor of crop yield; he may find that certain soil compositions predict high yield and other compositions predict low crop yield. But in addition, he wants to "upset the prediction" for the "poor" soil compositions so that they will generate high crop yield instead of their predicted low yield. In order to do this, he must identify what causes the differential yield--is it entire soil composition, or is it some aspect of composition, or is there some other explanation?

Ascertaining causality

One of the major goals of scientific inquiry is to identify causal relationships because one of the aims of the human endeavor is to "make things better" (by one's own definition) by upsetting the prediction that exists and causing a more desirable

outcome. Researchers may be able to predict that some children
will learn more slowly than others, or that certain people will
contract cancer, or that certain fields will yield fewer crops.
But these researchers want to upset those predictions: those
children who are expected to learn less would learn as well;
those people who are expected to get cancer would not contract
the disease; and those fields which are expected to yield few
crops would yield more. In order to upset prediction, the causal
factors must be identified and then changed through manipulation.
The process of identifying a causal factor, or causer, might be
something like this: First, a situation is described, and the researcher then gets some ideas about what causers are operating.
He then formulates a hypothesis (directional, if he has enough
information) regarding what factors predict the outcome he is interested in. He relies on intuition and theoretical and logical
tenets to design a study to test his hypothesis; he wants this
study to eliminate as many competing explanations of the results
as possible. He conducts the study and tests his hypothesis. If
his directional hypothesis is supported, he has evidence about
which factors predict the outcome of interest. He then formulates
a directional hypothesis regarding which factors cause the outcome; and he must now design a study in which he will eliminate
further competing explanations and in which he will manipulate
the hypothesized "causers" to see if a change in the causers
brings about the expected change in the outcome. The ultimate
test for having discovered a causer rests upon successive repli-

cations of the manipulated causer, all resulting in the hypothe-

sized change in the criterion.

Causality cannot be ascertained in a single study. There

will always be too many competing explainers of the results.

Campbell and Stanley (1963) have a well-organized categorization

of possible competing explainers. All of these possible competing

explainers must be eliminated, either by the design of the study,

or by logical analysis. The design of the study cannot rule out

all of the possible explainers, so logical analysis must always

be referred to in order to ultimately ascertain causality. For

some inexplicable reason, researchers often indicate they are not

trying to ascertain causal relationships, but a little question-

ing as to their purposes usually reveals a causative purpose.

Ennis (1973) has some valuable thoughts on causality which go be-

yond the scope of the present text.

Directional research hypotheses

The utility of a directional research hypothesis becomes

even more apparent when trying to detect where change occurs due

to what causers. A new teaching method is developed not simply to

change the rate of learning, but to increase the rate. A new diet

is developed not just to change the weight of overweight people,

but to reduce the weight. A new traffic regulation is put into

effect not merely to change the accelerating rate of accidents,

but to reduce the trend. If researchers are going to ultimately

upset prediction, then they must know the direction in which the causative factors elicit their effect.

Random assignment

The major tool for letting the design of the study eliminate possible competing explainers has been random assignment. One should note, though, that random assignment <u>does</u> <u>not</u> guarantee that the samples are equivalent. Random assignment only maximizes the probability that the samples will be equivalent. Very seldom are two random samples equal on a given variable, let alone equal on all variables. Furthermore, the populations to which most re- searchers want to generalize are so large or entirely theoretical that a random sample of that population is physically impossible. Most often the subjects are randomly sampled from some accessible population which may in many ways differ from the population to which the researcher wishes to generalize. Few studies in the literature employ a random sample from the population to which the researcher has generalized. These statements are not intended as a condemnation of the literature, but as a condemnation of researchers relying upon random sampling as a cure all.

On the other hand, simply arriving on the scene and mea- suring all the variables simultaneously does not allow causal statements either. Again, causality can only be ascertained by manipulating a variable and always observing the predicted re- sult. Variables in the real world are highly correlated and therefore a change in one variable <u>may</u> change one or more other predictor variables. But just because two variables are correlated does not mean that the one variable causes the other. Both variables could be caused by one or more other variables. Again, only a tight logical analysis can tease out the causative variables. Manipulation of the proposed causative variables is a necessary step in determination of causality.

One group--post only

The recent emphasis on criterion referenced testing and on the explicit stating of objectives implies that more researchers will be testing hypotheses about a single population mean. Consider a project utilizing methods to reduce alienation. One of their objectives might be: After six weeks of participation, the alienation mean score of the children in the project will be less than five. Now if the project director is interested only in how the project works for the few children in the project, he simply needs to look at the sample alienation mean to see if it is less than five. But a more reasonable desire is to infer to the adequacy of the project, with the intent of adopting it in other schools. With this desire, the project director wants to infer to a population of children. The Research Hypothesis in this case would be: "After six weeks of instruction, the Alienation mean score in the population will be less than five." To answer any research hypothesis in multiple linear regression, full and restricted models must be constructed.

The Research Hypothesis dictates a Full Model which must allow the Alienation mean to manifest itself:

Model 8.1: $Y_1 = a_0 U + E_1$

    where: Y_1 = Alienation scores.

Readers by now should recognize this model as "the unit vector model," yielding no differential predictability (R^2 = .00). The one regression coefficient that must be determined is a_0, and this will be the sample mean. The Statistical Hypothesis implies the restriction $a_0 = 5$. Forcing this restriction on the Full Model results in the following algebraic gyrations:

    Full Model: $Y_1 = a_0 U + E_1$

    Restriction: $a_0 = 5$

    Restricted Model: $Y_1 = 5U + E_2$

But since U = 1 for all subjects, 5U is a constant, and subtracting that constant from both sides yields the final form of the Restricted Model:

Model 8.2: $(Y_1 - 5) = E_2$

The same general F test formula is applicable to all hypotheses, providing that the unit vector is in both the full and restricted models. If this is not the case, and the present situation is not, then an alternative formula for the F test must be used (Bottenberg and Ward, 1963):

Equation 8.1: $F_{(m_1-m_2,\ N-m_1)} = \dfrac{(ESS_r - ESS_f)/(m_1 - m_2)}{(ESS_f)/(N-m_1)}$

    where: ESS_r = the error sum of squares in the Restricted Model;

               ESS_f = error sum of squares in the Full Model;

               m_1 = number of linearly independent vectors in the Full Model (number of pieces of information); and

m_2 = number of linearly independent vectors in the
Restricted Model (number of pieces of information).

Equation 8.1 is more general than is the F test in terms of

R^2, although the latter is more conceptually appealing. If one

were to apply Equation 8.1 to any of the previous examples, the

identical F would result.

The sum of the squared elements in E_1 in Model 8.1 will be

the ESS_f. The sum of the squared elements in E_2 (or Y_1 - 5U) in

Model 8.2 will be the ESS_r. Note that the Full Model utilizes one

piece of information (the unit vector), whereas the Restricted

Model utilizes no information; therefore, m_1 = 1 and m_2 = 0. The

difference between m_1 and m_2 is one, being equal to the number of

restrictions made, and also being the degrees of freedom numer-

ator for the F test. Table 8.1 contains the intermediate values

for the solution of this kind of hypothesis on 24 subjects. The

significance of the resultant F of 101.5 must be judged by re-

ferring to tabled values, and since this was a directional hy-

pothesis, one must use the 90th percentile of F if his alpha was

.05 and the sample results are in the hypothesized direction. If

the Alienation sample mean was greater than 5, there would have

been no need to go through the statistical gyrations; it would

have sufficed to report "not significant," and then suggest drop-

ping the project.

Hypotheses about a proportion could also be tested in this

manner. The criterion vector in this case would be a dichotomous

Table 8.1 Numerical solution for regression testing of an hypothesized population mean.

Y_1	$=$	a_0U	$+$	E_1	E_1^2	E_2 $(Y_1 - 5)$	E_2^2 $(Y_1 - 5)^2$
1				1	-1.5	2.25	-4	16
1				1	-1.5	2.25	-4	16
1				1	-1.5	2.25	-4	16
1				1	-1.5	2.25	-4	16
4				1	1.5	2.25	-1	1
3				1	.5	.25	-2	4
3				1	.5	.25	-2	4
2				1	- .5	.25	-3	9
1				1	-1.5	2.25	-4	16
4				1	1.5	2.25	-1	1
2	$= 2.5$			1	- .5	.25	-3	9
1				1	1.5	2.25	-4	16
3				1	.5	.25	-2	4
2				1	- .5	.25	-3	9
3				1	.5	.25	-2	4
1				1	-1.5	2.25	-4	16
5				1	2.5	6.25	0	0
3				1	.5	.25	-2	4
2				1	- .5	.25	-3	9
3				1	.5	.25	-2	4
3				1	.5	.25	-2	4
3				1	.5	.25	-2	4
4				1	1.5	2.25	-1	1
4				1	1.5	2.25	-1	1

$$\Sigma = 34 \qquad\qquad\qquad \Sigma = 184$$
$$\mathrm{ESS}_f = 34 \qquad\qquad\qquad \mathrm{ESS}_r = 184$$

$$\underline{F}_{(m_1-m_2,\ N-m_1)} = \frac{(\mathrm{ESS}_r - \mathrm{ESS}_f)/(m_1-m_2)}{(\mathrm{ESS}_f)/(N-m_1)}$$

$$\underline{F}_{(1-0,\ 24-1)} = \frac{(184-34)/1}{(34)/23}$$

$$\underline{F}_{(1,23)} = 101.5$$

Generalized Research Hypothesis 8.1

Directional Hypothesis: For a given population, the criterion
mean is greater than some specified
value a.

Non-directional Hypothesis: For a given population, the criterion mean is different than some
specified value a.

Statistical Hypothesis: For a given population, the criterion
mean is equal to some specified value a

Full Model: $Y_1 = a_0U + E_1$

Restrictions: $a_0 = a$

Restricted Model: $Y_1 = aU + E_2$; or $(Y_1 - aU) = E_2$

where:

Y_1 = criterion;
U = 1 for all subjects; and
a_0 is a least squares weighting coefficient calculated
to minimize the sum of the squared values in the
error vector.

Degrees of freedom numerator = (m_1-m_2) = $(1-0)$ = 1

Degrees of freedom denominator = $(N-m_1)$ = $(N-1)$

where:

N = number of subjects.

---

vector rather than a continuous vector as in the alienation example. The Generalized Research Hypothesis for testing a single
population value appears in GRH 8.1

Researchers having access to computing facilities can perform the required analysis quickly, as one computer run can provide all the component values of the F test. The substitution of

---

Applied Research Hypothesis 8.1

Directional Research Hypothesis: For a given population, the
                                 population mean of X_2 is
                                 greater than the national norm
                                 of 20.

Statistical Hypothesis: For a given population, the population
                        mean of X_2 is equal to the national
                        norm of 20.

Full Model: $X_2 = a_0U + E_1$

Restricted Model: $(X_2 - 20) = E_2$

alpha = .05

ESS_f = 7771 ESS_r = 8723

$\underline{F}_{(1,59)}$ = 7.21

Interpretation: Since the sample mean of X_2 is 23.98, the 90th
                percentile of $\underline{F}$ can be referred to. The neces-
                sary $\underline{F}$ value is 2.75, and since the calculated
                $\underline{F}$ is larger, the Statistical Hypothesis can be
                rejected, and the Research Hypothesis can be
                accepted: For the given population, the popu-
                lation mean of X_2 is greater than the national
                norm of 20.

---

the numerical values into Equation 8.1 must be done by hand, but

that is a small price to pay for the utilization of the flexible

multiple linear regression technique.

Applied Research Hypothesis 8.1 is provided for the reader

to verify his skills in testing a single population mean (or

single population proportion, if the criterion is dichotomous)

hypothesis.

One group--pre and post

The researcher is often interested in the gain from the beginning of the treatment to the end of the treatment. Hence, not
only a post measure is obtained, but also a pre-assessment on the
same measure. There are three ways to analyze the data, with the
last method being the preferred.

(1) One group--pre and post with gain as the criterion. The
pre-measure can be subtracted from the post-measure for each subject, yielding a gain, difference, or change score. This gain
score can be used as a criterion in the following Research Hypothesis: "The mean gain from pre to post is greater than zero
(or any specified value)." The structure of this Research Hypothesis is the same as GRH 8.1, with the pre-post difference as
the criterion.

(2) One group--two repeated measures. Since the pre- and
post-measure are of the same variable, one could treat both measures as a criterion, and use a pre-vector and a post-vector as
the two predictor variables (to see if knowledge of whether the
score was pre or post helps predict the score). (A pre-vector
would be a vector which identifies those scores in the criterion
which are pre-scores by containing a one if the score is a pre-
score, zero otherwise. A post-vector identifies the post-scores
in the criterion in the same manner. These are the same as the
time vectors (T_1 and T_2) in GRH 8.2.) The Research Hypothesis in
this case would be: "The post mean is higher than the pre mean."

The number of scores in the criterion would be equal to twice the number of subjects since every subject would have both a pre and post score. GRH 6.1 in Chapter Six would be used to test this Research Hypothesis, with the denominator degrees of freedom equal to [(2N) - 1], where N is equal to the number of subjects.

(3) One group--repeated measures, person vectors. The second analysis (above) is not as powerful as it might be, since some very important information about the criterion scores has been omitted. Indeed, one of the assumptions of statistical analysis-- independence of observations--is violated when person vectors are not included. Since every subject is tested twice, the researcher knows that one pre-score and one post-score come from a given subject. One other pre-score and one other post-score come from another subject, and so on. This information is quite valuable information, as important as individual differences are in most areas. That is to say, if a person scores low on the pre-test, he is expected to score relatively low on the post-test. Con- versely, if a subject scores high on the pre-test, he is ex- pected to score high on the post-test. The inclusion of person vectors, as illustrated in Table 8.2, drains the denominator degrees of freedom, but the increase in the amount of variance accounted for is usually of such a large magnitude that it off- sets the loss in degrees of freedom.

Of the three designs discussed in this section, the inclu- sion of the person vectors utilizes best the information in the study. Another way of stating the above is to say that in most

cases, the inclusion of the person vectors will yield a lower
probability value than will the other two designs, (1) or (2)
above. Design (3) is widely used and is of sufficient value to
present it as Generalized Research Hypothesis 8.2.

Additional repeated measures might be obtained, necessi-
tating simply more time vectors (T_i) in the GRH 8.2 Full Model.
Specific hypotheses would dictate the restrictions made on that
Full Model, and one can apply the multiple comparison notions
discussed in Chapter Six to the present design.

Table 8.2 Vector conceptualization of repeated measures
design.

$$Y_1 = a_0 U + a_1 T_1 + a_2 T_2 + p_1 P_1 + p_2 P_2 + p_3 P_3 + p_4 P_4 + p_5 P_5 + p_6 P_6 + p_7 P_7 + E_1$$

	U	T_1	T_2	P_1	P_2	P_3	P_4	P_5	P_6	P_7	E_1
Sam's pre	1	1	0	1	0	0	0	0	0	0	?
Joe's pre	1	1	0	0	1	0	0	0	0	0	?
Sue's pre	1	1	0	0	0	1	0	0	0	0	?
Pat's pre	1	1	0	0	0	0	1	0	0	0	?
Sal's pre	1	1	0	0	0	0	0	1	0	0	?
Hal's pre	1	1	0	0	0	0	0	0	1	0	?
Tom's pre	1	1	0	0	0	0	0	0	0	1	?
Sam's post	1	0	1	1	0	0	0	0	0	0	?
Joe's post	1	0	1	0	1	0	0	0	0	0	?
Sue's post	1	0	1	0	0	1	0	0	0	0	?
Pat's post	1	0	1	0	0	0	1	0	0	0	?
Sal's post	1	0	1	0	0	0	0	1	0	0	?
Hal's post	1	0	1	0	0	0	0	0	1	0	?
Tom's post	1	0	1	0	0	0	0	0	0	1	?

Two groups--post only

The introduction of a control group often allows the elimination of many competing explainers. The simplest design entail-
ing both a control and an experimental group is the one discussed

Generalized Research Hypothesis 8.2

Directional Research Hypothesis: For a given population, the post mean is higher than the pre mean, over and above expected individual differences.

Non-directional Research Hypothesis: For a given population, the post mean is different than the pre mean, over and above expected individual differences.

Statistical Hypothesis: For a given population, the post mean is equal to the pre mean, over and above expected individual differences.

Full Model: $Y_1 = a_0U + a_1T_1 + a_2T_2 + p_1P_1 + p_2P_2 + \cdots p_nP_n + E_1$

Restrictions: $a_1 = a_2$

Restricted Model: $Y_1 = a_0U + p_1P_1 + p_2P_2 + \cdots p_nP_n + E_2$

where:

Y_1 = criterion of <u>both</u> pre- and post-scores;
T_1 = 1 if pre-score, 0 otherwise;
T_2 = 1 if post-score, 0 otherwise;
P_1 = 1 if score from subject 1, 0 otherwise;
P_2 = 1 if score from subject 2, 0 otherwise;
.
.
.
P_n = 1 if score from subject n, 0 otherwise; and
a_1, a_2, p_1, p_2, $\cdots p_n$ are least squares weighting coefficients calculated so as to minimize the sum of the squared values in the error vectors.

Degrees of freedom numerator = $(m_1-m_2) = [(n+1) - n] = 1$

Degrees of freedom denominator = $(N-m_1) = [(n+n) - (n+1)] =$

$(n-1)$

where:

n = number of subjects

N = number of elements in each of the vectors--in all
other GRH, there was only one observation for each
subject, and N = n. For repeated measures analyses,
there will be as many elements in each vector as
there are subjects times how many times they were
measured.

Generalized Research Hypothesis 8.3

Directional Research Hypothesis: For a given population, the
post mean of the experimental
method is higher than the post
mean of the control method.

Non-directional Research Hypothesis: For a given population,
the post mean of the ex-
perimental method is not
equal to the post mean of
the control method.

Statistical Hypothesis: For a given population, the post mean
of the experimental method is equal to
the post mean of the control method.

Full Model: $Y_1 = a_1 G_1 + a_2 G_2 + E_1$

Restrictions: $a_1 = a_2$

Restricted Model: $Y_1 = a_0 U + E_2$

where:

Y_1 = criterion;
U = 1 for all subjects;
G_1 = 1 if subject given the experimental method, 0 otherwise;
G_2 = 1 if subject given the control method, 0 otherwise;
and
a_0, a_1, and a_2 are least squares weighting coefficients
calculated so as to minimize the sum of the squared
values in the error vectors.

Degrees of freedom numerator = $(m_1 - m_2)$ = $(2-1)$ = 1

Degrees of freedom denominator = $(N - m_1)$ = $(N-2)$

where:

N = number of subjects.

in this section. Only a post-measure is obtained, and for change
to be attributable to the experimental method, the directional
research hypothesis in GRH 8.3 must be supported. GRH 8.3 is exactly the same structure as GRH 6.1, the "two-group" structure.
Notice, though, in this context the definite importance of stating and obtaining a directional difference in favor of the experimental method.

Lack of knowledge as to the standing of the subjects before
treatment forces a qualified interpretation. Random assignment
does not guarantee that the two groups are initially equal, and
a higher post score by the treatment group could be a function of
that group being initially better. Or, the kinds of subjects
placed in the treatment condition may have responded better to
the experimental situation than did those subjects assigned to
the control situation. The next design (two groups--pre and post)
allows the researcher to eliminate this possible competing explainer.

Two groups--pre and post

The assessment of initial standing via a pre-test is one way
of obtaining some knowledge of initial standing. There are four
different ways of analyzing such data, the first three being
slightly easier to handle conceptually and operationally, the
last being more powerful statistically.

(1) Two groups--pre and post with gain as the criterion.
The pre-score can be subtracted from the post-score to yield a

---

Applied Research Hypothesis 8.2

Directional Research Hypothesis: For a given population, X_{12} achieves more gain (X_2-X_3) than does X_{13}.

Statistical Hypothesis: For a given population, X_{12} achieves gains equal to that of X_{13}.

Full Model: $(X_2-X_3) = a_0U + a_{12}X_{12} + a_{13}X_{13} + E_1$

Restrictions: $a_{12} = a_{13}$

Restricted Model: $(X_2-X_3) = a_0U + E_2$

alpha = .05

$R_f^2 = .00$ $R_r^2 = .00$

$\underline{F}_{(1,58)} = .009$ Directional p = .54

Interpretation: Since the mean gains are in the opposite direction than hypothesized, support for the
Research Hypothesis has not been established,
and the Statistical Hypothesis cannot be rejected. The same conclusion is reached by comparing p to alpha.

---

gain, difference, or change score, which then becomes the cri-

terion. The Research Hypothesis again fits into the GRH 6.1

framework: "The treatment method is better than the control

method in terms of the criterion Y_1 (gain from pre to post)."

Applied Research Hypothesis 8.2 gives the reader a chance to ap-

ply the two group hypothesis to data regarding gain scores.

Thus, the control group may indicate a significant gain,

whereas the experimental group shows a real gain over and above

that of the control group. The net difference between the two

Applied Research Hypothesis 8.3

Directional Research Hypothesis: For a given population, X_{12} is better than X_{13} on the criterion of X_2, over and above any differences observed on the pre-scores, X_3.

Statistical Hypothesis: For a given population, X_{12} is equal to X_{13} on the criterion of X_2, over and above any differences observed on the pre-scores, X_3.

Full Model: $X_2 = a_0U + a_{12}X_{12} + a_{13}X_{13} + a_3X_3 + E_1$

Restrictions: $a_{12} = a_{13}$

Restricted Model: $X_2 = a_0U + a_3X_3 + E_2$

alpha = .05

$R_f^2 = .928$ $R_r^2 = .927$

$\underline{F}_{(1,57)} = .81$ Directional p = .82

Interpretation: Since the weighting coefficient for X_{12} in the Full Model is less than the weighting coefficient for X_{13}, the Research Hypothesis cannot be supported, and the Statistical Hypothesis cannot be rejected. The same conclusion is reached by comparing p to alpha.

groups may be thought of as the final difference between the two groups corrected for their initial difference. Such a correction, though, involves the assumption that each unit of difference in initial standing will produce a constant difference in final standing (McNemar, 1962, p. 373). This assumption can be stated statistically as a correlation of 1.0 between the pre- and post-scores. Although this assumption would probably never be valid in an applied setting, testing gain scores is recommended over those

procedures previously discussed in this chapter. Most statisticians recommend the difference method over that of matching,
mainly because of the difficulty in selecting good matched samples, as shall be discussed shortly.

(2) Two groups--pre and post, with pre used as a covariate.
If one does not want to assume that the correlation between pre
and post if 1.0, then the pre-score can be used as a covariate.
The Research Hypothesis in this case would be: "The treatment
method is better than the control method on the post scores, over
and above any differences observed on the pre scores." This research hypothesis fits the framework of GRH 5.1, with the pre-
test as a covariate and the treatment and control dichotomous
vectors as the only other vectors in the Full Model. Applied
Research Hypothesis 8.3 gives one a chance to test such an hypothesis.

(3) Two groups--pre-test(s) different from post-test. There
are many research situations wherein it is unwise to administer
the same test pre as will be administered post. In this situation, behaviors related to the post-test can be assessed pre
(prior to the treatment) and used as covariates. The researcher
can assess as many behaviors as he has reason to suspect might
be different between his groups. The Research Hypothesis in this
case would be: "The treatment method is better than the control
method on the post-test Y_1, over and above the possible contaminating variables of A_1, B_1, C_1, etc." This research hypothesis
fits the framework given previously in GRH 6.4.

(4) Two groups--pre and post--interaction between method and time, with person vectors. Don't let the long and probably confusing heading scare you. What is now going to be done is simply to combine the notions of interaction and repeated measures (with directionality, of course). Figure 8.1 contains three possible empirical outcomes which would satisfy the following Research Hypothesis:

Research Hypothesis: The relative effectiveness of the treatment method as compared to the control method will be greater at time 2 (post) than at time 1 (pre), over and above the expected individual differences.

Statistical Hypothesis: The relative effectiveness of the treatment method as compared to the control method will be the same at time 2 (post) as at time 1 (pre), over and above the expected individual differences.

The Full Model reflecting the Research Hypothesis is:

Model 8.3: $Y_1 = a_0U + a_1(T_1{}^*X_1) + a_2(T_1{}^*X_2) + a_3(T_2{}^*X_1) +$

$$a_4(T_2{}^*X_2) + p_1P_1 + p_2P_2 + \cdots p_nP_n + E_3$$

The restriction implied by the Research and Statistical Hypothesis is:

$$(a_1-a_2) = (a_3-a_4)$$

resulting in the following Restricted Model:

Model 8.4: $Y_1 = a_0U + a_1T_1 + a_2T_2 + a_3X_1 + a_4X_2 + p_1P_1 + p_2P_2$

$$+ \cdots p_nP_n + E_4$$

where: Y_1 = the criterion variable containing both pre
             and post scores;
         T_1 = 1 if score from pre, 0 otherwise;
         T_2 = 1 if score from post, 0 otherwise;
         X_1 = 1 if score from treatment, 0 otherwise;
         X_2 = 1 if score from control, 0 otherwise;
         P_1 = 1 if score from person 1, 0 otherwise;
         $\vdots$
         P_n = 1 if score from person n, 0 otherwise.

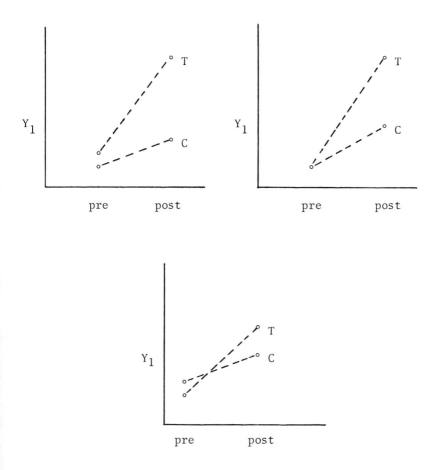

Figure 8.1. Three possible empirical outcomes wherein the treatment (T) increases more than the control (C).

To ascertain the number of linearly independent vectors in the Restricted Model, first consider the unit vector. Since one of the time vectors can be obtained by subtracting the other time vector from the unit vector, only one of the time vectors will be linearly independent. Likewise, only one of the treatment vectors will be linearly independent. Now if all of the person vectors

Applied Research Hypothesis 8.4

Directional Research Hypothesis: For a given population, the relative effectiveness of X_{12} over X_{13} is less on post (X_2) than on pre (X_3), over and above individual differences $(X_{21} \ldots X_{80})$.

Statistical Hypothesis: For a given population, differences (if any) between X_{12} and X_{13} are the same on post (X_2) as on pre (X_3), over and above individual differences.

Full Model: $X_{(2 \text{ and } 3)} = a_0 U + a_{17} X_{17} + a_{18} X_{18} + a_{19} X_{19} + a_{20} X_{20} + P_{21} X_{21} + \cdots P_{80} X_{80} + E_1$

where:

$X_{17} = X_{12} * X_{15}$
$X_{18} = X_{12} * X_{16}$
$X_{19} = X_{13} * X_{15}$
$X_{20} = X_{13} * X_{16}$

Restrictions: $(a_{17} - a_{18}) = (a_{19} - a_{20})$

Restricted Model: $X_{(2 \text{ and } 3)} = a_0 U + a_{12} X_{12} + a_{13} X_{13} + a_{15} X_{15} + a_{16} X_{16} + P_{21} X_{21} + \cdots$
$P_{80} X_{80} + E_2$

alpha = .05

$R_f^2 = .8830$ $R_r^2 = .8829$

$\underline{F}_{(1,58)} = .016$ Directional p = .4502

Interpretation: The difference between a_{20} and a_{19} in the Full Model is less than the difference between a_{18} and a_{17}, therefore the directional probability can be referred to. Since the directional probability is not less than the predetermined alpha, the Statistical Hypothesis cannot be rejected, and hence the Research Hypothesis cannot be accepted.

representing persons in the treatment group are added together, the treatment vector will be obtained. Hence one of those person vectors is linearly dependent. Likewise, one of the control person vectors is linearly dependent. Therefore only n-2 of the person vectors are linearly independent. Adding n-2 to the three previously determined linearly independent vectors results in (n+1) linearly independent vectors. Therefore, $df_n = (m_1-m_2) = [(n+2) - (n+1)] = 1$ and $df_d = (N-m_1) = [(n+n) - (n+2)] = n-2$.

Notice that the experimental group does not have to be equal to the control group on the pre-measure. The experimental group can be higher than the control group, or lower--the concern is in the relative effectiveness on the post-measure as compared to the superiority on the pre-measure. Applied Research Hypothesis 8.4 provides the models appropriate to the Appendix C data for this kind of hypothesis. Namboodiri (1972) reviews recent developments regarding repeated measures designs.

Matched groups

This section and the following section are included solely to familiarize students with the process of matching, so common in early empirical research. The pitfalls of matching have become more apparent, and the availability of computers has brought analyses such as GRH 6.4 to the forefront.

There are two somewhat different ways to match: Match individual pairs; or match groups on the basis of group indices (means, standard deviations, etc.).

Procedures for matching groups is discussed adequately by
McNemar (1962). Basically, subjects are pre-tested on a number of
supposedly relevant variables, and then assigned to groups such
that the resultant groups are "similar" (not necessarily equivalent) on the several supposedly relevant variables. Often, the
sole requirement of similarity is that of equal means. An additional requirement that is sometimes used is that of similar
standard deviations. Unfortunately, only visual comparisons are
utilized to ascertain similarity. A totally indefensible procedure that is sometimes used in place of the above is that of
visually comparing the extreme scores of the various groups. As
most researchers know, extreme scores tell little about the bulk
of the scores.

After the two matched groups are obtained, the format of
the research hypothesis would be similar to that of GRH 8.3. That
is, the matching variables are not used in the analysis of the
data, only for the delineation of the two groups. The matching
variable could be used as a "blocking variable." That is, the
matching variable could be used as a factor, not of interest to
the researcher, but as a means to reduce the within variability,
thus resulting in a smaller error component. It is relatively
cumbersome to use more than one blocking variable in this manner.

The most important problems related to matching groups will
now be discussed. First, the relevancy of the matching variables
is often of questionable value. If there is no relationship be-

tween the matching variable and the criterion, then the results
may be generalized to subjects that differ on the matching variable, although the experimenter who has needlessly matched does
not know that this generalization can be made.

A second problem involves the difficulty in equating the
groups on the matching variables. To come up with two groups
having equal means on a number of matching variables is a very
difficult task. Of course, the more matching variables one is
interested in, the more unlikely it is that one can come up with
equivalent groups. This problem will be further delineated in the
following discussion concerning matched pairs of individuals.

Matched pairs

A more thorough procedure for matching involves matching
individuals. Thus, for every person in the treatment group, there
is a person like him in the control group. The extent to which
the control subject is like the treatment subject is again a
function of the number and relevancy of the matching variables.
Sex may be an important matching variable in one study, whereas
in another study it may not be relevant. Whether or not a variable should be considered as a matching variable is an empirical
question which can only be estimated ahead of time on theoretical
grounds by the researcher. Matching pairs is an attempt to approach internal validity, as discussed by Campbell and Stanley
(1963).

Since for every person in the experimental group there is a

like person in the control group, person vectors can be utilized
as in GRH 8.2. Notice that here, as in the matching of groups,
the matching variables are not used in the analysis of the data.
The assumption in the matched pairs design is that the person in
the experimental group is exactly like the "like" person in the
control group. That this assumption is not valid should be readily apparent.

There are three major problems with the technique of matched
pairs. First, since the selection of one kind of individual demands a like person, the degrees of freedom are drastically reduced. For an "uncorrelated" or "independent" t test, the degrees
of freedom is N-2, where N is the total number of people in the
two groups. For the matched pairs analysis (referred to as "correlated" or "dependent" t test by some), the degrees of freedom
is the number of pairs minus one. Thus, if one decides to match,
the gain that one obtains from matching must be enough to over-
ride the loss in degrees of freedom (exactly half as many degrees
of freedom). If the matching variables are chosen judiciously,
the loss in degrees of freedom will not be of any consequence.

A second and more important problem with matched pairs concerns the availability of "identical" subjects. With individual
differences being as large and variable as they are, it is often
difficult to find an adequate match. Empirical studies using this
procedure often demand an initial sampling pool of 300 or 400 to
get even a crude match for 30 pairs of subjects.

The third problem with matched pairs is related to the above problem. The search for a matching individual often eliminates the extreme subjects, those who are deviant on only one measure. Conversely, those deviant matches that are found are often a function of a large error component, and on a second testing this "match" may indeed turn out to be not a good match after all.

Point change measuring multiple subjects

Some research questions involve the detection of a change at a certain point in time as a function of some stimulus change or input. An ideal situation would be to measure a number of subjects in a number of testing periods both before and after the stimulus change. The number of time periods before the stimulus change would not have to be equal to the number of measured time periods after the stimulus change, although the following example does have an equal number.

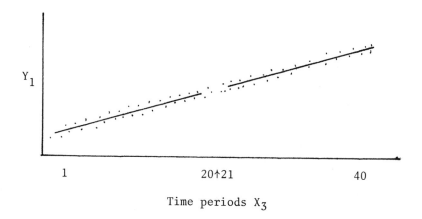

Figure 8.2. A situation where the stimulus input (identified by the arrow between Time periods 20 and 21) has not caused a change.

Figure 8.2 depicts a situation wherein the mean of the last 20 time periods is much greater than the mean of the first 20 time periods. But the careful researcher surely would not want to attribute a causal influence to the stimulus input which occurred between Time periods 20 and 21. A regression model which allows the two lines depicted in Figure 8.2 would be:

Model 8.5: $Y_1 = a_0U + a_1U_1 + a_2X_1 + a_3U_2 + a_4X_2 + E_5$

where: Y_1 = the criterion to be predicted;
U_1 = 1 if observation from Time period 1-20; 0 otherwise;
X_1 = the Time period if observation from Time period 1-20, 0 otherwise;
U_2 = 1 if observation from Time period 21-40, 0 otherwise; and
X_2 = the Time period if observation from Time period 21-40, 0 otherwise.

A very reasonable question to ask of the data in Figure 8.2 is, "How much error would be introduced if only one line were used to depict the data?" Another way of expressing this question is to ask if the slopes of the two lines are the same, and if the Y-intercepts of the two lines are the same. In Model 8.5, the symbols a_2 and a_4 represent the two slopes, so one is essentially asking if it is reasonable to force a_2 equal to a_4. Also in Model 8.5, the Y-intercepts are a function of a_1 and a_3. If a_1 is forced to be equal to a_3, then one is forcing the two lines to have the same Y-intercepts. If one simultaneously forces the two slopes to be the same ($a_2 = a_4$, both equal to a_5, a common weight) and the two Y-intercepts to be the same ($a_1 = a_3$, both equal to a_6, a common weight), then one is indeed asking if one line can do as adequate a job as the two lines. Forcing these two restric-

tions on Model 8.5, and with minor algebraic manipulation, one
obtains:

$$Y_1 = a_0 U + a_6 U_1 + a_5 X_1 + a_6 U_2 + a_5 X_2 + E_6$$

$$Y_1 = a_0 U + a_6 (U_1 + U_2) + a_5 (X_1 + X_2) + E_6$$

$$Y_1 = a_0 U + a_6 U + a_5 (X_3) + E_6$$

$$Y_1 = (a_0 + a_6) U + a_5 X_3 + E_6$$

Model 8.6: $Y_1 = a_7 U + a_5 X_3 + E_6$

where: X_3 = a continuous vector of Time periods.

The Y-intercept of this one line will be represented by a_7
and the slope will be represented by a_5, the coefficient of the
continuous vector of Time periods. If both of the restrictions
are indeed true for the sample, then the elements of E_6 will be
the same as the elements of E_5; the errors in prediction using
the one line will be comparable to the errors in prediction using
the two separate lines.

If the amount of variance in the criterion scores that is
being accounted for in the Restricted Model (Model 8.6) is significantly less than that in the Full Model, then one line does not
do as good a job as do two lines, and there may well be an effect
due to the stimulus change. Reasons for a cautious interpretation
will be discussed in the next section on curvilinear relationships.

In another situation, the researcher may expect that slopes
of the lines (the increase in Performance level) will stay the
same before and after the stimulus change, but that following the
stimulus change, there is a jump in Performance level so that the

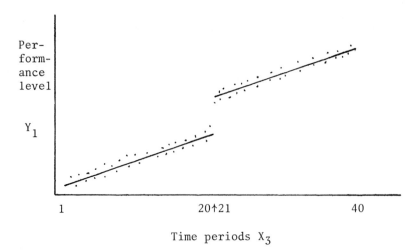

Figure 8.3. A case where the slope of the line after stimulus change is the same as before stimulus
change.

Y-intercepts are not equal, as shown in Figure 8.3.

In this case, the Full Model would allow for different Y-

intercepts but only one slope for the two lines:

Model 8.7: $Y_1 = a_0U + a_1U_1 + a_2U_2 + a_3X_3 + E_7$

where: Y_1 = Performance level;
U_1 = 1 if observation from Time period 1-20, 0
otherwise;
U_2 = 1 if observation from Time period 21-40, 0
otherwise; and
X_3 = Time period.

Since the researcher expects the Y-intercepts to be differ-

ent, the Statistical Hypothesis would state that the Y-intercepts

are equal ($a_1 = a_2$). Forcing the restriction onto the Full Model

would yield:

$$Y_1 = a_0U + a_1U_1 + a_1U_2 + a_3X_3 + E_8$$

Gathering vectors together which have the same weighting coefficient yields:

$$Y_1 = a_0U + a_1(U_1+U_2) + a_3X_3 + E_8$$

The Restricted Model would thus be Model 8.8.

Model 8.8: $Y_1 = a_4U + a_3X_3 + E_8$

If the restriction ($a_1 = a_2$) is not viable and the magnitude
of a_2 is greater than that of a_1, then one can attribute a causal
heightening effect to the stimulus change.

Curvilinear relationships which account for the data better than stimulus change

The interpretations in the above discussions have been somewhat cautious because the models were assuming and acting as if
there were rectilinear relationships. The authors of this text
suspect that there are quite a few curvilinear relationships
existing in the real world, but these relationships have not been
discovered because researchers have not looked for them. If there
is an underlying curvilinear relationship, one may be making a
costly mistake by trying to map the data with two straight lines.

Figure 8.4 depicts a set of data where the performance
scores after the stimulus change continue to follow the curvilinear trend that started before the stimulus change. Visually,
it appears that a single second degree curved line can account
for the data, rather than two second degree curved lines, or two
straight lines.

The two dashed lines in Figure 8.4 would possibly lead to

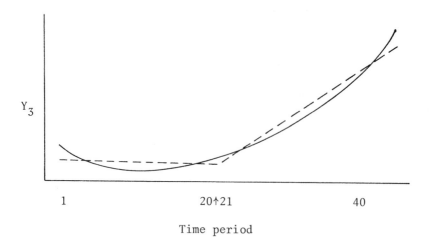

Figure 8.4. A case where a single second degree curved line
depicts the data as well as do two second degree curved lines, and better than two straight
lines.

the conclusion that the stimulus change after Time period 20 had

an effect (of raising the rate of change in Y_3). However, if a

single second degree curve fits all the data well (yielding a

high R^2), then what was happening systematically before the stim-

ulus change is still happening after the stimulus change. A re-

searcher would be hard pressed to argue that the stimulus change

in Figure 8.4 upset the normal course of events.

The regression approach to verifying the visual impression

would be as follows. Just as in the case with rectilinear lines,

one needs to generate a model containing two separate second

degree curved lines.

Model 8.9: $Y_1 = a_0U + a_1Z_1 + b_2Z_2 + s_3Z_3 + a_4Z_4 + b_5Z_5 +$

$$s_6Z_6 + E_9$$

where: Y_1 = the criterion performance score;
Z_1 = 1 if observation is from Time period 1-20, zero otherwise;
Z_2 = the Time period if observation is from Time period 1-20, zero otherwise;
Z_3 = the squared value of the element in Z_2;
Z_4 = 1 if observation is from Time period 21-40, zero otherwise;
Z_5 = the Time period if observation is from Time period 21-40, zero otherwise; and
Z_6 = the squared value of the element in Z_5.

To test the effectiveness of a single line, all that one needs to do is to equate corresponding components of the two separate lines. Since a_1 and a_4 are the two Y-intercepts, b_2 and b_5 are weights for the two linear components, and s_3 and s_6 are weights for the two second degree components, the restrictions are: $a_1 = a_4$, both equal to a_7, a common weight; $b_2 = b_5$, both equal to b_8, a common weight; $s_3 = s_6$, both equal to s_9, a common weight. Inserting these restrictions into Model 8.9, yields:

$$Y_1 = a_0U + a_7Z_1 + a_7Z_4 + b_8Z_2 + b_8Z_5 + s_9Z_3 + s_9Z_6 + E_{10}$$

$$Y_1 = a_0U + a_7(Z_1+Z_4) + b_8(Z_2+Z_5) + s_9(Z_3+Z_6) + E_{10}$$

$$Y_1 = (a_0+a_7)(U) + b_8(Z_7) + s_9(Z_8) + E_{10}$$

Model 8.10: $Y_1 = a_{10}U + b_8Z_7 + s_9Z_8 + E_{10}$

where: Z_7 is the continuous vector of Time periods, and Z_8 is the squared value of the corresponding element in Z_7.

The unit vector allows for a non-zero intercept, the Z_7 vector allows for a linear component, and the squared elements of vector Z_7 appearing in vector Z_8 allow for a second degree

curvilinear component. Model 8.10 thus allows for a single second
degree curve. With reference to data such as in Figure 8.4,
Model 8.10 will yield as good a fit as will Model 8.9. Thus, the
kind of relationship between time and the criterion which was
occurring prior to the stimulus change is also occurring after
the stimulus change.

Whether the researcher chooses a curvilinear or rectilinear
fit to the data is a function of both theoretical notions and
empirical data. A statistician cannot tell a researcher when to
investigate curvilinearity. What the authors of this text have
done is to show the applicability of the multiple regression
procedure to curvilinear types of problems (and to encourage
such application whenever the researcher thinks that such might
be fruitful).

Functional change in individual organism research

The preceding discussion has many ramifications for individual organism research. Kelly, Newman, and McNeil (1973) have
documented a more extensive application of the multiple regression approach to repeated measures of a single organism. The
following paragraphs should only serve to whet the appetites of
empirically oriented single organism researchers. See Chapters
Nine and Ten for a more extensive discussion.

It appears that the "eyeballing" of data emanating from
standard individual organism designs can be bolstered with probability statements of empirical reproducibility. Suppose that some

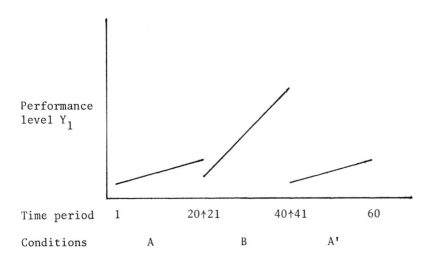

Figure 8.5. One possible set of data emanating from single organism research.

functional relationship is established under a set of conditions which shall be called "A". Then a new set of conditions "B" is introduced to the organism. Finally, the A conditions are re-introduced in a period called "A'". A set of data points depict the functional relationship between time period and performance level for each condition. Figure 8.5 illustrates one possible kind of outcome.

As with any collection of data, a number of research questions can be tested. For instance, one might ask if the A baseline function has been achieved during the A' condition. This question is answered by testing Model 8.11 vs. Model 8.12 on only the data from A and A'.

Model 8.11: $Y_1 = a_0 U + a_1 X_1 + a_2 X_2 + a_3 X_3 + a_4 X_4 + E_{11}$

where: Y_1 = criterion of Performance level;
        X_1 = 1 if criterion from Time period 1-20, zero
                otherwise (Condition A);
        X_2 = Time period if observation from Time period
                1-20, zero otherwise;
        X_3 = 1 if observation from Time period 41-60, zero
                otherwise (Condition A'); and
        X_4 = Time period minus 40 if observation from Time
                period 41-60, zero otherwise.

Restrictions: $a_1 = a_3$, both equal to a_5, a common weight; and

               $a_2 = a_4$, both equal to a_6, a common weight.

Forcing these restrictions onto the above Full Model results in

the following Restricted Model:

Model 8.12: $Y_1 = (a_0 + a_5)U + a_6X_5 + E_{12}$

        where: X_5 is the vector containing the (relative) time
                at which Performance level was observed.

Note that the adjustment made on the time scores in vector

X_4 has the effect of superimposing the data obtained during the

A' condition onto the data obtained during the A condition. If

there is a significant decrease in the amount of criterion vari-

ance accounted for by Model 8.12, then one cannot assume that the

same (linear) functional relationship exists under the A and A'

conditions. If there is no significant decrease in the amount of

variance accounted for, then the Restricted Model is as valid as

the Full Model, implying that the difference in the two func-

tional relationships is no more than what would be expected

through random fluctuations. If a non-linear baseline is ex-

pected under A', such a condition could be tested by including in

the model the appropriate non-linear terms.

    A possibly more interesting change question involves the

comparison of the functional relationship under condition B with

the (average) functional relationship under conditions A and A'.

If the B condition is effective, then the rate of increase ought

to be faster than under baseline conditions (A and A'). The

superimposed starting point for all conditions might reasonably

be expected to be the same (an empirically testable situation,

of course). The Research Hypothesis of interest could be: "As-

suming a common starting point for all conditions, the B condi-

tion yields a faster rate of increase on the criterion Y_1 than

does the average of conditions A and A'." The Full Model to test

this Research Hypothesis would be:

Model 8.13: $Y_1 = a_0U + a_2X_2 + a_4X_4 + E_{13}$

where: Y_1 = criterion of Performance level;
X_2 = Time period if observation from condition A,
Time period minus 40 if observation from
condition A', zero otherwise; and
X_4 = Time period minus 20 if observation from
condition B, zero otherwise.

The restriction implied by the Research Hypothesis is: $a_2 = a_4$,

both equal to a_8, a common weight. Forcing this restriction into

the above Full Model yields the following Restricted Model:

Model 8.14: $Y_1 = a_0U + a_8X_5 + E_{14}$

where: X_5 is a vector containing the relative time at
which Performance level was observed.

The Research Hypothesis is supported only if $a_4 > a_2$ in the

Full Model, and if the directional probability is less than the

predetermined alpha. The careful reader will begin to notice the

underlying similarities in the testing of various hypotheses.

9

Miscellaneous Questions about Research which Regression Helps Answer

Chapter Nine, in an interview format, covers a number of somewhat unrelated topics. Many of the questions in this chapter can best be answered through logic and a strong theoretic background in the research content area. But a good grasp of multiple linear regression provides some added insight. Many researchers have come to the authors over the past years with these very questions, and it is to their research zeal that the character of "Budding Researcher" is dedicated.

Samples and Populations

BUDDING RESEARCHER: When I am reporting my research, what do I need to say about the population to which my results apply?

DR. M. L. REGRESSION: The population to which you want to infer should be clearly indicated. The scope and meaningfulness of the population is entirely up to you as the researcher, and to the constraints under which you perform your research. But it is up to you to justify that the sample you investigated represents that population. If the population is a theoretical one, then you must justify that there is no reason to believe that the sample is discrepant from those subjects in the theoretical population.

348

There are no indices to indicate when a representative sample has not been obtained, although the sample and its selection must pass the scrutiny of your colleagues. The ultimate test is, of course, the repeated finding of the same results in samples from that population.

BUD: Can I achieve a representative sample by random sampling?

DR. MLR: Random sampling does not guarantee that the sample will be representative, but only that the chances are minimized that the (random) sample will be highly unrepresentative. Any one random sample could be quite unrepresentative of the population.

Random sampling demands that every entity in the population have an equal opportunity to appear in the sample. Very few pieces of research that appear in the behavioral science journals strictly meet this criterion. It is more often the case that only a limited number of entities are available for testing, only a certain percentage of questionnaires are returned, or the research is performed at a certain University in a certain class. Researchers must remember that the observed findings (unless replicated) may be a function of the particular sample, and may not be generalizable to the (different) population.

Collection of data on more and more subjects often will not overcome the above difficulty. If the sample is unrepresentative in the first place, then measuring more (unrepresentative) subjects simply solidifies the discrepancy. Similarly, striving for

100% return on questionnaires through second and third mailings
(and other associated harassments) will probably generate different kinds of responses than would the first questionnaire if it
had been responded to voluntarily.

BUD: How large a sample should I use?

DR. MLR: The desirability of large sample sizes has been an
illusion for many years. As you've already learned, the number of
subjects in a study enters into the calculation of the degrees of
freedom denominator (N - pieces of information in the full model).
The $\underline{F}$ test of significance is adjusted for the number of subjects
in the sense that, with a fixed R_f^2 and R_r^2, it is harder to get
statistical significance with a smaller number of subjects. With
reference to the $\underline{F}$ denominator degrees of freedom, the minimum
sample size is one more subject than predictor variables in the
full model. For if there were only as many subjects as predictor
variables in the full model, the denominator degrees of freedom
would be zero (a statistician's bad dream). And if you were
analyzing fewer subjects than predictor variables, the denominator degrees of freedom would be negative (a statistician's night-
mare). If only as many subjects are used as predictor variables,
the R^2 would always be 1.0, exemplifying the notion of producing
a spurious result by overfitting the data.

Now I'm not saying you should $\underline{always}$ use a small number of
subjects. What I'm trying to communicate is that you need only
one more subject than the number of predictor variables in order
to perform the $\underline{F}$ test of significance. Now if a researcher has a

good grasp of the functional relationship he is studying, he
needs only a few subjects. But I suspect that most relationships
being studied in the behavioral sciences right now and in the
near future will not yield extremely high R^2 values; thus larger
sample sizes are needed in order to be fairly certain that one is
getting a good estimate of the relationship in the population.
Fortunately, there is no recognized rule of thumb about the number of subjects to use. The less tight a grasp the researcher has
on the functional relationship, the larger the recommended sample
size. The wider the range of values on the continuous predictor
variables, the larger the recommended sample size. The more reliance on data snooping rather than on theory and past research,
the larger the recommended sample size. On the other hand, you
must never let a small sample size eliminate an analysis. You
must realize that small sample sizes reduce statistical power--
the probability of finding significance if the research hypothesis is true in the population. Indeed, increasing the sample
size will lower the probability value, while change in sample
size has no systematic effect on the R^2 value (after the effects
of overfitting have been eliminated--say 20 subjects per predictor variable). Increasing the sample size merely results in
the sample R^2 being a better estimate of the population R^2. That
is to say, increasing the sample size tends to yield a sample R^2
that is not closer to one or to zero, but to the population R^2,
whatever it is. This seems to be a compelling reason for relying upon R^2 for decisions, rather than solely on the probability

value.

All findings require replication, and unexpected findings on small samples simply indicate less chance of being replicated. The proof is always in the prediction, and whether or not a finding from a small sample has merit rests in empirical replication.

Criterion and Predictor Variables

BUD: What variable do I use as my criterion?

DR. MLR: Strangely enough, that question is often asked of us statisticians, but it can never be specifically answered by anyone but the researcher. The criterion variable is the variable which measures the phenomenon of interest to the researcher. The researcher is intrigued as to why subjects vary on this variable, and he sets out to account for variability on this variable. Often the phenomenon is measurable by more than one variable, although the MLR approach can handle only one criterion variable at a time. You could use the same predictor model to predict two different criterion variables separately. A friend of mine named Spaner played around with that idea and published his thoughts on it in 1970. And if you're really interested in using multiple criterion variables simultaneously, a good introduction to multivariate procedures is in a book that a guy called Veldman wrote in 1967.

Another notion about variables is that a particular variable may be a criterion variable for one researcher and a predictor variable for another researcher. In developmental studies, a criterion of 10 day behavior may also be a meaningful predictor of

30 day criterion behavior (McNeil, 1974).

BUD: How do I know what variables to use as predictors?

DR. MLR: Most importantly, the nature of the predictor vari-

ables you use should emanate from a combination of past research

and your theoretical structure. Given this as a premise, what

specifically do you want to know about predictors?

BUD: How should I measure the concepts I've decided to use

as predictors?

DR. MLR: Two researchers may agree on using a given con-

struct as a predictor, but they might use two quite different

measures of that construct. You've got to remember that the R^2

and resulting interpretations are the result of the actual data,

and not of the construct which you hope the variable is measur-

ing. The scores that are obtained are hopefully good approxima-

tions of the construct to which you would like to infer. But, for

instance, you must always keep in mind that the measure of IQ

that is used in a statistical analysis is most likely not synon-

ymous with the construct of IQ. Until someone is able to "see the

construct of IQ," we must act as if the operational measure is at

best a (good) approximation. The goodness of that approximation

is, in measurement terms, referred to as validity. You should

also be aware of the fact that the measures need not be of an

interval scale, as written by my friends Keith McNeil and Francis

Kelly in 1970.

I'd like to make the point that the researcher doesn't need

to follow any of the "guidelines" about using "recognized" tests
or measurements. The predictor variables can be measured in any
way the researcher sees fit. If a high R^2 is obtained, more credit to him. There is nothing sacred about any of the prevalent
measures in any of the behavioral sciences. Indeed, the many
years of attention paid to certain "well-known tests" may have
stagnated the growth of knowledge about relationships in that
field. On the other hand, low R^2 values may not indicate that the
criterion can't be accounted for by the constructs measured in
the predictor set, but that the measures of the constructs were
poor ones and need to be rescaled or totally remeasured.

BUD: Should all of my chosen predictor variables be uncorrelated?

DR. MLR: That's another of those guidelines I'd rather you
ignored. Most multiple correlation literature encourages the use
of uncorrelated predictor variables. Analysis of variance designs
are set up such that the predictor variables are uncorrelated.
But the real world isn't fashioned that way. More importantly,
the ultimate criterion of the value of a predictor variable is
its effect on the R^2. If the R^2 is significantly increased by the
inclusion of a correlated predictor variable, then by definition
that variable is a good variable (in that set of predictor variables).

When you use both linear and polynomial terms in a regression model, those variables are very highly correlated, but you
know how valuable and even necessary they may be. Additional

group membership variables, beyond those for which you hope to

establish differences, may also increase the R^2 even though they

will also be correlated. If you want to get further into that, my

friends McNeil and Spaner wrote an article in 1971 which you

could read. There is a special kind of predictor variable--one

that is uncorrelated with the criterion but highly correlated

with other predictor variables--and it has been given the name

"suppressor variable." This kind of variable, by whatever name

you want to call it, is a predictor variable that was of no value

in the bivariate way of looking at the criterion but is indeed

valuable in the multivariate model. Since most past research is

of a bivariate nature, you might want to have less faith that

past research will give you many insights into productive multi-

variate hypotheses.

Some of the suppressor variable situations will merely be

rescaling problems. Others will depict intricate and inseparable

relationships between distinct constructs. These will be the

variables that "complexly" measure the "complex behavior" so

often discussed in the behavioral sciences.

BUD: How many predictor variables should I use in my analy-

sis?

DR. MLR: You've got to use as many predictor variables as it

takes to result in an R^2 that satisfies you. And I hope you're

not going to be satisfied until you get an R^2 close to 1.00. You

have an infinite number of predictor variables at your disposal.

You choose to measure a limited number of variables from an in-

finite number of possibilities. And once those limited number
have been measured, the possibilities of polynomials and interactions are unlimited. Since you are the researcher, it is up to
you to choose from the unlimited set of transformed variables
those that are to be used in the analysis. This decision is a
very difficult one, for the value of a variable is in its ability
to increase R^2, not in whether it is a "linear interaction" term,
or a "third degree" term, etc. But if you choose to defend the
inclusion of predictor variables on some other grounds than practical utility, remember that the originally measured variable
(the "linear" term) must be defended just as much as the "second
degree" term, or any other term for that matter. And since man is
probably too frail an entity to be able to posit reasons for including one variable out of an infinite set, the final decision
to include a predictor variable should thus be deferred to pragmatic reasoning--whether or not the variable increases R^2.

When I first introduce curvilinear models to eager young
researchers like yourself, most of these researchers excitedly
include the second degree term in their prediction models. Then
when third degree models are introduced, the researchers excitedly include third degree terms. When it becomes apparent that
any power can be used, then they want to know, "How high a power
should I include in my model?" I can't be emphatic enough about
this: Previous research and theoretical views determine the research hypothesis and hence predictor variables. Lacking solid
previous research or a tight theoretical view, predictor vari-

ables should be used which will increase the R^2.

Some researchers have been led to believe that if a third power is used, the second power and linear term must also be used. But I'm saying that the researcher does not have to include lower powered terms. Likewise, the inclusion of an interaction term does not demand that the variables being interacted also appear as separate terms. The inclusion or omission of a variable should always be thought out by the researcher. One must remember that interpretation of a variable's effects are always within the context of the other variables in the regression model. If a variable is not included in a model, one is assuming that the effects of that variable, within the context of the other variables, is negligible. These notions are quite important when one considers the drain on the degrees of freedom by including these sometimes undesired variables and the additional cost involved in measuring them and/or analyzing them.

The Dollar Cost of Prediction

BUD: You've been encouraging the inclusion of many predictor variables, but each additional predictor I use costs money, time, and effort.

DR. MLR: You're right. The cost may be in the collection of the data, or it may be in the computer time to transform the data or to obtain the results. The F test you use to compare models places the same cost on each predictor variable. In reality, some variables cost a lot more than others, while some may be essentially free. Measures that require a lot of testing time or that require a lot of intensive follow-up would cost a lot. Some variables are readily available in school files or in published data banks. All of the transformed variables cost essentially nothing, except for the small amount of computer time needed to make the actual transformation.

Researchers often implicitly take these concerns into consideration in determining what variables to measure. You should not overlook a potentially important variable just because it costs a lot--you should have more interest in accounting for your criterion! On the other hand, you should pay more attention to transformed variables for two reasons. Firstly, the transformed variables may increase the R^2. Secondly, the cost of transforming variables is essentially zero.

The business of considering the cost of predictor variables is a difficult and complex problem. Recently, a method of assigning cost figures to the predictor variables and then using this information in determining the "best" model was proposed by a friend of mine, John Pohlmann, in 1973. Until cost values can be more validly assessed, Pohlmann's method remains ahead of its time.

Attenuation, Replication, and Cross-validation

BUD: I've got another concern about my criterion variable. I've read that an unreliable criterion causes something called "attenuation."

DR. MLR: Attenuation is the reduction in predictability due
to the unreliability of the criterion. Formulas exist for "correction due to attenuation" and are often used. It seems a little
ironic to boast about the degree of predictability you would have
if you had a perfectly reliable criterion. It is often the case
that the researcher does not have perfectly reliable criteria;
and furthermore, when you try to predict a criterion, you must
have some faith in the measurement of that criterion. Prediction
of a criterion is an exercise in validity, and increasing the
"statistical" reliability of the criterion may decrease the validity of the functional relationship.

The R^2 you get from your sample is most likely an overesti-
mate of the population R^2 because the weighting coefficients in a
regression model are a function of those sampled subjects. Several adjustment formulas exist for reducing the R^2 by taking into
account the number of subjects and the number of pieces of information in the model (Newman, 1973; Klein and Newman, 1974).
But these formulas are "average statistical adjustments." I prefer to replicate my findings (perform the same study on new subjects to see if I get significance again) or to cross-validate by
applying the weighting coefficients from my original sample to a
new sample of subjects from the same population. Suppose that for
a given sample, the following results were obtained:

Model 9.1: $Y_1 = 6*U + 1.3*X_1 + 2.6*X_2 + 3.3*X_3 + (-66.1)*X_4 + E_1$

The weighting coefficients above are unique to the sample, but if
the R^2 is relatively high, they should be applicable to other

samples from the same population with little loss in R^2. In order
to check this out, you would have to obtain another sample of
subjects, measure them on the same criterion and predictor variables, and then multiply the original sample weights times the
appropriate predictor variables for the new sample to obtain a
predicted criterion for each subject. The correlation between the
observed criterion for each subject in the new sample and the
predicted criterion would give you an idea as to the "goodness"
of those initially discovered weighting coefficients. In this
process called cross-validation, you are investigating the replicability of the <u>magnitude</u> as well as the <u>direction</u> of the weighting coefficients.

Now, if you'd like to practice this notion to get a better
understanding of it, here's something you might try on the data
in Appendix C. Take the first 30 subjects and solve the regression weights (using the computer of course) for this model:

Model 9.2: $X_2 = a_{12}X_{12} + a_{13}X_{13} + a_{14}X_{14} + E_2$

You'll get a solution like this one:

$X_2 = (-69.82533)*U + 4.62768*X_{12} + 0.0*X_{13} + .92690*X_{14} + E_2$

Then take the last 30 subjects and generate a new variable, which
would be a predicted criterion, by using the weights you obtained
on the first 30 subjects. You'll need the transformation statement:

$X(15) = 4.62768*X(12) + .92690*X(14) - 69.82533$

Then look at the correlation between the criterion (X_2) and the
predicted criterion (X_{15}). That correlation (.98) indicates the

degree of stability of the function as expressed by the variables
and their weighting coefficients.

Cross-validation on a study which is based on a strong theoretical background should yield equally high R^2 values for the
original and later samples; but cross-validation on a study based
on data snooping procedures can be expected to yield a lower R^2
on the second sample than on the first. Small samples will even
cross-validate well if the "true" functional relationship has
been found. (You could cross-validate on a very small sample,
Einstein's Law ($E = MC^2$) or Newton's Law ($D = 1/2GT^2$) as illustrated at the end of this chapter.)

Sequencing of Hypotheses

BUD: Must I state and test just one hypothesis at a time, or
is there any way to combine the testing of several hypotheses?

DR. MLR: The MLR approach makes it quite easy to posit research hypotheses and to build models and F tests to test those
hypotheses. Most regression computer programs allow for a number
of hypotheses to be tested on one computer run. The problem then
arises as to finding the most efficient and logically defensible
sequence of testing those hypotheses. There are a number of ways
of approaching this problem--from statistical, computer, and researcher points of view.

BUD: What sequencing has been developed by statisticians?

DR. MLR: The statistician's major methods are interaction
and multiple comparisons. Researchers are admonished to: "test

for interaction before investigating main effects; if interaction is found, then look only at simple effects." In 1974 a friend of mine, John Williams, indicated two concerns he had about this procedure. First, a test of significance is being performed in hopes of not finding significance. And secondly, the error rate--chance of accepting one of the research hypotheses falsely--is greater than the stated alpha. The stated alpha is appropriate for each hypothesis, but if, say, alpha equals .05 for each of three hypotheses, the probability of finding at least one hypothesis significant is .14 (considerably higher than .05). Furthermore, I think a researcher may be interested in the main effects even if interaction exists. The point is that not all researchers need to take the same route.

Most multiple comparison procedures demand that the researcher compute the one-way F first, and then investigate an orthogonal set of hypotheses. But the researcher may be interested in only one comparison or in several non-orthogonal comparisons. If the comparisons have been hypothesized before the data is analyzed, then the researcher is not capitalizing on vagaries in the data. Furthermore, there is no need for all of the (orthogonal) multiple comparisons to be calculated and interpreted by the researcher. That seems to me to be like getting answers to questions you didn't ask.

If all orthogonal comparisons are desired, then a more efficient way to define vectors than (1,0) is available. Kerlinger

and Pedhazur (1973) present the various alternatives. My position
is that one is not interested in those kinds of questions often
enough to justify the learning of those procedures. As Kerlinger
and Pedhazur (1973) indicate, the same $\underline{F}$ values will be obtained
under the various systems.

BUD: If I don't need to follow the multiple comparison
route, what is the procedure developed from the computer point of
view?

DR. MLR: I'm not recommending this procedure for hypothesis
testing purposes, but I'd like you to be aware of it for hypothesis generating purposes. Most computer installations have a
stepwise regression program. Very few specific hypotheses can be
tested with such a program. The hypothesis testing sequence is
controlled only very slightly by the researcher, although some
stepwise programs allow for an "importance" indicator for each
variable. But specific models generally cannot be constructed and
tested with stepwise programs.

The researcher reads into the computer his criterion variable and all the predictor variables in which he has an interest.
The idea is to find which variables form the "best" predictor
model for that criterion. Forward selection procedures find the
one variable in the predictor set which is most highly correlated
with the criterion. This variable, along with the unit vector, is
referred to as the best model at step 1. Then the program searches the variables to find which predictor, in combination with the
two already "in the system," will yield the highest R^2. Once

found, these three variables become the best model at step 2. The process continues until all variables are in the system, or until no variable can be found such that the increase in R^2 is significant at a level specified either by the program or by the user. The stepwise procedure was developed for continuous variables. Few applications have involved interactions or non-linear terms. Few have involved dichotomous variables for two reasons. First, the inclusion of linearly dependent vectors will cause most stepwise programs to not run. Second, if a variable is to be represented by three mutually exclusive dichotomous vectors, what interpretation would one place on the situation wherein one of those dichotomous vectors was in the system, but the other two were not?

That friend of mine, John Williams whom I spoke of earlier, has developed a procedure, referred to as "setwise regression" to be used in such a case. Instead of using single variables, the researcher is allowed to define sets of variables. While sets can be defined on a logical basis, the use of a setwise procedure would seem mandatory when binary coded variables are included and there are more than two categories involved. The setwise procedure drops one set at a time in a stepwise fashion.

Some problems arise with the stepwise approach. There are a great many hypotheses being tested, none of which has been specified by the researcher. The resulting "best model" will most likely be overfit quite drastically, and hence replication of that model is relatively unlikely. Also, the stepwise procedure

may stop too soon, in that two variables considered simultaneously might significantly increase the R^2, whereas neither one of
them may separately significantly increase the R^2. Some stepwise
versions allow variables already in the system to be deleted if
they aren't making a contribution in a much larger model. Versions that do not allow for this additional flexibility may end
up with a quite inferior "best model."

Some of these problems with forward stepwise regression are
resolved with backwards stepwise regression. In this approach,
all variables are considered as the best model at step 1. Then
the variables (or sets, as in setwise) are evaluated one at a
time to find which one will, when omitted from the system, reduce
the R^2 the least. That variable is then omitted, and the model
utilizing all the variables except that one is the best model at
step 2. The process is repeated for step 3, and so on until
either all variables have been omitted, or until omitting any of
the variables would result in a significant loss in predictability. The backward procedure is preferable although many hypotheses are tested (resulting in a lot of computer time and a large
probability of making at least one type I error). The resulting
model may not replicate because of overfitting. At any rate, the
particular hypothesis you want to test may or may not appear--
and if it does appear, you may have difficulty recognizing it.

Applications of the stepwise procedure have typically ignored curvilinear variables, as well as interaction variables.
Also, too much attention has been placed on the sequence in

which variables either enter or leave the system. This sequencing
is a function of all the intercorrelations between the predictors
as well as how highly the predictors correlate with the criterion. Small changes in one or two correlations could quite drastically change the sequencing. You might want to read Kunce
(1971) and a rejoinder by McNeil and Lewis (1972) for an example
of these notions.

Basically, stepwise regression programs perform an hypothesis formulation function, whereas the regression procedures which
are discussed in this text are concerned with hypothesis testing
(and hypothesis formulation). Both have their place, although
hypothesis formulation must be followed by hypothesis testing.

BUD: What method for sequencing hypothesis testing can you
recommend from the researcher's point of view?

DR. MLR: Build your hypotheses upon past research (yours or
someone else's) and your theoretical views. If you have enough
information and enough confidence in your theory to state directional research hypotheses, then state them in the order of your
interest and expectations and test them in that order. You can
expect statistical support for such hypotheses and you should be
able to progress toward causal interpretation.

BUD: What about those areas of my theory where past research
has given me no clues about what functional relationships to expect? I can't state directional research hypotheses about those
relationships, can I?

DR. MLR: You've made a good point. I've always been quite

emphatic about the researcher stating his research hypothesis.
The research hypothesis should be a statement of expected outcome
based upon past relevant research, theoretical relationships, and
synthesis of relevant constructs. But most past research has been
bivariate in nature. That is, only one predictor variable was
used to account for criterion variance, and the amount of criterion variance accounted for is usually quite small. Therefore,
theory development has been held back, and consequently there is
little knowledge from which one can synthesize. Data snooping is
one way of finding functional relationships. As long as one constantly remembers to replicate the findings on a new sample of
data, I see nothing wrong with "seeing what works." Using a stepwise regression program is one way to discover predictor variables; accidental (serendipitous) findings are another way.
Guidelines as to how to obtain serendipitous findings are non-
existent, but you should be willing to cross-validate a variable
that appears to be valuable in a pilot study. Many important
variables in the sciences have been discovered accidentally, although their value was repeatedly tested (cross-validated) before
they were widely accepted. Mosteller and Tukey (1968) aptly summarized the researcher's reaction to results found through data
snooping when they said, "Here we must stop our calculations with
indications and be careful to think of our results only as hints
as to what to study next, rather than as established results."

To give you an idea of what data snooping is all about, I'll
tell you a "data snooping analogy." The bread is analogous to a

meaningful discovery, and the telling of friends about the news is analogous to publishing one's results for the benefit of the profession.

You have been asked by your spouse to go down to the milk discount store to buy some milk. Now, if you do exactly as you are asked, you will go directly to the milk counter, pick up the milk, pay for the milk, and return home.

Let's suppose that you snoop around while you're there and you accidentally notice that this particular store has bread at an exceptionally low price. In fact, at the price that it is listed, you might label this "the best bread buy in town." If you are the least bit economically minded, you will take note of this price and purchase a loaf or two, especially if you are aware of procedures whereby bread can be stored for a long period of time.

You are not "sure" that this low price will prevail for any period of time, but you would certainly consider informing some of your friends about the "best bread buy in town." You would probably indicate your reservations by saying something like, "Yesterday I accidentally discovered that the milk discount store that I trade at had a good price on bread. You might want to go down there and see if they still have this good bargain; I really don't know if they still have that good buy." That is, you are somewhat reluctant to yell too loud about the bargain until the bargain can be verified. The longer that the store retains this low price, the oftener and louder you will announce your discovery to more of your friends. If the bargain remains long enough and enough people are able to verify your finding, then the "best bread buy in town" becomes a lawful fact.

That lawful fact might not have been discovered if you had not snooped around while at the milk discount store looking for milk. An auxiliary bargain discovered while looking for something completely different must be verified on a second trip before much faith can be placed in the stability of that bargain.

As a researcher, please take the analogy to heart. Snooping around in your data is not antithetical to good research. If you find some meaningful relationship while snooping around, then you are obligated to replicate your results before you say too much about them. Your research hypotheses should be based upon theory which is often loosely defined as past research. There is often little past research upon which to base your hypotheses. Furthermore, the past research is often of poor quality and bivariate in nature and therefore possibly misleading. When one snoops around in his data, that study must be considered as a pilot (past research) upon which hypotheses are developed for the replication sample.

It could be argued that researchers have drastically held back scientific advancement by not snooping around in their data. It seems very unfortunate to collect large amounts of data and then to look at that data with blinders on. Let's take those blinders off--but at the same time remember to leave our replicators on!

BUD: Can I use multiple regression procedures to help in my

data snooping?

DR. MLR: Data snooping is easily accomplished within the MLR approach. All of your measured variables (about which you have hypotheses, or hunches, or you just want to see what's going on) can be included in a full regression model. If a lot of predictor variables are being used, a large number of subjects should be used so that the resulting functional relationship is not due entirely to the idiosyncracies of the sample at hand.

The R^2 for the Full Model using all measured predictor variables could be compared against the unit vector model to see if any significant criterion variance is being accounted for. If you want to limit your predictor set to, say, one-half the number of predictors you started out with and you want to find the best ones, you might use one of the stepwise regression approaches.

Data snooping can also be thought of in terms of trying to increase R^2 over the value you have obtained based on your previous hypothesis testing. (This snooping will then lead to further hypothesis testing.) Suppose a particular researcher would like to obtain 90% predictability. The R^2 of his Full Model is only .70. What can be done now? More snooping is called for, but where? By now you should be thinking of possible interactions and polynomial forms. Not only X_1, but X_1^2, X_1^3, X_1^4, and one could even give the predictor the fifth degree. Not only X_2, but also X_2^2, X_2^3, X_2^4, etc. Not only X_1*X_2, but also $X_1*X_2^2$, $X_1^3*X_2^2$. Not only X_3^2, X_3^3, but also $X_1^{2.3}*X_3^{2.3469}$; ad infinitum. Yes, there are an infinite number of transformed variables that you could get from any set of measured variables. But an infinite set of predictors can't be used because you can't have the required number of subjects (infinity plus one). Ideally, theory and past research should have guided us and assisted us in selecting the "appropriate" set to use. The primary guideline, though, should be pragmatic predictability. When R^2 is up to .90, then this researcher will be happy. (Astronauts going to the moon desired R^2 around .99, and those desiring to return to Earth required a somewhat higher R^2.) The actual minimum R^2 value is up to the individual researcher and will depend upon the state of the art in his content area and the amount of impact he desires to make in that area.

Snooping may be a relatively costly approach. Computer time is needed to generate the various transformed variables. Furthermore, the amount of computer time required to solve for a regression model increases drastically when the number of predictor variables increases. Some of this computer time can be saved by splitting the sample in half and snooping only on the one half-- saving the second half for the required cross-validation as discussed earlier. Only those transformations which were found useful in the "snooping sample" would have to be computed in the "cross-validation sample." The cross-validation sample would require additional computer time, but cross-validation is a research step which is desired anyway.

Some have suggested that factor analysis could be used first to reduce the set of predictor variables to a small set of uncorrelated predictors (Connett, Houston, and Shaw, 1972). This procedure would have the advantage of reducing the number of predictors, but the possibility of obtaining the desired level of R^2 might be lost (see Newman, 1972, for other problems). Most interesting criterion variables are, by the admonition of researchers investigating them, complexly caused. It is too naive to believe that real-world causers are both rectilinear and uncorrelated.

What does the researcher have after he has snooped and subsequently cross-validated? He has a highly predictable and highly reliable way of accounting for variability in his criterion of interest. He may likely not understand why the predictors are combined in the way they are. Indeed it could be argued that scientific advances occur only when data hits you between the eyes and says, "Hey, I'm something you didn't expect to find. I'm reliable and predictable. Use my functional relationship because I work."

Data snooping may lead to serendipitous findings, and when I realize how many aspects of our world have been found serendipitously, I become amazed (e.g., gravity, pasturization, vulcanization of rubber, popcorn). The concepts of transformed variables and cross-validation within the data snooping framework discussed above should give more researchers the opportunity to advance science.

BUD: Thanks so much for answering my questions, but I get the feeling that a lot of what you say is wishful statistical thinking and that few researchers will ever get a chance to practice what you preach.

DR. MLR: Quite the contrary; and to put a lot of these notions into perspective, let me tell you the story, related to me
in 1970 by K. McNeil, of how a budding researcher like yourself
used MLR to discover a well-known law.

Meeting the Goals of Research With Multiple Linear Regression

Multiple linear regression is discussed as it relates to
several goals of research: Predictability, Parsimony, Replication
and Validity Generalization. These goals are presented with the
development of a well established physical law. The emphasis is
upon the percent of variance accounted for in the criterion under
investigation, rather than on statistical significance from random events. Additional remarks concerning curvilinear relationships and data snooping are also presented. Dingman's Canons of
Reproducibility are discussed within the framework of multiple
linear regression and the goals of research.

The following material represents one way which Sir Isaac
Newton might have developed his law of gravity. The material is
presented with the intent of showing the advantage of using the
multiple linear regression technique as one's statistical tool.
(See Kelly, Beggs, McNeil, Eichelberger, and Lyon (1969) for a
more thorough discussion of the procedure.)

Statement of the problem

It seems that for years some of our most competent researchers have been looking for the functional relationship between the amount of time that an object has been falling in space, and the distance the object has fallen. Stating the problem symbolically, d=f(t), and verbally, "distance is what function of time?" I believe that I have finally discovered this functional relationship between time and distance. This means that if I know the amount of time an object has been falling, then I can tell you how far it has fallen.

One of the surprising findings that my research strategy has led me to is that I need know nothing about the object, if I can assume that the object is falling in a vacuum, where there is no resistance to its fall. The only additional variable that I need to know is what I call the "gravitational constant"--computed from the forces being exerted by the earth, the sun, the moon, and other heavenly bodies. For any one time and place, this variable has the same value for any object.

Goal I: Predictability

Review of the literature

To give a little background, let me review previous research. Galileo (1632) obtained a lot of data on falling objects but only investigated the linear relationship between time and distance. Galileo and even before him, Aristotle, utilized the multiple linear regression procedure in testing their hypotheses, but these great thinkers did not realize the value and the flexibility of this procedure. They only utilized a very restricted form of the technique, the form which ascertains the rectilinear

relationship between the two variables.

Galileo utilized the following model:

Model 1: $D_1 = a_0 U + a_1 T_1 + E_1$

   where: D_1 = the vector containing the distances the objects
               have traveled;
            T_1 = the vector containing the amount of time the
               objects have traveled;
            U = the unit vector which allows the regression
               constant to be non-zero;
            E_1 = the vector containing the difference between
               the actual distance and the predicted distance,
               that predicted from the pool of predictor vari-
               ables on the right-hand side of the equation;
               and
            a_0 and a_1 are weighting coefficients of best fit,
               determined from the sample data. Best is here
               defined in the least squares sense, minimizing
               the sum of the squared values in E_1.

Model 1 produced a significant fit of the data, but the R^2

value was discouragingly low ($R^2 = .40$). This R^2 value is the

squared value of the correlation between the observed and pre-

dicted values, and can be interpreted as the proportion of vari-

ance in the criterion distance measures which is predicted, or

accounted for, by the predictor variables in the sample (in the

case of Model 1, the regression constant and the continuous vari-

able of time).

A linear model which allows for curvilinear relationship

Upon looking at a bivariate plot of the data in Figure 1, it

becomes obvious that there is not simply a linear relationship--

but a curvilinear relationship between time and distance. And,

since the rate of acceleration of the curve continuously in-

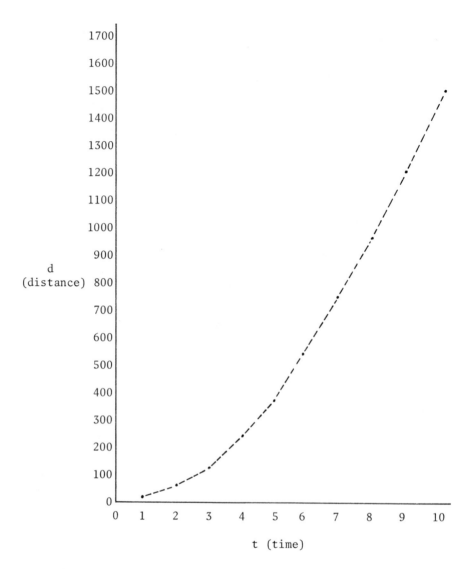

Figure 1. Bivariate plot of data suggesting a possible cur-
         vilinear relationship between time and distance.

creases, there must be a second degree curvilinear relationship.

The multiple linear regression model which allows a second-degree

relationship to exist is:

Model 2: $D_1 = a_0U + a_1T_1 + a_2T_2 + E_2$

All of the variables in Model 1 appear in Model 2, with the addition of T_2. The ith element of T_2 is the squared value of the corresponding value in T_1 (thus, if the object took 4 seconds to fall the designated distance, then T_2 would be 4^2 or 16).

The variable T_2 allows the second degree curve in the data to manifest itself, if in fact there is a second degree curve in the data. I then fit Model 2 to the data for a 100-pound stone. What surprised me was not that I obtained a significant fit of the data, but that the R^2 value was .99. Distance could be almost exactly predicted from knowledge of the regression constant, time, and time squared.

Goal II: Parsimony

A more parsimonious model

"Parsimony" means to explain something in the simplest way possible. With respect to the problem at hand, Model 2 is incorporating three pieces of information in predicting the criterion of distance. The notion of parsimony demands that we investigate the predictability of the criterion by dropping out either T_1, T_2, or U, or some combination of these three vectors. If T_2 is dropped out of the model (actually I am hypothesizing that the weight (a_2) associated with T_2 is equal to zero), I end up with the model utilized by Galileo. I tested the predictability (R^2) of Galileo's model (Model 1) against the R^2 of Model 2 and found a significant decrease (p < .01) in the amount of variance being

accounted for in Galileo's model. Model 2 was accounting for 99%
of the variance, while Galileo's model was accounting for some
40%. I would like to parenthetically add here that many researchers had interpreted Galileo's results incorrectly. Galileo had
tested the predictability of his model against one which contained no differential information:

Model 3: $D_1 = a_0U + E_3$

The test indicated that there was a significant decrease in
predictability, and in particular, indicated that the rectilinear
relationship between time and distance was something other than
zero. As the Model 2 R^2 of .99 indicated, models can be built
which more accurately predict the criterion of distance. The
point is that we should not rest on our haunches until we get
close to accounting for 100% of the variance.

A second model more parsimonious than Model 1 involves hypothesizing that the weight associated with T_1 is equal to zero.
Again this reduces to dropping that particular variable out of
the pool of predictor variables:

Model 4: $D_1 = a_0U + a_2T_2 + E_4$

This last model doesn't contain the variable of time at all,
except in its squared form, and the surprising thing is that
Model 4 was as predictive as the original full model, Model 2. In
fact, the R^2 of Model 4 was also .99, exactly equal to that of
Model 2. What was also true, but fully anticipated, was that the
weight for the unit vector was equal to zero, i.e., the "Y-intercept" is zero. This in effect substantiates our impression that

the distance covered by a falling stone in zero minutes is exactly zero feet. The final values for the more parsimonious model
(Model 4) were: $D_1 = (0*U) + (16*T_2)$, or $a_0 = 0$ and $a_2 = 16$. The
goal of parsimony has been satisfied in that the criterion has
been predicted with a small amount of predictor information, in
this case only one bit of predictor information is needed, that
of the square of time.

Goal III: Replicability

Replication of the functional relationship

It was indicated previously that the regression weights were
optimum weights for that particular sample data. We are concerned, of course, about predicting the distance measure for not
just this one sample of stones, but for any stone from a clearly
defined population of stones. My first concern was to check the
replicability of the obtained regression weights for another set
of data on 100-pound stones.

One process of checking replicability simply involves taking
the obtained regression weights (from Model 4) and applying them
to the new data. This process produces a predicted distance and
the predicted distances are then correlated with the actual observed distances. In this particular instance, every new time
score was simply squared and then correlated with the associated
distance score. The resulting (Pearson product moment) correlation was .98, yielding an R^2 of .96, which again is quite satisfactory. The regression weights found for Model 4 were checked

for replicability on several other samples of data with 100-pound

stones, and all R^2 values were above .95.

Goal IV: Validity Generalization

Generalization of the functional relationship

There are several aspects to the replication problem. In the

above discussion, we discussed how we replicated the functional

relationship in a single population. I was also interested in

generalizing this functional relationship to other populations,

such as populations of other weights. I discovered that the

weight of the objects turned out to be not important. Here is how

I obtained evidence to make this statement. I wondered about the

generalizability of the results on the 100-pound weight to other

weights. I obtained data on a 50-pound weight and proceeded to

apply the regression weights from the 100-pound weight in Model

4. To my surprise, the R^2 resulting from this replicability study

was comparable to the aforementioned replicability studies. That

is, the functional relationship between time and distance was the

same for both 100- and 50-pound weights.

It was not a great intellectual leap to investigate the

possibility that the functional relationship was similar for all

weights! To check this hypothesis, I simply included weight mea-

sures in the predictor pool of variables in Model 4:

Model 5: $D_1 = a_0U + a_2T_2 + a_3T_3 + E_5$

where: T_3 is the weight of the objects being measured.

Of course, to make this model viable, I had to measure objects of

various weights. When Model 5 was applied to a set of data on
stones of differing weights, the regression weight a_3 turned out
to have a numerical value of zero. Thus, when the predictive efficiency of Model 5 was compared against the predictive efficiency of Model 4, no predictive information was lost. The hypothesis was supported that the functional relationship between
time and distance was similar for objects of any weight (more
specifically, for the range of weight values we used in the analysis). To put it another way, weight is not needed in defining
the functional relationship between distance and time. The functional relationship has been generalized across all weight levels.

A return to the gravitational constant

I still haven't discussed this "gravitational constant."
This is the most difficult discussion of all, but it should be
included because it increases the generalizability of my findings
and also illustrates the flexibility of the multiple linear regression approach.

All the data that has been presented so far was obtained in
my experimental labs. But objects move in other places besides my
lab. In particular, it has been known for a long time that objects move faster when they are at lower altitudes. Also, the
functional relationship which I found did not replicate on observations made by Galileo in outer space. Therefore, the functional relationship between time and distance may be modified by

the gravitational field in which the object is measured. That is, various observations provided clues as to what kinds of variables might be important to investigate.

I shall not go into the actual calculation of this gravitational constant, except to indicate that it can be quite accurately measured. The gravitational constant will be symbolized as "G_1". I subjected data which included objects measured within various gravitational fields to Model 6 (an extension of Model 4) to check the functional relationship within the various gravitational fields:

Model 6: $D_1 = a_0U + a_2T_2 + a_4G_1 + E_6$

Model 6 yielded an R^2 value of .50, indicating that 50% of the criterion variance was predicted by the predictor set. To be sure, this is a large amount but not as high as we had been accustomed to. I was careful not to make the same mistake that Galileo had made years before, that of stopping with an extremely parsimonious model, which does only a fair job of accounting for the criterion variance. Generally, one reduces the amount of predictability when the variable pool is reduced. Conversely, the inclusion of another predictor variable will generally increase the predictability, the question being: Is the increase in predictability significant?

Curvilinear interaction

The bivariate plot (Figure 2) between time and distance for the various "G_1" levels was visually inspected, and it became

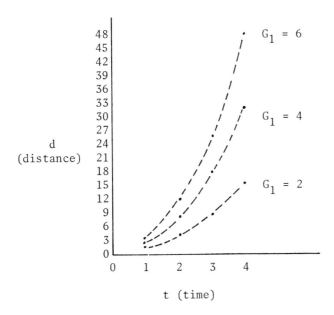

Figure 2. Bivariate plot of data indicating the curvilinear
interaction between time and gravity in the prediction of distance.

obvious that there was likely an interaction between the gravita-

tional constant "G_1" and the square of time "T_2" (Figure 3). A

model was accordingly constructed which allowed this hypothesized

interaction to exist: where the vector (G_1*T_2) represents a sin-

gle number which is the product of the gravitational field and

the square of time:

Model 7: $D_1 = a_0U + a_2T_2 + a_4G_1 + a_5(G_1*T_2) + E_7$

In accord with anticipations, the resulting R^2 was .99, in-

dicating that Model 7 was indeed a good reflection of the func-

tional relationship. I carefully noted that a_0, a_2, and a_4 all

had numerical values extremely close to zero, indicating that the

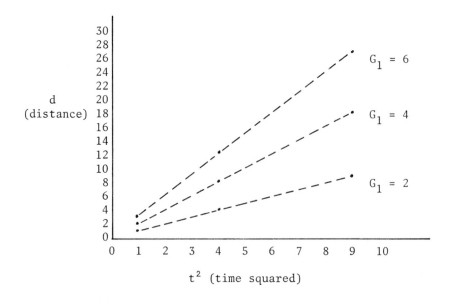

Figure 3. Bivariate plot of data which led to the notion
that G_1 and t^2 interact in the prediction of d.

variables associated with these regression weights were all somewhat useless in obtaining the high degree of predictability. Thus
the hypotheses that the parameter value of these weights is, in
fact, equal to zero was tested. By setting a_0, a_2, and a_4 in
Model 7 equal to zero, the following model is derived:

Model 8: $D_1 = a_5(G_1*T_2) + E_8$

When the R^2 of Model 8 was tested against the R^2 of Model 7,
a significant decrease in the amount of variance being accounted
for was not found. In fact, the R^2 of Model 8 was .98, well within the error expected from the measurement errors in our data.

The Goal of Predictability was met since the interaction
model predicted a high percentage of the criterion variance, al-

though a good linear relationship did not exist in the prediction
of distance. The "simplified" interaction model (Model 8) predicted just as satisfactorily as Model 7; therefore, the "simplified" interaction model was accepted as a better model (because
it simultaneously met the Goals of Predictability and Parsimony).

After running successful replications on other sample data,
Model 8 was deemed to be "lawful" for the phenomenon being studied (because it met the Goals of Predictability, Parsimony, Replication, and to some extent Validity Generalization). The functional relationship between time and distance is thus that function indicated by Model 8.

From my multiple linear regression analyses of Model 8, I
note that the numerical value of a_5 is .5. Verbally then, D_1 is
equal to one-half the product of G_1 times T_2, symbolically: $D_1 =
1/2(G_1 * T_2)$. If we call distance "d", the gravitational constant
"g", and time "t", we have the following symbolic equation: d =
$1/2gt^2$. Let's call that Newton's Law.

Additional remarks by the present author

It is hoped that behavioral science researchers will pay
attention to the methodology that has been presented. It seems
inappropriate for these researchers to continually argue that the
behaviors they look at are inherently complex; the behaviors may
indeed be complex, but how many have even attempted to investigate curvilinear interaction as Newton did? It is time for us to
realize that complex behavior demands complex methodology as well

as complex theory.

There is a problem as to what variables one should choose, but that is a content question which cannot be answered by a methodologist. Transformations and intricate relationships should be sought, rather than discarded as completely confounding the issue (Glass, 1968). Interaction has traditionally been viewed as a "bad" effect, rather than as an effect which will assist in explaining the phenomenon at hand (Kelly et al., 1969).

The notion of a curvilinear relationship is not new to the methodological scene, but we must certainly admit that few investigators make a habit of checking the curvilinear predictability. Indeed, some statistical books discourage the inclusion of highly correlated predictor vectors in regression models (all polynomials are highly correlated with the original vector). Whether a researcher should include a vector which allows for curvilinearity is partly a function of past research findings and partly a function of the nature of his data. Even though the addition of another predictor variable cannot decrease the R^2, it is not the case that such a vector will always statistically increase the R^2 value (McNeil and Spaner, 1971), and such investigations may prove quite rewarding.

Inspecting the sample data is considered to be a poor research strategy by many present day researchers. But if the goal of research is to develop generalizable scientific findings, then it can be argued that a researcher is essentially committing a crime when he does not "squeeze" the data for all it is worth.

When such squeezing is done, though, the researcher is committed to replicating his findings on an independent sample. The stability, and hence value, of the finding must rest on the replicated study rather than on the original study alone.

Dingman (1969) presented three canons of reproducibility for evaluating the robustness of scientific findings. It seems that replicability is the same as Canon of Reproducibility I: "The data matrix must be so homogeneous that if an experiment or multivariate procedure is carried out on a sub-sample, then other sub-samples will give similar results." The discussion of validity generalization in the present paper seems to be synonymous with Dingman's Canon of Reproducibility II: "The results of similar experiments or multivariate procedures on different samples of experimental subjects must give equivalent results." Dingman's third Canon of Reproducibility demands multiple criteria, an area which the present paper unfortunately does not cover. Perhaps future developments with procedures such as canonical correlation will provide an objective function which will provide information for the three canons in designs where there are multiple criteria as well as multiple predictors.

References

Dingman, H.F. Scientific method and reproducibility of results. Multivariate Behavioral Research, 1969, 4, 517-522.

Galileo, G. Dialogues Concerning Two New Sciences. circa 1632.

Glass, G.V. Correlations with products of variables: Statistical formulation and implications for methodology. American

Educational Research Journal, 1968, 5, 712-727.

Kelly, F.J., Beggs, D.L., McNeil, K.A., Eichelberger, T. and
    Lyon, J. Research Design in the Behavioral Sciences: Multi-
    ple Regression Approach. Carbondale, Illinois: Southern
    Illinois University Press, 1969.

McNeil, K.A. and Spaner, S.D. Brief report: Highly correlated
    predictor variables in multiple regression models.
    Multivariate Behavioral Research, 1971, 6, 117-125.

10

Miscellaneous Applications

Applications to Evaluation Situations

There has been an increased emphasis placed on evaluation in recent years. This is a welcomed trend, as too often new techniques or procedures have been adopted without a critical check on their effectiveness. In the following sections the notions of evaluation are related to the designs of previous chapters, particularly Chapter Six. Next some notions of cost efficiency are presented, restricting weighting coefficients to values other than zero, teacher effectiveness, the prediction of under- and over-achieving students, and some ideas about the place of regression analysis in the context of one proposed evaluation model.

There is not total agreement as to whether evaluation and research represent the same activities. The following paragraphs are presented, not with the intent of settling the question, but with indications that the designs in Chapter Six can be used for evaluation purposes.

Criterion-referenced testing

The lack of either adequate or available control groups in many practical evaluation situations has often led to the speci-

fication of criterion levels to be reached. When the value of the
technique or procedure is to be generalized beyond the specific
sample, inferential techniques such as in GRH 8.1 are called for.
Only one group of subjects is tested, but some rationale exists
so as to posit a criterion level that they should have reached.
Whether that rationale is logical may be questioned, but at least
the criterion level is specified. That the technique or procedure
caused the subjects to reach criterion is usually equivocal, and
can only be ascertained through comparisons to a control group.

Control groups

Control groups provide a tighter control over some competing
explainers of the data, but are not always feasible or available
in practical applications. When the researcher has authority over
the selection of the control group, designs such as GRH 6.1, GRH
6.2, and GRH 6.3 can be used. Often, though, control groups are
"given" to the researcher, and hence similarity cannot always be
assumed. In these cases, statistically controlling for the pre-
test, or a number of entering behaviors, is a reasonable solution
utilizing the notions in GRH 5.1 and GRH 6.4 with k = 2 groups.

Interaction between pre-test and treatment

Treating possible contaminating pre-variables as covariates
may not always be sufficient. The traditional analysis of covariance assumption is that the relationship between the covariate
and the criterion is the same for all treatments. This is the as-

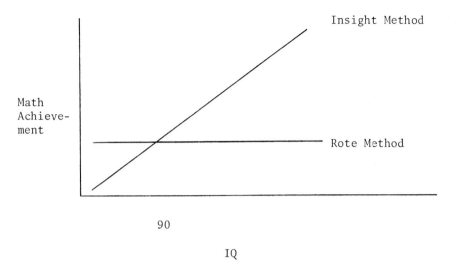

Figure 10.1. Interaction between type of method and IQ
upon the criterion of Math Achievement.

sumption of no interaction (parallel slopes). You should be fa-

miliar enough and comfortable enough with regression by now to

realize that one could test this assumption. The discovery of

interaction should not be treated as a negative finding, but as

an important finding in and of itself. With reference to Figure

10.1, the Insight Method is better, overall, than the Rote Method

overall. But clearly the low IQ subjects do better with the Rote

Method. Assuming the two methods to be of equal cost, the recom-

mendation from the evaluator would be to use the Rote Method with

subjects whose IQ falls below 90, whereas the Insight Method

should be used for subjects who have an IQ of 90 or higher.

Now assume that no subjects with an IQ of less than 90 were

placed in the Rote Method. The best guess through extrapolation

is the solid line in Figure 10.1 for the Rote Method. But not
having observed subjects with IQ's less than 90, one cannot say
for sure what their resultant Math Achievement would be. One must
always be careful when extrapolating outside the range of observed values. Obtaining a predicted score for unobserved values
within the observed range is a much safer procedure, especially
if one has obtained a relatively high R^2.

An article by Novick and Jackson (1970) proposed that Bayesian statistics could provide better solutions to prediction problems because they did not have to use common Y-intercepts and
common slopes. Although many applied regression analyses (and all
covariance analyses) in the past used common slopes, one must
realize that common slopes do not have to be used. The Bayesian
prediction models may be more fruitful than the regression approach, but as of yet the Bayesian approach has not been compared
to regression models using varying slopes. (See Newman, Lewis,
and McNeil (1973) for a more complete discussion.)

Evaluating programs that have cost differences

Very seldom will two programs being compared have exactly
the same cost. Cost here is meant to be not only the announced
dollar cost, but all associated costs such as teacher in-service,
teacher rejection, changes in janitorial policy, increased use of
library books, increased vandalism on the part of dissident students, etc. It is a difficult task to list all possible costs,
and extremely difficult to change all of these costs into a sin-

gle unit such as dollars. The ultimate decision of adopting a
given program should, though, consider all of these costs.

One way to consider the differential cost of two programs
in an evaluation plan is to demand more from the program that is
more costly. Assuming that most of the important costs have been
considered, one could arbitrarily demand that the more costly
program be more than three units better than the cheaper program
before adoption of the more costly program. In times of austerity
a criterion of five units better might be set. In times of unlimited monetary resources from parents who want the best for
their children, the most effective program (more than zero units
better) would be the program adopted. Even though the required
number of units is arbitrarily set, the units are decided before
the data was obtained, making the final decision determined by
the data and not by any unique impressions from the data. Adoption of more stringent alpha levels would have a similar effect,
but stating a particular unit value allows one to make a more
conclusive statement if significance is found.

Generalized Research Hypothesis 10.1 presents the "a unit
question," and Applied Research Hypothesis 10.1 presents a "3
unit question."

Non-zero restrictions can also be made on a continuous predictor variable as indicated in Figure 10.2. Suppose a community
has indicated a willingness to pay more taxes if it can be reasonably predicted that an increase for each $1 spent on each
child in math class will result in more than 2 units mean in-

Generalized Research Hypothesis 10.1

Directional Research Hypothesis: For a given population, Method A is more than "a" units better than Method B on the criterion Y_1.

Non-directional Research Hypothesis: For a given population, Method A is other than "a" units better than Method B on the criterion Y_1.

Statistical Hypothesis: For a given population, Method A is "a" units better than Method B on the criterion Y_1.

Full Model: $Y_1 = a_0U + a_1X_1 + a_2X_2 + E_1$

Restrictions: $a_1 = a_2 + $ "a"

Restricted Model: $Y_1 = a_0U + (a_2+\text{"a"})X_1 + a_2X_2 + E_2$

$$Y_1 = a_0U + a_2X_1 + \text{"a"}X_1 + a_2X_2 + E_2$$

$$Y_1 = a_0U + a_2(X_1+X_2) + \text{"a"}X_1 + E_2$$

$$Y_1 = a_0U + a_2U + \text{"a"}X_1 + E_2$$

$$Y_1 = (a_0+a_2)U + \text{"a"}X_1 + E_2$$

$$Y_1 = a_3U + \text{"a"}X_1 + E_2$$

But since "a"X_1 is not an unknown quantity, it can be subtracted from both sides of the equality, yielding:

$$(Y_1-\text{"a"}X_1) = a_3U + E_2$$

where:

Y_1 = the criterion;
U = 1 for all subjects;
X_1 = 1 if Method A, zero otherwise;
X_2 = 1 if Method B, zero otherwise; and
a_0, a_1, a_2, and a_3 are least squares weighting coefficients calculated so as to minimize the sum of the squared values in the error vectors.

In order to use the general F test, the criterion must be the same in both the full and restricted models, which is not the case here. These two models can be evaluated by the F formula in Chapter Eight, which utilizes the error sum of squares.

Degrees of freedom numerator = $(m_1 - m_2)$ = $(2-1)$ = 1

Degrees of freedom denominator = $(N - m_1)$ = $(N-2)$

where:

N = number of subjects

Applied Research Hypothesis 10.1

Directional Research Hypothesis: For a given population, X_6 is more than 3 units better than X_7 on the criterion X_2.

Statistical Hypothesis: For a given population, X_6 is 3 units better than X_7 on the criterion of X_2.

Full Model: $X_2 = a_0 U + a_6 X_6 + a_7 X_7 + E_1$

Restricted Model: $(X_2 - 3 * X_6) = a_0 U + E_2$

alpha = .05

ESS_f = 7565 ESS_r = 7572

F = .05   Directional p ≈ .50

Interpretation: Although the weighting coefficients are in the hypothesized direction, the directional probability is not lower than alpha; therefore, fail to reject the Statistical Hypothesis.

---

crease on the Math Test. Assuming that data is available concerning dollars spent per child in math class (X_3) and resultant scores on the Math Test (X_2), and assuming a linear increase over the range of dollars spent, Applied Research Hypothesis 10.2

Applied Research Hypothesis 10.2

Directional Research Hypothesis: For a given population, for
each additional unit increase
in X_3, the increase in the
criterion X_2 will be more than
two units.

Statistical Hypothesis: For a given population, for each additional unit increase in X_3, the in-
crease in the criterion X_2 will be
equal to two units.

Full Model: $X_2 = a_0 U + a_3 X_3 + E_1$

Restriction: $a_3 = 2$

Restricted Model: $X_2 = a_0 U + 2 X_3 + E_2$

But since $2X_3$ is not an unknown quantity, it must be subtracted from both sides of the equation, yielding as the

Restricted Model: $(X_2 - 2X_3) = a_0 U + E_2$

alpha = .05

$ESS_f = 559$ $ESS_r = 566$

$\underline{F} = .68$ Directional $p \simeq .20$

Interpretation: Although the weighting coefficient a_3 in the
Full Model is in the hypothesized direction,
the directional probability is not less than
alpha; therefore, the statistical hypothesis
must be rejected.

---

might be appropriate.

Restricting weighting coefficients to values other than zero

is important enough to present Generalized Research Hypothesis

10.2. Any of the previously discussed Generalized Research Hy-

potheses could have had a weight restricted to some constant

Generalized Research Hypothesis 10.2

Directional Research Hypothesis: For a given population, for each additional unit increase in the predictor X_1, there will be more than k units increase in the criterion Y_1.

Non-directional Research Hypothesis: For a given population, for each additional unit increase in the predictor X_1, there will be other than k units increase in the criterion Y_1.

Statistical Hypothesis: For a given population, for each additional unit increase in the predictor X_1, there will be k units increase on the criterion Y_1.

Full Model: $Y_1 = a_0U + a_1X_1 + E_1$

Restrictions: $a_1 = k$

Restricted Model: $(Y_1 - kX_1) = a_0U + E_2$

   where:

   Y_1 = criterion;
   U = 1 for all subjects;
   X_1 = predictor score; and
   a_0 and a_1 are least squares weighting coefficients calculated so as to minimize the sum of the squared elements in the error vectors E_1 and E_2.

Degrees of freedom numerator = $(m_1 - m_2)$ = $(2-1)$ = 1

Degrees of freedom denominator = $(N - m_1)$ = $(N-2)$

   where:

   N = number of subjects

NOTE: The $\underline{F}$ test utilizing error sums of squares, discussed in Chapter Eight, would have to be used with this hypothesis, as with any which specifies a non-zero weight.

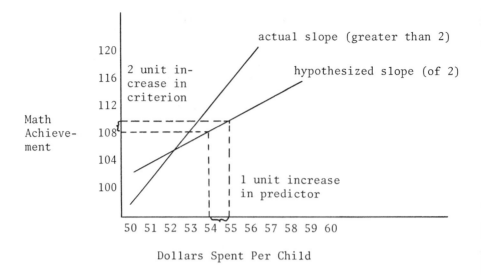

Figure 10.2. Data which might yield a slope greater than 2.

value other than zero. Many times zero is a meaningful restriction, although the authors of this text feel that more often a
weight might have been restricted to some non-zero value if the
researcher had felt comfortable in doing so. Restricting a
weighting coefficient to zero is only one out of an infinite number of ways of restricting that weighting coefficient. It is
important to have a rationale for restricting a weighting coefficient to some value, but the rationale must be as thoroughly
thought out and defensible for restricting the coefficient to
zero as to any other numerical value.

   If all the costs of programs can be assessed, then one
might want to incorporate the actual cost into the analysis. The
gain per dollar cost could be used as a criterion when evalu-

ating the relative effectiveness of two programs. That is, the
gain or criterion worth of the program could be divided by the
dollar outlay or cost. The two "program vectors" would be the
predictor vectors, resulting in a model structure exactly like
GRH 6.1. The decision as to which program is better would be a
statistical decision. The decision as to which program to imple-
ment would be an administrative decision based upon amount of
money available as compared to the amount of educational gain de-
manded (or loss tolerated) by the community.

Teacher and student evaluation

Accountability has entered the classroom, and many school
systems are interested in comparing one teacher to all the other
teachers in the system at the same grade level. Students are not
always randomly placed into classrooms; hence teachers are faced
with students who have quite different entering characteristics
than students in other classrooms. Therefore, covariates are a
necessity. Perhaps previous achievement, motivation, IQ, and a
host of other variables should be considered as possible contaminating variables. The criterion variable(s) should be ones that
are relevant to the objectives of that particular school system.
The possible contaminating variables and the criterion should be
proposed and discussed by all the teachers. A list of the agreed-
upon contaminating variables and criteria should then be passed
on to the administration for their consideration. Once the variables have been established by the administration, then each

teacher knows the basis upon which an evaluation will be made. Once the criterion variable and contaminating covariables are agreed upon, then the covariance notions incorporated in GRH 6.4 would be appropriate.

A slightly different route might be useful in the evaluation of individual student growth. One can use the standard error of estimate to ascertain whether the student is achieving "above that of students like him" (the overachiever), or whether the student is achieving "below that of students like him" (the so-called underachiever). Figure 10.3 depicts a simple example of the above notions.

Those students in Figure 10.3 who are above the top dashed line have achieved more than other students who had the same entering IQ. Those students below the bottom dashed line are

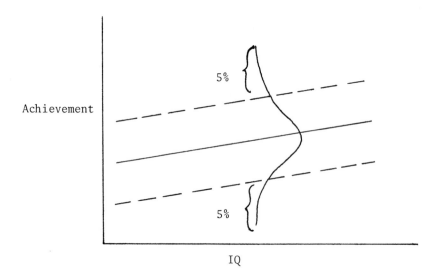

Figure 10.3. Identification of overachievers and underachievers.

achieving less than other students who had the same entering IQ.

Note that overachievers can be both high and low IQ students.

Likewise, underachievers can be both high and low IQ students.

The distance between the dashed line and the solid line of best

fit is dependent upon how many students the particular school

system wants to identify and either reward (the overachievers)

or remediate (the underachievers). If the achievement scores are

normally distributed about the line of best fit, then the func-

tional relationship that is being used is a good one. If the

previous sentence seems cryptic, reread the discussion in Chapter

Seven on homogeneity of variance. Thus the unit normal curve can

be referred to after the percentage of identified students has

been set. If 5% of the students are to be rewarded, and 5% re-

mediated, then the two dashed lines would be 1.64 standard de-

viation units away from the line of best fit. The standard de-

viation of the discrepancies between the actual achievement and

the predicted achievement is the standard deviation of the errors

in prediction. This standard error of estimate can be found from

the following formula:

Formula 10.1: $S_{est} = S_y \sqrt{1-R^2}$

R^2 is the proportion of criterion variance accounted for.

$(1-R^2)$ will then be the proportion of criterion variance unac-

counted for. The square root of variance is the standard devia-

tion, so $\sqrt{1-R^2}$ is the proportion of criterion standard deviation

that is not accounted for, or the proportion of criterion stan-

dard deviation that is error. Multiplying this proportion by the

actual criterion standard deviation (S_y) gives the desired quantity, the standard deviation of the error in predicting the criterion from the predictor variable(s) (the standard error of
estimate).

Note that if R^2 is low, say 0, then S_{est} = S_y, or all of the
criterion variance is error variance. The larger the R^2, the
smaller S_{est} becomes in relation to S_y. When R^2 = 1, all of the
criterion variance is accounted for, all of the errors in prediction = 0, the standard deviation of these errors would be
zero, and S_{est} = 0.

The regression program referred to in the Appendix can print
the criterion, the predicted criterion, and the error in prediction for all students (described in McNeil and McNeil, 1975). The
program can be modified to print the names of the students who
are the underachievers and the overachievers, along with their
scores. Suppressing the names and scores of those students who
are within the two dashed lines seems reasonable, as these discrepancies between the actual criterion and the predicted criterion could well be simply random fluctuation, rather than something systematic that needs rewarding or remediating.

The multiple linear regression procedure allows for a more
numerous set of predictors than that depicted in Figure 10.1. The
notions are the same though, although figures cannot be drawn to
depict, say, five continuous predictors. The identified underachiever in this situation is achieving much below students with
the same profile of five entering scores. Although another stu-

dent may not exist in the school exactly like him on these five
entering variables, the use of continuous variables allows the
inference made in the preceding sentence.

A few precautionary remarks are in order here. Students may
be designated as overachievers or underachievers when in fact
they are not. That is, chance may have placed them in one of the
two extremes, or one bad or good test score might have influenced
their placement. Erroneous placement is especially crucial for
the identified underachievers. Therefore, it is suggested that
counselors extensively examine the student's folder before attempting any remediating. Statistical and computer techniques can
save time in identifying potential students to work with, but man
should not abdicate all of his decision making and responsibility.
One mispunch by the keypunch operator can change Johnny from an
overachiever to an underachiever. More importantly, all of the
relevant variables may not have been taken into consideration.
Who would ever have predicted that Johnny would not be doing well
this year because Sally no longer is his girl?

The presentation in this section is employing the standard
error of estimate for the identification of individuals. Some
statistical authors use similar procedures and concepts to state
confidence intervals around the observed sample statistic, whether it be the mean, differences between means, or line of best
fit. The emphasis in this text has been the testing of hypotheses
and the increasing of R^2, rather than the identification of population values.

Regression models for one evaluation model

The place of regression analysis in the evaluation setting

has been well documented by Webster and Eichelberger (1972). They

present the CIPP model and then develop regression models for an

actual application in the educational setting. The CIPP model

delineates four kinds of evaluation:

> Context evaluation serves planning decisions by identi-
> fying unmet needs, unused opportunities, and underlying
> problems which prevent the meeting of needs or the use of
> opportunities; input evaluation serves structuring deci-
> sions by projecting and analyzing alternative procedural
> designs; process evaluation serves implementing decisions
> by monitoring project operations; and product evaluation
> serves recycling decisions by determining the degree to
> which objectives have been achieved and by providing in-
> sights with regard to the causes of the obtained results.

Each of these four kinds of evaluation are discussed in more

detail by Webster and Eichelberger, and then regression models

are presented with alternative lines of sequencing of hypotheses

depending upon whether there is previous significance or lack of

significance. The possible inferences that can be drawn from re-

sults are discussed, and decisions that might be made as a result

of these inferences are delineated.

The important aspect of this article is that it indicates a

close connection between research, evaluation, and statistical

procedures (particularly regression analysis). There will have to

be arbitrary administrative decisions in the real world--the

point that Webster and Eichelberger are making is that before the

administrative decision could be "yes," statistical significance

must be established. But statistical significance may not be enough evidence for the administrative decision. Statistical significance identifies the likelihood of random occurrence, but an administrator might be justified in not taking action until at least, say, 80% of the variance in the criterion is accounted for ($R^2 > .80$). Statistical significance is a necessary, but not sufficient condition for practical significance.

Policy Capturing

The following fable was written by Christal (1968). It is an enjoyable introduction to an interesting application of multiple regression--that of capturing policies, and is presented in its entirety.

Selecting a harem: A short fable

Once upon a time, there was an Oriental king who was concerned as to how he might make a name for himself in history. "I know," he said, "I'll select a harem larger than King Solomon's."

So the word went out, and soon thousands of young girls were arriving from the various provinces to seek the king's approval.

Early one morning the king began his selection process. As each girl filed by, he looked her over carefully and then expressed his judgment.

"Excellent!" he would say. "This one is very pleasing to my eye." Or perhaps he would hum and haw with indecision. Many times he would show his disapproval in no uncertain terms. "Never!" he would say. "Pass on! Pass on!"

In each instance, the Court Recorder attempted to quantify the king's degree of approval by checking the appropriate level on a 9-point scale which had been devised especially for the occasion by the Chief of the Royal Psychometricians.

By suppertime the king had considered some 300 girls. His eyes and imagination were beginning to tire.

"Most High First Counselor," he said, "you've been watching me all day, and by now you should know my likes and my dislikes. I've decided to leave the selection of my harem in your hands. But take care! If your choices do not please me, it will be your head!"

After the king retired, the Most High First Counselor summoned the Chief of the Royal Psychometricians. "I'm passing the job on to you," he said. "If you fail to please the king, your head will roll along with mine."

The Chief of the Royal Psychometricians called his staff together and explained the situation.

"We must not fail," he said, "or it will be all of our heads."

"How shall we proceed?" asked one of the young staff members who was fresh out of the Royal Academy.

"Well," responded the Chief, "we know how the king rated the first 300 girls. Right?"

"Right!"

"And we can see everything the king saw when he looked at the girls. Right!"

"Right!"

"Then all we have to do is to uncover the girly characteristics considered by the king and determine how he weighted them
in his judgment. This is a natural for the Multiple Linear Regression Model." (See Bottenberg and Ward, 1963).

"But how do we know which characteristics he considered?"
asked the neophyte.

"We don't, you fool! Didn't they teach you anything in that
school? That's what the regression model is for. If a girly characteristic adds to our ability to predict the king's ratings, we
may assume he gave it consideration. Now let's get on with the
business."

"How about height?" asked one of the staff members. "Does
the king like short girls or tall girls?"

"Neither," replied another. "I would guess that the relationship between height and the king's preference is curvilinear.
Some girls are too tall, while others are too short."

"Well," responded the neophyte, "if the relationship is
curvilinear, then we cannot use the linear regression model. If
we were to plot the curve between height and acceptability, I
think we would find it to be parabolic."

"They really didn't teach you very much in that school, did
they?" commented the Chief. "What is the general equation for a
parabola?"

"$aX^2 + bX + c$," responded the neophyte.

"Bravo!" declared the Chief. "Now let X be a vector of

heights. If we square each value in the height vector, we generate a new vector, X^2. Now if we introduce these two predictors in
the regression model, what will be the form of the resulting
equation?"

"aX^2 + bX plus the regression constant c," replied the young
man.

"Simple, isn't it?" responded the Chief. "You see, there's
no problem in fitting curvilinear relationships with the linear
regression model as long as the proper power terms are introduced
as predictors. The linear restriction is on the weighting system,
not on the form of the predictors." (See Figure 1.)

"How about eye color?" asked one of the other staff members
who was eager to move on. "I'm sure the king looked at the color
of each girl's eyes."

"Fine," said the Chief, "we will consider eye color in our
equations. Since eye color is not an ordered variable, we must
introduce a separate categorically coded predictor for each
color."

"What the Chief means," whispered one of the staff members
to the neophyte, "is that for a variable associated with a particular eye color, each girl will be assigned a value of 1 if her
eyes are that color, and a value of 0 if her eyes are not that
color."

"It's been my observation," said one of the group, address-
ing the Chief, "that the king likes blue eyes on blondes, but not
on brunettes."

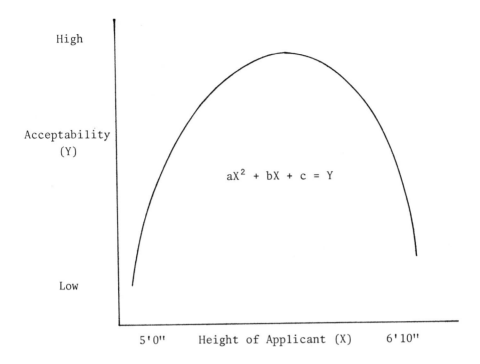

Figure 1. Relationship between Height of Applicant and
judged Acceptability.

"That's easily handled," responded the Chief. "First we will introduce categorical predictors for each hair color; then we can cross-multiply eye-color and hair-color variables in order to generate the appropriate interaction predictors." (See Table 1.)

"I thought," said one of the group, "that the regression model assumes the predictors to be normally distributed, and also that their joint-distribution is normal. We certainly can't meet these assumptions using powered terms, interaction terms, and categorically coded predictors."

"You're right," said the Chief, "if you're thinking about

Table 1. Examples of categorically coded and interaction
predictors.

Applicant No.	X_1 Blue Eyes	X_2 Brown Eyes	X_3 Brown Hair	X_4 Blonde Hair	X_5 (X_1*X_3)	X_6 (X_1*X_4)	X_7 (X_2*X_3)
1	1	0	1	0	1	0	0
2	0	1	1	0	0	0	1
3	0	1	1	0	0	0	1
4	1	0	0	1	0	1	0
5	0	1	0	1	0	0	0
6	1	0	0	1	0	1	0
.	.	.	.	.	.	.	.
.	.	.	.	.	.	.	.
.	.	.	.	.	.	.	.
.	.	.	.	.	.	.	.
N	1	0	1	0	1	0	0

the multi-normal model. But we're going to use the fixed-X model, which does not involve those assumptions. We would be stupid to restrict ourselves to normally distributed predictors. It would force us to omit most of the variables which we know the king considered."

"But," objected the staff member, "if we use the fixed-X model, we cannot generalize beyond the computing sample."

"Who can't?" responded the Chief. "Let's not assume our equation will fail to hold up just because our predictors are not normally distributed."

"Well, I'm from Outer Missourivich," said the staff member.

"Very well," replied the Chief, "if it will make you feel better, we will develop our equation on the first 150 girls rated by the king, and then check how well the equation predicts his judgments of the remaining 150 girls."

And so went the conference into the wee hours of the morning. Over a hundred predictors were eventually defined, each representing a girly characteristic which might have influenced the
king's judgment. The time had now come for the acid test. Could
they produce an equation which would simulate the king?

Some of the royal guards had to be called in to help measure
and evaluate each girl on the predictor variables. It was a mad-
house with 20 guards checking eye colors, weighing, and measuring
the 300 girls.

By late afternoon, the raw data had been accumulated. All
that night and throughout the next day and night, one could hear
the constant clicking of the abaci beads coming from the royal
computing shop. Then came the answer:

"We got an R^2 of .87, and it held up in the cross-applica-
tion sample," reported a messenger to the Chief of the Royal
Psychometricians, who was with his staff in the coffee room anx-
iously awaiting the results.

"Hmmm," said the Chief, "that's pretty good. But it's not
good enough for me to risk my head on it. There must be some
variable we failed to consider."

"Maybe the king likes girls who look like his mother," offered the neophyte. "Men often do."

"You're a genius," said the Chief. "It's certainly worth a
try."

"How can we quantify that?" asked a staff member. "You can't
measure it with a yardstick."

"We'll establish a rating board," responded the Chief. "Each board member will judge how much each girl looks like the king's mother. We will use an average of their ratings for each girl as our new predictor."

When the new variable was introduced into the king's policy equation, the R^2 jumped to .94. Everyone now felt confident that they had an equation which would truly simulate the king. The rest was routine. By the end of the week, all of the 8,000 girls in the applicant pool had been evaluated by the final policy equation, and those with the highest composite scores were se-lected.

The king was very pleased with the results, and as a reward, he gave the Most High First Counselor and the Chief of the Royal Psychometricians their choice of the leftovers.

Using multiple linear regression to capture policies

The above article was concerned with a fictitious example of policy capturing. Some real-life examples might also be useful at this point. Suppose the policy capturing problem involves the hiring of new teachers in a school district. Now how does the person who is responsible for hiring new teachers usually go about hiring those teachers? Applications are sent in and often interviews are conducted for those applicants who "looked good" on paper. The person doing the hiring usually focuses on one or two (sometimes a few more) good or bad traits of the applicant. In most employment situations, there are many relevant variables

in the application and in the interview. Human beings, though,
are usually not capable of incorporating all of this information
in order to make a rational and consistent decision. This is not
to say that one or two variables may not be the crucial variables
but the person doing the hiring should have the opportunity to
incorporate as many variables as are relevant. Furthermore, the
people doing the hiring are often not able to delineate "the
policy" (assuming one exists) used to determine which teachers
were to be hired. This is not a condemnation against hiring personnel, it is simply a frailty of mankind. Multiple linear regression can be used to determine if there is a policy, and if
there is one, capture it, and once captured evaluate it in terms
of the magnitude of R^2. Once this policy has been captured, it
can be applied consistently to all new applicants, and thus reduce the amount of time needed to screen applicants. Furthermore,
prospective teachers can learn the rules of the game--the desired
characteristics of teachers in that district. Persons could then
become aware of the crucial variables even before acquiring the
training for that profession.

The variables to focus on in a policy capturing application
depend upon the variables being used by the hiring personnel. Unfortunately, those people cannot always delineate those variables. A researcher in his own field may often have a good idea
of the important variables, but often there are many more variables relevant to that situation than are being used. One suggested procedure is to develop folders of teachers who applied

for positions the previous year, or to even develop fictitious
applicants. Then present these folders to those doing the hiring
(the "experts" in the field) and have the experts rate the folders in terms of goodness--the likelihood of this person being
hired. The advantages of fabricated data include:

1. One does not have to construct new tests which one thinks
   might be important;

2. One can include variables which might not be obtainable
   without a large investment;

3. One can include scores on ambiguous concepts.

The variables should cover the range that would be expected
in the real world. A lot of combinations of the variables should
occur, such as a teacher with 10 years experience being 50 years
old, another being 40 years old, and another being 35 years old.

In the construction of these folders, one should try to omit
irrelevant information. The problem is that one does not always
know what is or is not irrelevant information. There should be
enough "irrelevant" information in the folders to make the folders realistic to the raters. If the folders contain fabricated
data, the folder should appear realistic to the rater; i.e., do
not include an applicant 18 years old with 20 years of teaching
experience.

In estimating the criterion score, one could rank the applicants, or one can assign a score on some predetermined scale. The
latter procedure seems preferable because it is entirely reasonable to expect two applicant profiles to result in equally good
teachers.

After the judgments have been made, one could then put meaningful numerical values on as many variables as one wanted. This is not to say that the experts used these particular variables, but that these variables were available to the experts. It would be possible to overlook some variable or some combination of variables such as interactions or polynomial terms that the experts have in fact used. Indeed, if one does not put into the regression model the variables that the experts have used, then the policy cannot be captured. The weighting coefficients for extraneous variables (ones not used by the experts) will be close to zero, and the deletion of those variables will not cause a significant loss in predictability.

One could compare two raters. One could also compare one of the raters to all of the other raters, and if that one rater is different, then one might want to throw out that rater. One would probably, after he has ascertained the agreeing raters, use the pooled rating of those raters using the same policy. Assume that in a given situation there are four raters. Two raters agree amongst themselves, but differ from two other raters (who are using the same policy). Which policy does one then use? This would be an arbitrary decision, unless the one set of raters had more authority than the other set. It could be in a given situation of seven raters that six raters agreed amongst themselves, and were using a policy quite different from the seventh rater. But if that seventh rater is the boss, then his policy would be the one to use. That is to say, just because several raters agree amongst themselves, the policy to be used has not necessarily been captured. One would want to capture the policy of the decision-maker(s).

The policy capturing notions can also be applied to situations involving salary, firing, grade assignment, rating performance in athletic contests such as gymnastics and diving, and certification.

Since the policy capturing procedure is done empirically, the raters do not have to verbalize their policy. Indeed, they may not even be able to verbalize their policy nor the variables used in the policy. As long as the rater is consistent, then the regression procedure can be used to capture that policy.

It is important to note that, as Bottenberg and Ward (1963, p. 122) indicate, "the regression equation may adequately simulate the judgments but it is not necessarily descriptive of the thought processes that produced those judgments." That is to say, a correlate of the policy (possibly the policy) that raters have been implicitly using has been captured. Why raters use that particular policy is another question. There is no reason to believe that they knew they were using that policy. They might actually be using a completely different policy that happens to produce an end result that is perfectly in accord with the policy that has been captured. Viewpoints has carried a number of applied policy capturing articles, abstracts of which can be found in Appendix E.

An applied policy capturing example with the data in Appendix C

Now an applied example of teacher applicants will be discussed. Use the data in Appendix C, and variables X_3, X_4, X_5, and
X_6. Attach names to those four variables, such as X_3 = hours of
education courses; X_4 = motivation; X_5 = age; X_6 = 1 if female, 0
if male. For each of the 60 subjects, look at X_3, X_4, X_5, and X_6
for that subject, and on the basis of those four variables make
an estimate of the goodness of that person's teaching on a scale
from 1 to 99. (You are creating a variable.) Record these scores
in the same order that the subjects appear. Do not return to previous subjects. Treat each subject as completely different, although attempt to be systematic in terms of which variables you
are weighing heavily and which ones you are giving only passing
consideration. In other words, don't be completely random in
estimating the "goodness of teaching."

Now go back and try to capture your policy intuitively in
the sense of saying, "What values did I actually use in assigning weights to the various variables? Did I actually multiply X_1
by 3 and subtract a third of X_2, adding a sixth of X_3 and subtract 6 times X_4, and add in a constant value of 3 for the unit
vector?" Estimate what your policy was, and then consistently apply those weighting coefficients to all 60 subjects to come up
with a "systematic goodness score." Now punch these two score on
the data cards. Punch the "estimated goodness" score in columns
37 and 38, and the "systematic goodness" score in columns 39-43,

with column 42 containing the units, and column 43 containing

tenths. This format allows for as large a number as 9,999.9.

Next, give the punched data to someone else reading the text. It

is then the task of each person to try to capture the policy (of

someone else) for both criterion variables, using X_3, X_4, X_5, and

X_6 to predict first the "estimated" criterion and then the "sys-

tematic" criterion. With the "systematic" criterion, one would

expect an extremely high R^2. The R^2 will be less than perfect

only because of rounding error (reporting to only one significant

digit), or because of either a miscalculation or mispunch. The R^2

will be somewhat less than 1.0 for the "estimated" criterion. The

lower the R^2, the less systematic you were in using your implicit

policy.

The difference in R^2 of the "systematic" policy and the

"estimated" policy indicates the amount of error variance intro-

duced by not explicitly stating the policy. Now note that the

estimation of the goodness of teaching could not have been made

on any more than these five variables (the unit vector, X_3, X_4,

X_5, and X_6). Therefore, the discrepancy between the R^2 of 1.00

and that of the "estimated" criterion is due to either unsystem-

atic policy or to having interacted some of these variables (e.

g., $X_6 * X_3$; $X_3 * X_4 * X_6$) or using the predictor variables in a non-

linear fashion. (Some students might like to use a systematic

policy containing interactions or polynomial terms. That is, use

only $(6 * X_3 * X_4)$ in the systematic calculation of the criterion

score, and then generate $(X_3 * X_4)$ and include this generated variable in the model so that it can be utilized in the prediction

equation, and obtain its weight of 6.)

Clinical Prediction

Clinical prediction often involves the observation of one

subject over a long period of time, obtaining a lot of informa-

tion about that one subject, and then finally making a predic-

tion. Clinical prediction therefore is quite opposite to statis-

tical prediction wherein one obtains a parsimonious amount of in-

formation on a number of subjects in order to make a prediction.

There has been a long-standing discussion between clinical

people and statistical people as to which procedure is better. A

recent article by Kunce (1971) attempted to indicate the frail-

ties of the (stepwise) regression procedure in the prediction of

clinical behavior. Kunce (1971) was attempting to justify his

a priori selection of weighting coefficients. For those readers

interested in clinical research, reading of the Kunce article is

suggested.

A rejoinder to that article was written by McNeil and Lewis

(1972). The important point of their article was that clinical

prediction may be better than statistical prediction, but that

neither Kunce's data nor his procedures verify the advantage of

the clinical prediction model. The last paragraph of the McNeil

and Lewis (1972) paper supports the notion that statistical anal-

ysis is only a tool in that statistical analysis does not lead to

understanding of the data. Those researchers who desire under-

standing as a goal of research must turn to something beyond
statistical analysis (faith?).

Most of the problems that Kunce (1971) discusses about the
regression procedure deal with the stepwise procedure rather than
the hypothesis-testing procedure as discussed in the present
text. Many researchers are unfortunately not aware of the hypothesis-testing aspect of regression.

The multiple linear regression procedure can be valuable to
the clinical researcher, whether he be investigating a small number of subjects, or only one subject. A later section in this
chapter discusses procedures which can be applied to single organisms. Most of the literature using a single organism is found
in the behavior modification area, although the notions and models can be applied to any content area. Two other sections in
this chapter (Trend and Time Series; Developmental Applications)
present variations of the single subject design.

Single Organism Research Applications

Most statistical books suggest that analyzing repeated measures from a single subject is not legitimate because of nonindependence of observations (correlated error). The position taken
in this text is that time series data are real phenomena, and
that the consideration of trend vectors eliminates the problem of
nonindependence. Other statistical authors are proposing statistical analysis of such data (Shine, 1973; Glass, Willson, and
Gottman, 1974; Ward and Jennings, 1973, pp. 224-228; Kerlinger

and Pedhazur, 1973). If the reader is still concerned about analyzing such data, we suggest that replications of the results be
routinely planned, so as to further substantiate the results.

A recent article by Kelly, Newman, and McNeil (1973) illustrates some ways multiple linear regression can be used to answer
meaningful questions with single organism research designs. Although some of the material is redundant with earlier material,
the entire paper is reproduced here, with the permission of the
authors, so as to retain the flow. Researchers should remember
that these are only a few examples, and that the research hypothesis must be stated first. Once the research hypothesis is stated, the models used to test that hypothesis are easily constructed.

Suggested Inferential Statistical Models
For Research in Behavior Modification

For a number of reasons, research in operant psychology usually has been summarized using descriptive statistics. We suspect
the chief reason inferential statistics has been given little
attention is that early laboratory research yielded such clear
cut distinctions that one did not have to resort to tests of
statistical significance. Another reason might also be attributed
to poor advice from statisticians regarding limitations of single
subject data.

Presently operant psychological techniques have diffused to
applied behavior modification where complex behaviors are manipulated in settings where control of extraneous variables is dif-

ficult to achieve. As a consequence, data often fail to exhibit the clear magnitude of effects observed in the data derived from laboratory manipulation. Under these circumstances, often the data no longer hit the reader between the eyes, even though the expected trend seems to be present.

We believe researchers who use behavioral modification procedures should consider wider use of inferential statistics, especially when some doubt exists regarding the outcome of an experimental manipulation. Given this belief, this paper presents a number of inferential statistical models that may aid the operant investigator in presenting his data. These models are essentially specific applications of the generalized analysis of variance using multiple regression procedures to partial variance.

None of the procedures presented are particularly new nor startling, yet we believe their use has not been fully appreciated as research tools.

The following designs are illustrated:

(1) single organism curve fitting of response latency

(2) multiple organism curve fitting

(3) evaluating A-B-A' observed mean differences

(4) evaluating A-B-A' observed curve differences

Example 1: Single organism curve fitting of response latency

In a study of reinforcement schedules, (see Figures 1 and 2), if the experimenter was interested in the latency of the response (criterion variable) over sets of five response blocks within the

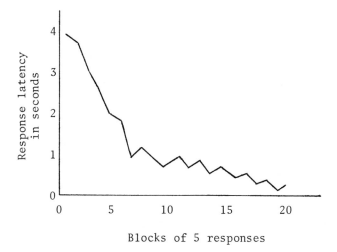

Figure 1. An observed response latency relationship on a
Fixed Ratio 100 Schedule for Day 1.

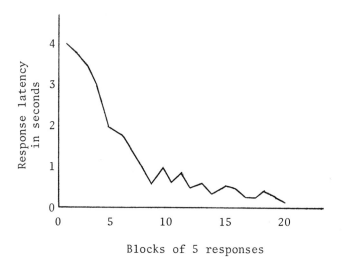

Figure 2. An observed response latency relationship on a
Fixed Ratio 100 Schedule for Day 2.

schedule (independent variable) the following are only a few of

the many research questions that can be asked regarding the design:

A. What kind of functional relationship between response
   blocks and response latency exists on day 1?

B. Is there a significantly different functional relationship between response blocks and response latency for
   day 1 and day 2?

C. Is there a significant difference between the initial
   response latency on day 1 and the initial response
   latency on day 2?

All of these questions can be answered with the use of multiple linear regression analysis by defining vectors and models
as outlined in more detail in Kelly, Beggs, McNeil, Eichelberger,
and Lyon (1969). If one expects a second degree functional relationship in question A above, the following models can test this
expectation.

Model 1: $Y_1 = a_0 U + b_1 X_1 + c_1 Z_1 + E_1$

Model 2: $Y_1 = a_0 U + b_1 X_1 + E_2$

> where: Y_1 = the criterion latency score;
> U = a 1 for all sessions;
> X_1 = number of the response block within the
> schedule;
> Z_1 = the squared values of X_1; and
> a_0, b_1, and c_1 are least squares weighting coeffi-
> cients computed so as to minimize the squared
> components in the error vectors E_1 and E_2.

The proportions of variance (R^2) accounted for by the above
two models can be compared by using the generalized F test used
in all regression analyses. If the F test is significant at the
specified alpha level, then the full model (Model 1) is preferred

over the restricted model (Model 2), indicating that there is a second degree curvature to the data. If the $\underline{F}$ test is not significant, then the restricted model is preferred over the full model because the restricted model is a more parsimonious accounting of the criterion. Parsimony in regression methodology is measured by the number of predictor variables. Other kinds of curves could be investigated, such as third degree, fourth degree, and logarithmic.

The second question, "Is there a significantly different functional relationship between latency and response block on day 1 and day 2?," can be tested statistically by determining if the graphs for the two days have the same Y-intercepts, same linear component, and same second degree curvature. A full model must be developed which will allow for differences on the three parameters, and then this model must be compared with a model which forces a common value for the three parameters. Models 3 and 4 below are those two models:

Model 3: $Y_1 = a_1U_1 + b_1X_1 + c_1Z_1 + a_2U_2 + b_2X_2 + c_2Z_2 + E_3$

Model 4: $Y_1 = a_0U + b_3X_3 + c_3Z_3 + E_4$

> where: Y_1 = the criterion latency score;
> U_1 = 1 if the criterion from day 1, 0 otherwise;
> U_2 = 1 if the criterion from day 2, 0 otherwise;
> X_1 = number of the response block within the schedule if criterion is from day 1, 0 if criterion from day 2;
> X_2 = number of the response block within the schedule if criterion is from day 2, 0 if criterion from day 1;
> $Z_1 = X_1*X_1$ = the squared elements of X_1;
> $Z_2 = X_2*X_2$ = the squared elements of X_2;
> U = 1 for all observations;

X_3 = number of the response block within the schedule, (disregarding day);

Z_3 = the squared number of response block within the
schedule, (disregarding day); and

a_0, a_1, a_2, b_1, b_2, b_3, c_1, c_2, and c_3 are least
squares weighting coefficients determined so as
to minimize the squared elements in the error
vectors E_3 and E_4.

The reader should note that Model 3 permits the reflection
of two second degree lines, one for day 1 and one for day 2. Model 4 forces the lines of best fit for day 1 and day 2 to be the
same. That is, the two days have common: intercept, linear component, and second degree component.

The R^2 of the above two models can again be tested with the
generalized F test.

If the R^2 of Model 3 is found to be not significantly different from the R^2 of Model 4, then one can conclude that the
functional relationship for day 1 and day 2 are the same. Indeed,
a stable state can be defined where the functional relationship
among a number of contiguous days are not statistically different. In the example given above, one may say a stable state is
observed when two contiguous days can be reflected by one line.
If stability for three or four sessions is desired to specify a
stable state, then one need only expand Model 4 to provide aU,
bX, and cZ components for each day under consideration. A comparison of this model to Model 4, using the F statistic, can provide an answer for the stable state. If the observed F is non-
significant, then a stable function can be said to exist. To ascertain whether that function required a second degree component

would necessitate a comparison such as Model 1 vs. Model 2.

If the R^2 of Model 3 is found to be significantly larger than the R^2 of Model 4, then the two second degree curves of best fit are not the same. Stated in another fashion, it could be concluded that the functional relationship is not the same for the two days. But because three parameters were simultaneously investigated, one could not, from the above analysis, discover which of the three parameters were different. If one was interested in specifically how the two days differed, then any one of the following three restrictions could be individually made and tested:

$a_1 = a_2$ (replaced by a_0 a common Y-intercept),

$b_1 = b_2$ (replaced by b_3 a common linear component),

$c_1 = c_2$ (replaced by c_3 a common second degree component),

resulting in the following models:

Model 5: $Y_1 = a_0U + b_1X_1 + b_2X_2 + c_1Z_1 + c_2Z_2 + E_5$

Model 6: $Y_1 = a_1U_1 + a_2U_2 + b_3X_3 + c_1Z_1 + c_2Z_2 + E_6$

Model 7: $Y_1 = a_1U_1 + a_2U_2 + b_1X_1 + b_2X_2 + c_3Z_3 + E_7$

By testing Model 3 against Model 5, one can determine if the model that allows for, among other things, different Y-intercepts, accounts for a greater proportion of variance than does a model which forces the Y-intercepts to be the same. If Model 3 does have a significantly larger R^2 than Model 5, then the data for the two days do have different Y-intercepts. Such a finding would imply that there is a significantly different initial response on days 1 and 2. If there is no significant difference between Models 3 and 5, then the Y-intercepts for the two days cannot be considered different. Note that the above conclusion does not say anything whatsoever about the linear and second degree slopes.

By testing Model 3 against Model 6, one can determine if the first degree function of day 1 is different than the first degree function of day 2.

A question similar to traditional statistics that one might ask of these data are, "Is there a linear interaction between day and trial blocks?" Model 8, allowing for interaction, would be compared against Model 9 which depicts a state of affairs of no interaction.

Model 8: $Y_1 = a_1U_1 + a_2U_2 + b_1X_1 + b_2X_2 + E_8$

Model 9: $Y_1 = a_1U_1 + a_2U_2 + b_3X_3 + E_9$

This test is investigating linear interaction, and if one suspected a second degree interaction (above and beyond the linear one) the appropriate models would be Models 3 and 7 which we have already compared.

The $\underline{F}$ tests just presented relate to inferences that can be made regarding one organism's population of responses under a particular schedule. The above analyses considered sample estimators of two potentially different populations (i.e., a_1 and a_2; b_1 and b_2; and c_1 and c_2). If the observed differences between the two sample estimators of population values are likely due to sampling variation (a nonsignificant $\underline{F}$), then one can assume that the two samples come from the same within-organism response population. One now has a good idea as to what this one organism will do under this one schedule.

Example 2: Multiple organism curve fitting

Once a stable curve is determined within the one organism,
one may wish to compare this functional relationship with another
organism with a similar history.

Suppose the curved relationship depicted in Figures 1 and 2
were found to be the same and all the partial regression weights
were non-zero, then Model 4 ($Y_1 = a_0 U + b_3 X_3 + c_3 Z_3 + E_4$) would
reflect the relationship for that particular organism.

Given another organism with a similar history, one may use
the procedures in Example 1 to express the functional relationship for this individual.

A legitimate question may be: Is the functional relationship
within organism 1 significantly different from the functional relationship within organism 2?

In order to answer this question, a model (Model 10) can be
constructed to incorporate the two separate equations and then
restricted down to another model (Model 11) which imposes common
weights for the two organisms.

Model 10: $Y_1 = a_1 U_1 + b_1 X_1 + c_1 Z_1 + a_2 U_2 + b_2 X_2 + c_2 Z_2 + E_{10}$

where: Y_1 = the criterion latency score;
        U_1 = 1 if the criterion is for organism 1, zero
             otherwise;
        X_1 = number of the response block within the sched-
             ule if the criterion is from organism 1, zero
             otherwise;
        Z_1 = the squared elements in X_1;
        U_2 = 1 if the criterion is from organism 2, zero
             otherwise;

X_2 = number of the response block within the schedule if the criterion is from organism 2, zero
otherwise; and

Z_2 = the squared elements in X_2.

Model 11, which sets:

$a_1 = a_2$ (replaced by a_3, a common Y-intercept),

$b_1 = b_2$ (replaced by b_3, a common linear component),

$c_1 = c_2$ (replaced by c_3, a common second degree component),

would be:

Model 11: $Y_1 = a_0U + b_3X_3 + c_3Z_3 + E_{11}$

Model 11 is different from Model 4 because data for 2 organisms
are being considered.

If the differences between Model 10 and Model 11 are likely
to be due to sampling variation (nonsignificant $\underline{F}$ test), then one
can conclude with some degree of confidence that, for the schedule under consideration, the response characteristics for the two
organisms reflect a population that does not differ significantly.

If the differences between Model 10 and Model 11 are not
likely to be due to sampling variations (significant $\underline{F}$ test),
then one must determine which (one, two, or three) of the three
population estimators (a_0 and/or b_3 and/or c_3) are not equal. A
highly probable expectation might be that the shape of the curve
is stable over organisms, but the initial response latency varies
over organisms (some just get to work sooner). Model 12 can be
cast and tested against Model 10 to ascertain this situation.

Model 12: $Y_1 = a_1U_1 + a_2U_2 + b_3X_3 + c_3Z_3 + E_{12}$

Note, Model 12 differs from Model 11 only in the elements which

reflect the intercept ($a_1 \neq a_2$). Given a nonsignificant difference between Model 10 and Model 12, one can conclude with some
degree of confidence that the population estimators of the linear
and second degree slope are common for the two organisms. A significant difference between Model 12 and Model 11 would then indicate that the two Y-intercepts (initial response latency) were
different.

Example 3: Evaluating A-B-A' observed mean differences

A third type of example is the traditional problem of evaluating mean differences in the A-B-A' design. The discussion here
can be applied to single subjects or means of a group of subjects; care must only be exercised to be sure the generalization
is to the appropriate populations (within subject or across subjects).

Some of the questions one may be interested in answering
are:

(1) Was there a significant increase in response rate during
condition B over condition A?

(2) Was there a significant decrease in response rate under
A' condition as compared to the B condition?

(3) Are A and A' response rates significantly different?
Each of these questions could be tested by comparing the scores
in one condition to the scores in another via the t test. We prefer to utilize the data in the third condition to help obtain a
more stable estimate of the expected variability of scores in the

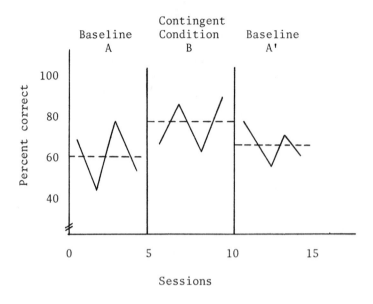

Figure 3. Percent correct for 16 sessions under three dif-
          ferent conditions in A-B-A' design (dashed
          lines represent means for each condition).

two conditions being tested. This approach assumes that the vari-

ability within all three conditions is relatively similar. A

latter example will discuss what may be done if the variability

within conditions is not similar.

Figure 3 is one typical example where the above questions

would be appropriate.

All the questions stated above can be answered by making

specific restrictions on Model 13.

Model 13: $Y_1 = a_1U_1 + a_2U_2 + a_3U_3 + E_{13}$

        where: Y_1 = response rate;
                 U_1 = 1 if the criterion from condition A, zero

> otherwise;
> U_2 = 1 if the criterion from condition B, zero
> otherwise; and
> U_3 = 1 if the criterion from condition A', zero
> otherwise.

Model 13 allows for three separate means, one for each condition. Since the three research questions are concerned with means and tacitly assume, by omission, that slopes are of no concern, slope functions _are_ not specified in the model. We shall return to questions about slopes in Example 4.

To answer question (1) "Was there a significant increase in response rates during condition B over condition A?" would require restrictions on the weighting coefficients in those two conditions: $a_1 = a_2 = a_4$. This single restriction would yield Model 14:

Model 14: $Y_2 = a_4U_4 + a_3U_3 + E_{14}$

> where: $U_4 = U_1 + U_2$

Note should be made that the above question is a directional question (B increase over A, rather than just a change from A to B) and, as such, an adjustment on the probability associated with the _F_ value is required.

If the comparison of Model 13 to Model 14 yields a significant _F_ value (directional test), and the weight a_2 is greater than a_1 (in Model 13), one can conclude with some degree of confidence (1-alpha) that response rate under condition B is greater than condition A.

Question (2) "Was there a significant decrease in response rate under A' condition as compared to condition B?" can be answered by restricting the weighting coefficients in these two conditions: $a_2 = a_3 = a_5$, a common weighting coefficient. This single restriction requires Model 15 to be:

Model 15: $Y_2 = a_1 U_1 + a_5 U_5 + E_{15}$

    where: $U_5 = U_2 + U_3$

If the $\underline{F}$ test yielded an adjusted probability value which was less than the set alpha level, then one can conclude the decrease was significant.

The last question associated with recovery of baseline, question (3) "Are A and A' response rates significantly different?" can be answered by comparing Model 16 and Model 13:

Model 16: $Y_2 = a_2 U_2 + a_6 U_6 + E_{16}$

    where: $U_6 = U_1 + U_3$

If the $\underline{F}$ value (non-directional since A' can be significantly lower or higher than A) is non-significant, then one can conclude with some degree of confidence that the baselines are not significantly different. In this case one would want to use a relatively high alpha level (say .60) because acceptance of the no difference hypothesis is here desired.

Inspection of Figure 3 might be instructive of the value of the preceding hypothesis testing. Condition B does seem to yield an increase in response rate and may be interpreted to be an obvious difference. But what about the decrease in response rate in A' from B? The eyeball detects a difference, yet not really unequivocal. Likewise, A and A' look a little different; is this a likely sample variation? We think these last two questions indicate the need for an inferential test in all instances.

Example 4: Evaluating A-B-A' observed curve differences

A fourth example we have encountered is an A-B-A'-B' study which involves a situation where a functional relationship exists between treatment condition, day in the treatment, and response rate.

Consider Figure 4. A group of children are observed for 10 days under a typical condition (A), then tokens are instituted for correct response on work material for a 10-day period (B), at which time tokens are withdrawn for 10 days (A') and then a return to tokens with an added stipulation that the tokens can be cashed in for preselected toys, etc., which the children "desire."

For illustrative purposes, let's specify a few questions which may seem reasonable. In the simple test of means, as provided in Example 3, we would ignore the negative slope under condition B. The similarity of variability that we assumed and probably observed in Example 3 does not hold true in Figure 4. The variability about the condition B mean is much greater than in the other three conditions.

A different set of questions than the ones asked below or the same questions asked in a different sequence would result in different models. One may ask the following sequence of questions:

(1) Has percent correct baseline been achieved? This can be operationally defined as a slope in condition A that is not significantly different from zero.

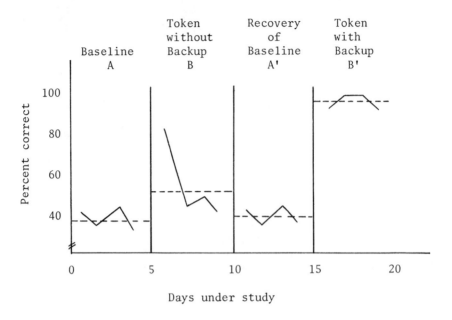

Figure 4. Percent correct by days within four experimental
conditions (dashed lines represents means for
each condition).

(2) Is there a significant negative slope relating day to
percent correct in condition B?

(3) Has percent correct baseline been recovered under A'
condition? This can be operationally defined as a non-
significant difference in the means of A and A'.

(4) Is percent correct significantly higher under condition
B' than under conditions A and A'?

These four questions can be answered by placing the appropriate restrictions on Model 17.

Model 17: $Y_3 = a_1U_1 + b_1X_1 + a_2U_2 + b_2X_2 + a_3U_3 + b_3X_3 + a_4U_4 +$
$b_4X_4 + E_{17}$

where: Y_3 = percent correct;
        U_1 = 1 if Y_3 is from condition A, zero otherwise;
        X_1 = day (1-10) associated with Y_3 under condition A, zero otherwise;
        U_2 = 1 if Y_3 is from condition B, zero otherwise;
        X_2 = day (1-10) associated with Y_3 under condition B, zero otherwise;
        U_3 = 1 if Y_3 is from condition A', zero otherwise;
        X_3 = day (1-10) associated with Y_3 under condition A', zero otherwise;
        U_4 = 1 if Y_3 is from condition B', zero otherwise; and
        X_4 = day (1-10) associated with condition B', zero otherwise.

The reader might note that Model 17 is simply an equation which allows for a slope and intercept for each of the four conditions. $Y_3 = a_1U_1 + b_1X_1$, for example, is Y = a + bX for condition A; $Y_3 = a_2U_2 + b_2X_2$, is Y = a + bX for condition B, etc.

To answer question (1), the weight b_1 can be set to zero; that is, we force the line under condition A to have a zero slope. Model 18 reflects this restriction.

Model 18: $Y_3 = a_1U_1 + a_2U_2 + b_2X_2 + a_3U_3 + b_3X_3 + a_4U_4 + b_4X_4$

$+ E_{18}$

We can test Model 18 (restricted) against Model 17 (full) and if there is no significant loss in the R^2 value, then we can conclude with some degree of confidence (two-tailed F test) that a baseline is achieved and Model 18 may be used as a new full model because it is the more parsimonious model (let us assume b_1 does equal zero).

Question (2) "Is there a significant negative slope relating day to percent correct in condition B?" can be answered by restricting Model 18 by setting b_2 (the slope coefficient for condition B) equal to zero. Model 19 represents this restriction.

Model 19: $Y_3 = a_1U_1 + a_2U_2 + a_3U_3 + b_3X_3 + a_4U_4 + b_4X_4 + E_{19}$

Note that both b_1 and b_2 are deleted from this model; b_1 because
we found it equal to zero and b_2 because we wish to test the
slope for condition B. In the case pictured in Figure 4, we
should find a significant loss in R^2_{19} over R^2_{18} because a distinct negative slope exists. This would be a one-tailed test
since we asked a directional question (negative slope). Given
this situation, we retain Model 18 as the full model because it
best describes the data.

The meaning of this finding is that it is ridiculous to ask,
"Is the mean of B greater than the mean of A?" There may be a
significant difference between these means, but the observed
state of affairs is that initially tokens increase percent correct, but over time the effect dissipates towards a return to
baseline.

Model 18 has a mean for condition A and an intercept and a
slope for the other three conditions. An inspection of Figure 4
should lead one to conclude that the means of A and A' are about
equal and that the slopes are equal to zero. To answer question
(3) "Has percent correct baseline been recovered under A' condition?" we must set $a_1 = a_3 = a_5$, a common weight and $b_3 = 0$. Note
two restrictions are being made, equal means and no slope for
condition A'.

Model 20 reflects these restrictions.

Model 20: $Y_3 = a_5U_5 + a_2U_2 + b_2X_2 + a_4U_4 + b_4X_4 + E_{20}$

where: U_5 = 1 if percent correct (Y_3) is from condition A
         or A', zero otherwise.

If R^2_{20} is not significantly different (two-tailed test)
from R^2_{18}, then one can conclude that the baseline was not significantly different from the original baseline (mean of A =
mean A'). Let us assume this is the case.

Question (4) "Is percent correct significantly higher under
condition B' than under conditions A and A'?" implies that there
is a zero slope for condition B'; however, Model 20 includes a
weighting coefficient for a slope (b_4). Before we answer question
(4), we must cast Model 21 to reflect an assumed slope of zero in
condition B'.

Model 21: $Y_3 = a_5 U_5 + a_2 U_2 + b_2 X_2 + a_4 U_4 + E_{21}$

Then we can answer question (4) by setting in Model 21 $a_4 = a_5 =$
a_6, a common mean which yields:

Model 22: $Y_3 = a_6 U_6 + a_2 U_2 + b_2 X_2 + E_{22}$

where: U_6 = 1 if Y_3 is from condition A, A', or B', zero
              otherwise; and
         a_6 = the mean of conditions A, A', and B'.

Given the data in Figure 4, we would expect R^2_{22} to be significantly smaller (one-tailed test) than R^2_{21}; therefore, we
could conclude that percent correct under condition B' is significantly larger than under baseline conditions.

These four questions were asked and tested using inferential
statistics. The particular set of questions were only a limited
set that could have been asked; they were, however, the four
questions the researcher was interested in investigating.

The regression models have been presented to illustrate clearly and simply how the multiple regression procedure can help researchers who are interested in behavior modification designs. The models that have been presented so far, specifically for the questions under Examples 3 and 4, do reflect the research question asked in as simple a manner as possible so that the concepts can be most easily understood. However, more precise models (repeated measures) can be written to answer these questions, but they have not been presented until this point because they tend to be a little more complex, even though the concepts are similar.

The following is a brief and oversimplified introduction to the concept of repeated measures. For question number 1 in Example 3, which was originally answered by testing Model 13 against Model 14, a repeated measures model will be written to more precisely answer that question. Another repeated measure model will be written to answer question number 1 in Example 4.

When experiments are designed so that all subjects are used in all treatments (K), then one has S*K observations, where S is the number of subjects and K is the number of treatments. This situation constitutes a repeated measures design. It is unlikely, when using this design, that there are S*K uncorrelated (independent) observations. Since the same subjects are in a number of treatments, there is likely to be a correlation between treatments due to subjects or S-treatment interaction. This problem can be considerably reduced by employing "person vectors" in the multiple regression equation (Kelly et al., 1969).

A person vector (P) is a vector that contains information of which person was responsible for that particular criterion score. By using such vectors, one can control for the variance in each of the K treatments due to individual differences (person differences). One can determine the amount of variance due to the treatments, independent of individual differences.

The following is a repeated measures model to test Example 3's question "Was there a significant increase in response rate in condition B over condition A?" (assuming that the same subjects are used in each of the conditions). This question can be tested with more precision by using Models 23 and 24 than previously done using Models 13 and 14. Models 23 and 24 assume that there are three subjects, and each is in both treatments.

Model 23: $Y_3 = a_1U_1 + a_2U_2 + a_3U_3 + a_4P_1 + a_5P_2 + a_6P_3 + E_{23}$

Model 24: $Y_3 = a_7U_7 + a_3U_3 + a_4P_1 + a_5P_2 + a_6P_3 + E_{24}$

where: Y_3 = response rate;
U_1 = 1 if the criterion is from condition A, zero otherwise;
U_2 = 1 if the criterion is from condition B, zero otherwise;
U_3 = 1 if the criterion is from condition A', zero otherwise;
$U_7 = U_1 + U_2$;
P_1 = 1 if the score on the criterion was made by Person number 1, zero otherwise;
P_2 = 1 if the score on the criterion was made by Person number 2, zero otherwise; and
P_3 = 1 if the score on the criterion was made by Person number 3, zero otherwise.

Since the research question is dealing with means, the full model (Model 23) allows each treatment to have its own mean. The re-

search question did not require information about slopes; therefore, no vectors dealing with slopes have been included.

By making the restriction $a_1 = a_2 = a_7$ the means of conditions A and B are being forced to be equal. Any differences between the two models' abilities to account for criterion variance
would have to be due to this restriction. If there is a significant difference between Models 23 and 24, using a one-tailed test,
and if $a_2 > a_1$, then one can state that the mean of B is significantly greater than the mean of A, at some alpha level above and
beyond any variability likely to be caused by individual differences. The part of the statement dealing with individual differences is what differentiates the results of testing Models 13 and
14 from testing Models 23 and 24. The testing of Models 23 and 24
is more precise since it takes into consideration variance that
can be accounted for due to individual differences.

The following is a repeated measures design to test Example
4's question "Has percent correct baseline been achieved in condition A?" If we operationally define and conceptually accept the
idea that, if the slope of the baseline data is not significantly
different from zero (a zero slope would indicate a stable percent
correct), then one could statistically test this question with
more precision than Models 17 and 18 by using Models 25 and 26.
This assumes that the same subjects are being used in all treatments, and in this particular case we assume three subjects.

Model 25: $Y_3 = a_1U_1 + b_1X_1 + a_2U_2 + b_2X_2 + a_3U_3 + b_3X_3 + a_4U_4 +$
$$b_4X_4 + p_1P_1 + p_2P_2 + p_3P_3 + E_{25}$$

Model 26: $Y_3 = a_1U_1 + a_2U_2 + b_2X_2 + a_3U_3 + b_3X_3 + a_4U_4 + b_4X_4 +$

$\qquad\qquad p_1P_1 + p_2P_2 + p_3P_3 + E_{26}$

where: Y_3 = percent correct;
$\qquad$ U_1 = 1 if Y_3 is from condition A, zero otherwise;
$\qquad$ U_2 = 1 if Y_3 is from condition B, zero otherwise;
$\qquad$ U_3 = 1 if Y_3 is from condition A', zero otherwise;
$\qquad$ U_4 = 1 if Y_3 is from condition B', zero otherwise;
$\qquad$ X_1 = day (1-10) associated with Y_3 under condition A, zero otherwise;
$\qquad$ X_2 = day (1-10) associated with Y_3 under condition B, zero otherwise;
$\qquad$ X_3 = day (1-10) associated with Y_3 under condition A', zero otherwise;
$\qquad$ X_4 = day (1-10) associated with Y_3 under condition B', zero otherwise;
$\qquad$ P_1 = 1 if the score on the criterion was made by Person number 1, zero otherwise;
$\qquad$ P_2 = 1 if the score on the criterion was made by Person number 2, zero otherwise; and
$\qquad$ P_3 = 1 if the score on the criterion was made by Person number 3, zero otherwise.

The inclusion of the person vectors in Models 25 and 26 increases the possibility of finding significance if indeed baseline has not been achieved. Certainly one would not want to act as if baseline had been achieved when in fact it hadn't. We strongly believe that the regression procedures as illustrated represents the proper case where the research questions dictate the statistical procedure.

A special note regarding time series

A statistical analysis of time measures has been a special and troublesome problem. Traditional methods, when applied to time designs, can yield results that are incorrect. The basic reason for this is that such data as time measures tend to fluctuate. If it is possible to legitimately assume that the vari-

ability is uniform over the whole criterion, there is no longer
a statistical problem (Boneau, 1960, pp. 148-162; Kerlinger,
1964, p. 319). In operant conditioning designs, the legitimacy of
making the assumption that the variability is stable over the
area of interest is more likely to be correct due to the procedures used in calculating baseline. The procedures generally require collection of enough data so that one can determine if the
responses are stable over the area of interest. The number of observations necessary depends upon the stability and adequacy of
the function being fit. At the lower limit, only one more observation than predictor pieces of information is needed in order to
calculate the statistical index. In essence, one needs only a
sufficient number of observations to provide a stable indicator
of the functional relationship, and this is primarily a non-
statistical decision.

References

Boneau, C.A. The effects of violations of assumptions underlying
the t-test. Psychological Bulletin, 1960, 57, 49-64.

Box, G.E. Some theorems on quadratic forms applied in the study
of analysis of variance problems. I. Effect of inequality of
variance in the one-way classification. Annals of Mathematical Statistics, 1954, 25, 290-302.

Draper, N.R. and Smith, H. Applied Regression Analysis. New York:
John Wiley, 1966.

Edwards, A.L. Experimental Design in Psychological Research, 3rd
Ed. New York: Holt, Rinehart, and Winston, 1968.

Kelly, F.J., Beggs, D.L., McNeil, K.A., Eichelberger, R.T., and
Lyon, J. Research Design in the Behavioral Sciences: Multiple Regression Approach, Carbondale, Illinois: Southern

Illinois University Press, 1969.

Kerlinger, F.N. Foundations of Behavioral Research. New York: Holt, Rinehart, and Winston, 1964.

Norton, D.W. An empirical investigation of some effects of non-normality and heterogeneity on the F distribution. Unpublished Doctoral Dissertation. State University of Iowa, 1952.

Trend and Time Series

The material presented in the previous section is not all that new to some disciplines. Economists have for many years been concerned with time series analysis. Their data usually represent the same entity repeatedly observed over a number of months or years. The amount of time separating the repeated measures is, of course, up to the researcher, and depends upon his interests. The time span could be days, hours, or even smaller units such as minutes or seconds.

The concern with data of this sort is not so much to find the causal factors, but to ascertain trends existing in the data. Usually, the only variables measured are the criterion and the time (month, day, etc.) at which the criterion was observed. One usually does not consider time by itself as a causative factor, but given the pattern or trend of the criterion scores over an extended period of time, one may be able to predict the criterion value at a given later time period.

Few time series can be reflected solely by a straight line; therefore, more than a linear trend exists in the data. Higher order polynomials can be tested to see if those kinds of trends

exist in the data. Often, though, the nature of the criterion being investigated can yield a hint as to what kinds of trends to
use. Suppose subjects are given ten trials on each of five successive days. One might anticipate a similar trend within each
day, superimposed on the overall trend such as in Figure 10.4.
One would then want to allow for an overall linear trend in the
data, as well as a common daily second degree trend. The requisite model allowing those trends would be:

Model 10.1: $Y_1 = a_0 U + a_1 X_3 + a_2 X_3^2 + a_3 X_1 + E_1$

where: Y_1 = Performance;
         X_1 = Trial number;
         X_2 = Day number; and
         $X_3 = X_1 - [(X_2-1) * 10]$ (Trial within day)

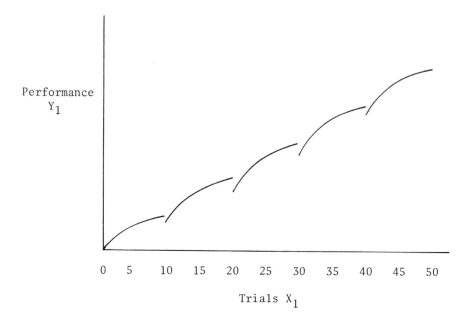

Figure 10.4. Similar trend within each Day, superimposed on
             overall linear trend.

Figure 10.5 presents some hypothetical economic data which
will allow some of these concepts to be clarified further. Fitting a straight line to the data yields an R^2 of .89, which is
significantly different from zero (using alpha = .01). Therefore,
there is a linear trend over months. Note, though, that the data
points are somewhat discrepant from the line of best fit, and
furthermore that the deviations of the data points from the line
of best fit are rather systematic. For the November, December,
and January months, the observed criterion values are much higher
than the linearly predicted values. For the months of May, June,
and July the observed criterion values are much less than the
linearly predicted values. The fact that the R^2 is somewhat less
than 1.00 indicates that the linear fit is not "the" fit to the
data. The systematic values of the errors in prediction indicate
that there are some other fits to be found--that the criterion is
not fluctuating solely in a linear fashion over time.

One way of attempting to discover the other trend(s) in the
data is to rotate the figure such that the straight line going
through the data points is horizontal. The problem now is to discover the remaining trend, after the linear trend has been ex-
tracted. (Or, to discover the trend in the data, over and above
the linear trend.) The data in Figure 10.6 appear to follow a
trigonometric function, over and above the linear function. That
supposition will be tested in Applied Research Hypothesis 10.3.

The careful reader will note that the data have been snooped
upon. Even though an extremely high R^2 has been obtained with the

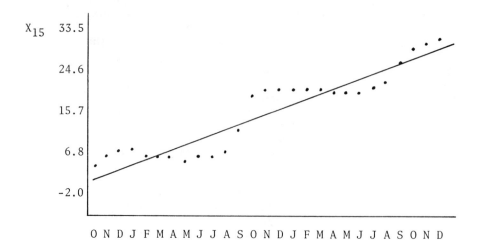

Figure 10.5. Sinusoidal relationship between Month (X_{16})
and the criterion (X_{15}), superimposed upon
the linear relationship.

Figure 10.6. Sinusoidal relationship between Month (X_{16})
and the criterion [$X_{15} - (.9724*X_{16})$] after
the linear trend has been taken out of the
criterion.

Applied Research Hypothesis 10.3

Directional Research Hypothesis: For a given population, there
is a sinusoidal trend over the
last months in the prediction
of X_{15} (with maximums occurring in December and minimums
in June), over and above the
linear trend.

Statistical Hypothesis: For a given population, there is not a
sinusoidal trend over the last months
in the prediction of X_{15} (with maximums
occurring in December and minimums in
June), over and above the linear trend.

Full Model: $X_{15} = a_0U + a_{16}X_{16} + a_{17}X_{17} + E_1$

where:

$X(17) = DSIN(X(16)*6.283/12.)$
$X(16)$ = the consecutive months numbered 1-30.
6.283 = twice the value of pi
12. = the period of the expected trend (12 months in a
year)

Restricted Model: $X_{15} = a_0U + a_{16}X_{16} + E_2$

alpha = .001

R_f^2 = .98 R_r^2 = .89

$\underline{F}_{(1,57)}$ = 492   Directional p < .0000001

Interpretation: Since the maximum parts of the sine curve are
during December and the minimum parts are dur-
ing June, as hypothesized, the directional
probability can be referred to. Since the directional probability is less than alpha, the
statistical hypothesis can be rejected in favor
of the research hypothesis.

Full Model in ARH 10.3, the researcher must obligate himself to
the replication process. Once he has replicated the functional

relationship, he is in a position to predict to future months.
Such a prediction assumes that the relevant influences remain the
same as in the empirically analyzed months. Unforeseen events
such as wars, changes in policy, and changes in consumer habits
will have their (unpredictable) effect on the magnitude of the
criterion.

Regression Analysis in Developmental Studies

Any given measure or scale used in the behavioral sciences
produces a score for an individual at a particular point in time,
even though that individual is probably in a state of change with
regard to that very measure. This state of affairs becomes particularly obvious when attempting a longitudinal study of a process
such as human development. Measures that must be pinpointed in
time are used to describe an ever changing phenomenon. Given that
this must be the case, how does the developmental researcher
proceed to select predictor and criterion variables? This section
discusses some of the issues involved in that selection.

The choice of criterion variables (as well as predictor
variables) must reflect the interests of the researcher. The use
of MLR should enable the researcher to state and test whatever
hypotheses she has (providing she can collect the necessary data)
so it is hoped that she will investigate her hypotheses regarding
all criteria of interest to her.

The ultimate criterion variable for human development may
well have to do with the adult that the child becomes. But there

are certainly an infinite number of criteria to be considered
along the way. These interim criteria, though, have the discon-
certing habit of becoming predictors for later criteria. For example, the ages at which different infants learn to pull themselves to a standing position may be a criterion at one point;
that variable may also be one of the predictors of the skill of
6-year-olds at skipping rope, or of the ability of 20-year-olds
to perform on a gymnastic balance beam.

The problem of criterion variables becomes especially difficult, though, when one considers how little is known on a descriptive level--let alone on a predictive basis--about the course
of normal development. It is difficult to decide upon criterion
variables when one is not sure what the process one is investigating is all about.

If the researcher can specify a criterion variable of interest to her, the task is then to select predictor variables
which will account for the variance observed in that criterion.
This choice will reflect the researcher's theoretic views about
development, especially the question of whether a child's structure and behavior are a result of something inherent in himself,
or are a product of his environment and reinforcement history, or
a result of both of these influences. Most current developmental
researchers follow some form of the theory that a combination of
heredity and environment influences development. It should be
pointed out, however, that although the word "interaction" is
often used to describe the relationship of heredity and environ-

ment, the statistical relationship investigated is generally additive rather than actually an interaction. The additive investigation would result in a model such as the following:

Model 10.2: $D_1 = a_0U + a_1H_1 + a_2V_1 + E_2$

where: D_1 = some aspect of development;
H_1 = some aspect(s) of the child's inherent structure; and
V_1 = some aspect(s) of the child's environment.

The addition (or substitution) in the above model of the interaction between heredity and environment (H_1*V_1) may well account for more of the variance in D_1, since it is likely true that a given environmental aspect will have an entirely different effect upon children with differing endowments. This approach has been suggested by Bell (1971), Brazelton (1969), and Korner (1967), among others. The resulting model would be either of the following:

Model 10.3: $D_1 = a_0U + a_1(H_1*V_1) + a_2H_1 + a_3V_1 + E_3$

Model 10.4: $D_1 = a_0U + a_1(H_1*V_1) + E_4$

The researcher's approach to predictor variables may be finding predictor variables which will increase the R^2 of predicting her criterion of interest, and/or testing the contribution to development of one predictor, over and above a given set of other predictors.

It may seem an obvious notion, but it should be remembered that most child development measures are very closely tied to age if the range of ages in the sample is sufficient. Age is then a competing explainer (unless the predictor of interest is age), and therefore in developmental studies either a measure of age should be included as a covariate, or all subjects must be very close together in age.

The task of a developmental researcher may be conceptualized as follows. First, the researcher needs a map or description of the process of development of the particular criterion variable she wants to study. This description may exist or she may have to provide it herself.

Second, the researcher wishes to discover what variables lead to (predict) differences between children at various levels of the developmental criterion. It will be most productive at this stage to investigate variables which are manipulatable along with non manipulatable variables.

Third, the researcher wishes to determine what variables (hopefully, some of the already found predictors) are causers of the differences in development. The task here is to determine which variables can be manipulated to produce different (hope- fully "better") development. It cannot be assumed that a variable which is found to be a predictor is a causer. For example, it may be found that the more a mother talks to an infant, the earlier the infant will speak two-word phrases. But upon manipulation of the predictor variable (getting a group of mothers to increase the amount of time they talk to their infants), it may be found that there is no change in the age of speaking two-word phrases. This is a serious problem with much of the research in the area of human development (as well as with any content area). It is

often assumed that a variable which is found to be a predictor of
the criterion of interest will result in change in the criterion
if the predictor is manipulated and recommendations are made to
parents and educators (and others) on that basis. It is necessary
to manipulate the predictor variable to determine if it is a
causer.

Missing data

The problem of some subjects having missing data is high-
lighted by the multiple linear regression procedure because of
the inclusion of multiple predictor variables. The possibility of
any one subject missing a score is geometrically increased when
one considers each additional predictor variable. The computerized regression procedure would interpret the blank (missing)
scores as zero and would end up giving a valid-appearing printout, but the results would not be valid, as one certainly did
not want each of the missing data points to be expressed as zero.

The easiest solution to the missing data problem is to try
to get complete data, thereby completely eliminating the missing
data problem. This is more likely to be the situation if the
study is carefully designed, the questionnaire is carefully constructed, the treatments are carefully developed and applied,
etc. Researchers should be cautioned, though, that forcing complete data oftentimes generates bad data. For example, mailing of
a second or third or fourth questionnaire, continually knocking
on someone's door, and demanding subjects to appear for an ex-

perimental situation will often generate negative attitudes in those subjects and hence invalid data. In these cases it would be better not to have the data than to have bad data.

Another solution to the missing data problem is to simply exclude those subjects who are missing data. But eliminating subjects because of missing data most likely redefines the population from which one has sampled, and hence to which one can generalize. Students in school who are missing some data over a two-week period may be the unhealthy kids, and/or the truant kids. Therefore, if one continually uses only complete data, the population to which one can legitimately generalize to may be healthy and nontruant students.

Another solution to the missing data problem would be to insert the mean value of the predictor variable. Insertion of a mean value assumes that the person with missing data is like the average subject with data. Another problem with inserting mean values is the reduction of the variance of the predictor variable resulting in a variable with lowered predictive value. The extreme case is where there is only one subject who has a score on the predictor variable. All other subjects are missing that predictor score. Insertion of the mean (the score of the one subject) would result in all subjects having the same score, and hence the variance of the variable would be zero, resulting in a vector with no predictive value. One way of getting around the problem of reduced variance is to insert a random score from the vector into the missing data locations. However, if the persons who are missing data are different from those with data, then the procedure of inserting random scores is, on the average, decreasing the relationship between that predictor variable and any other variable.

The best solution to this missing data problem may be to directly test to see if subjects missing data are different on the criterion from those who do not have missing data, using the procedures outlined in GRH 6.1. The Full Model would contain a (1,0) vector for persons who have complete (predictor) data, and a (1,0) vector for persons who have incomplete (predictor) data. A significant difference would indicate that those subjects who are missing data are different on the criterion from those subjects who have complete data.

Another procedure would be to predict the value of the missing score from the other predictor variables. That is, use as a criterion the predictor variable of concern and include only those subjects who have complete data on all the predictor variables. Then find the weighting coefficients for the functional relationship between the criterion (predictor variable of concern) and the remaining predictor variables. These weighting coefficients can then be applied to the subjects who have missing data on that particular predictor variable. This last procedure seems to be of benefit when there are a large number of variables relative to the (small) number of subjects. This procedure, though, does lead to a more systematic relationship between the predictors and the criterion than might really exist. Replication is therefore an important corollary of this procedure.

How to Beat the Horses

This chapter is concluded with some insight as to how you might earn some money at the race track. You might consider this your reward for sticking with the multiple regression procedure.

When a bet is made at the race track, the bettor, unless he enjoys throwing his money away, acts as if he is making a prediction. He is putting not only faith, but also money, behind some kinds of information that he has mixed together in some fashion. Most bettors cannot verbalize their prediction model. Some bettors don't even have a model, picking their choice randomly or through some kind of superstitious behavior. Some bettors have models that they apply differentially; for example, they let past earnings and track conditions play a large part in the model, unless the horse ran badly the last time out. If the horse did not place the last time out, then the only variables looked at are jockey weight and starting position. But give the horse one of the top jockeys in the country and that becomes the sole determiner--a winner for sure.

A little reflection should indicate that very seldom will a single variable, or even two variables, be of enough predictive value to be useful. What needs to be done is to develop a regression model with (probably) a large number of predictor variables. This model then should be applied uniformly to all horses.

Lest you get too excited, such a scheme has already been developed by someone who calls himself "The Wizard" and has been written up by James Morgan in Look (June 1, 1971). Although he investigated 120 possible predictor variables, the Wizard evidently did not use interactions and polynomial terms. Incorporating such variables in the prediction model would most likely increase predictability. If the criterion (winning a horse race) is a complex phenomenon, then a complex prediction model is needed to account for that complex phenomenon.

Attempting to use a prediction model at the race track assumes that winning a horse race is a predictable behavior. Random events such as the randomly assigned starting position and chuckholes may lower predictability, but not eliminate the possibility of obtaining a somewhat reasonable R^2. Doping of horses, "deals" made between owners, and "slowdown" instructions from the owner to the jockey may also be important variables, although in most cases not available to the average regression person.

Assuming that winning a horse race is basically a predictable behavior (lawful behavior might be a better phrase here), there are still other factors which automatically reduce the predictability. The track doesn't want to lose money, so it has built-in variables that force the R^2 toward zero. Good horses are required to carry extra weight, and if a horse has been consistently winning, he is raised to a better level of competition. If these efforts don't reduce meaningful predictability, then the odds come into play. For if there is clearly a probable winner in the field, most bets will be placed on that horse, reducing the odds and hence the payoff if the horse does in fact win.

The Wizard uses winning percentage as his criterion, although you may have some ideas about a better criterion to use. This chapter must be ended now as curvilinear models are being transported to the track. Keep regressing, and see you at the track!

11

The Strategy of Research as Viewed from the Multiple Linear Regression Approach

The hypotheses and their associated models presented since Chapter One have been of a rather restricted nature when compared with the research notions of Chapter One. In that first chapter, the arguments were made that behavior (whether of humans, rats, university administrations, plots of land, etc.) is complex and that a number of predictor variables may be needed to account for that behavior. In accounting for such behavior it has been shown that categorical information could be used (Chapters Six and Eight), that continuous variables could be used (Chapters Four and Five), and that transformations of originally scaled variables should be considered (Chapter Seven). It must be emphasized that the data and inherent functional relationships determine the variables necessary to obtain the desired R^2. Both dichotomous and continuous variables can be used in the same regression model.

Some of the generalized research hypotheses were presented in the previous chapters in part to show that traditional least squares solutions are all computational simplifications of the general least squares procedure. The regression formulation was shown for such traditional solutions as the Pearson correlation (GRH 4.2), Phi correlation (GRH 6.6), t test that the Point Bi-

serial correlation equals zero--also called the t test for the
difference between two independent means (GRH 6.5, 6.1), t test
for the difference between two dependent means (GRH 8.2), t test
for a single population mean (GRH 8.1), one-way analysis of variance (GRH 6.2), two-way analysis of variance (pp. 202-206), and
analysis of covariance (GRH 6.4). Every one was computed using
the same least squares procedure. To aid in the calculation of
these solutions by hand, simplified formulas have been developed
to fit special cases; and then these solutions became grouped according to their computational similarities (correlation, t
tests, $\underline{F}$ tests, etc.). Now that the computer is available to perform calculations, statistics should be put in their conceptual
place rather than their computational place. Any hypothesis can
be tested by the regression approach (as long as the researcher
is using the least squares approach with a single criterion). The
specific research hypothesis dictates the full and restricted
models and the testing of those models. Rather than spending his
time searching for the "appropriate" simplified statistical test
which will fit his design, the researcher should spend his time
explicitly stating his research hypothesis so that those full and
restricted models can be generated and tested.

The Most General Regression Model

As a way of organizing one's approach to the stating of research hypotheses, it may be helpful to point out that all of the
research hypotheses discussed in previous chapters can be stated

as "over and above" hypotheses. This is not to suggest that this is the best way to state any hypothesis, but to emphasize that each is a form of "the most generalized hypothesis."

All hypotheses can be either phrased or rephrased into a single structure. That structure contains a criterion variable and one or more variables which are being tested, and may contain one or more covariables. (The unit vector is nearly always a co-variable, but it could be a variable being tested, a covariable, or neither. This will be more fully discussed later in this chapter.) The single structure for expressing all research hypotheses is:

> "Is knowledge of variables X_1, X_2, ... X_k valuable, over and above knowledge of variables Z_1, Z_2, ... Z_g, in the prediction of the criterion Y_1?"

One regression model is specified by the Research Hypothesis as having full knowledge of all of the mentioned predictor variables:

Model 11.1: $Y_1 = a_1 X_1 + a_2 X_2 \ldots + a_k X_k + c_1 Z_1 + c_2 Z_2 \ldots$

$$+ c_g Z_g + E_1$$

where: Y_1 = the criterion variable;
X_1, X_2, ... X_k = the variables being tested;
Z_1, Z_2, ... Z_g = the "over and above" variables; and
a_1, a_2, ... a_k, c_1, c_2, ... c_g are least squares weighting coefficients calculated to minimize the squared elements of the error vector E_1.

The Full Model must be compared to a Restricted Model which has as predictors only the "over and above" information. Each of the k variables of interest are restricted from the Full Model by setting their weighting coefficients equal to some numerical value. There are an infinite number of numerical values that one could use (Werts and Linn, 1969). The actual restriction(s) need to be a function of past theory, empirical findings, and expectations of the researcher. In order to simplify the following discussion, each of the k variables of interest will be set equal to 0 (making k restrictions).

The restrictions would be:

$$a_1 = 0; \quad a_2 = 0; \quad \ldots \quad a_k = 0$$

Forcing those restrictions onto the Full Model would result in the following Restricted Model:

Model 11.2: $Y_1 = c_1 Z_1 + c_2 Z_2 \ldots + c_g Z_g + E_2$

The increase in R^2 in going from the Restricted Model to the Full Model would be due to the predictive information in the X_1, X_2, $\ldots$ X_k variables that is "over and above" the predictive information in the Z_1, Z_2, $\ldots$ Z_g variables in the Restricted Model.

The Research Hypothesis could also be stated as:

> "Are variables X_1, X_2, $\ldots$ X_k predictive of Y_1, holding constant the variables of Z_1, Z_2, $\ldots$ Z_g?"

or as

> "Are variables X_1, X_2, $\ldots$ X_k predictive of Y_1, while covarying the effects of variables Z_1, Z_2, $\ldots$ Z_g?"

And if the variables appearing in the Full Model but not in the Restricted Model are all mutually exclusive group membership vectors, then the Research Hypothesis could be stated as:

> "Are the k groups different on Y_1, after adjusting for differences on the variables of Z_1, Z_2, $\ldots$ Z_g?"

Since there are k variables to be assessed and g covariables, then m_1 = k + g. (Note that it is being assumed that all linearly dependent vectors have been omitted from the Full and Restricted Models.) The number of predictor pieces of information in the Restricted Model is g, therefore, m_2 = g. The numerator degrees of freedom for the $\underline{F}$ test then is:

$[m_1-m_2] = [(k+g) - (g)] = k$

The denominator degrees of freedom for the $\underline{F}$ test then is:

$[N-m_1] = [N-k-g]$

R^2 of the Full Model minus R^2 of the Restricted Model is the proportion of criterion variance accounted for by the additional k variables. Indeed, this average increase in R^2 becomes the numerator for the $\underline{F}$ test:

$(R_f^2 - R_r^2)/(m_1 - m_2) = (R_f^2 - R_r^2)/k$

In order for the $\underline{F}$ test (in terms of R^2--see Equation 4.6) to be computed on regression models, three requirements must be met:

1. The same criterion must appear in both models.

2. All predictor variables in the restricted model must appear in the full model either directly or as a linear combination of the full model predictor variables.

3. The unit vector must appear in both models.

A quick glance at Models 11.1 and 11.2 will verify that the first two requirements are met. The same criterion appears in both models, and the Z_1, Z_2, ... Z_g predictor vectors in the Restricted Model appear in the Full Model. On most applications, researchers will want to include the unit vector in both models (as one of the Z_i vectors) as a way of taking care of the arbitrary scaling of variables. Indeed, the unit vector so commonly appears in both models that most computerized regression programs provide it automatically. (The program depicted in Appendix A includes the unit vector automatically.) The unit vector has been included as a covariable so often that now it is assumed to be a covariable, unless otherwise indicated. It must be remembered, though, that the unit vector can also be tested (i.e., be one of the X_i vectors) as indicated in GRH 8.1. Furthermore, the unit vector need not appear in either model--the variables that appear in a given model are always a function of the question the researcher is asking.

Any number of X_i variables may be investigated, and any number of Z_i variables may be controlled (including none if that is desired). (The number of predictor variables must always be less than the number of subjects.) Furthermore, the variables may be dichotomous, or they may be continuous. The criterion variable in multiple regression is limited to only one variable, but it may be either dichotomous or continuous. Combinations of each of these options result in specific kinds of research hypotheses, some of which were discussed in previous chapters. (Multivariate regression models allowing for multiple criterion variables are beyond the scope of this text.)

To reemphasize the point made earlier, even though all hypotheses can be reworded in the "over and above" fashion, many hypotheses make more sense stated otherwise. For instance, "For a given population, Treatment A is better than Treatment B on the criterion" seems more communicative than does "For a given population, Treatment A is better than Treatment B on the criterion, over and above the overall mean."

Where is this discussion leading? Suppose Researcher A has theorized that variables X_1, X_2, X_3, and X_4 account for the variability in the criterion of interest. An obtained R^2 of .90 is used as evidence for that statement. Researcher B, though, realizes that all of the criterion variance has not been accounted for, and proposes that in addition, variable X_6 is necessary. Two researchers have posited two different states of affairs, and since the one is a restricted case of the other, the argument can be tested through the regression approach. Since Researcher B is proposing a new, untried variable, and since he is using more information than Researcher A, he is obligated to have a higher R^2. The difference in R^2 could be agreed to before data collection, or the two researchers could agree to use a test of significance for the difference in R^2, as in Generalized Research Hypothesis 11.1. With this GRH, the text has finally arrived at the degree of complexity that was outlined in Chapter One.

Researchers have in the past used a slight variation of the hypothesis in GRH 11.1. In testing to "see if the multiple correlation is significant," a researcher is actually testing the full model against one with zero predictability. The combined predic-

---

Generalized Research Hypothesis 11.1

Directional Research Hypothesis: For a given population, X_6 is positively predictive of the criterion, Y_1, over and above U, X_1, X_2, X_3, and X_4.

Non-directional Research Hypothesis: For a given population, X_6 is predictive of the criterion, Y_1, over and above U, X_1, X_2, X_3, and X_4.

Statistical Hypothesis: For a given population, X_6 is not predictive of the criterion, Y_1, over and above X_1, X_2, X_3, and X_4.

Full Model: $Y_1 = a_0U + a_1X_1 + a_2X_2 + a_3X_3 + a_4X_4 + a_6X_6 + E_1$

Restrictions: $a_6 = 0$

Restricted Model: $Y_1 = a_0U + a_1X_1 + a_2X_2 + a_3X_3 + a_4X_4 + E_2$

where:

Y_1 = the criterion;
U = 1 for all subjects;
X_1, X_2, X_3, X_4, X_6 = continuous or categorical information; and
a_0, a_1, a_2, a_3, a_4, and a_6 are least squares weighting coefficients calculated so as to minimize the sum of the squared values in the error vectors, E_1 and E_2.

Degrees of freedom numerator = $(m_1-m_2) = (6-5) = 1$

Degrees of freedom denominator = $(N-m_1) = (N-6)$

Note: If all predictors are continuous, $m_1 = 6$. If some reflect categorical information, the linear dependencies must be ascertained (e.g., if X_1 and X_2 represent males and females, then one of them is linearly dependent, and $m_1 = 5$ and $m_2 = 4$).

---

tive power of the predictors is being questioned. Since more than one restriction is being made, a directional hypothesis cannot be

Generalized Research Hypothesis 11.2

Non-directional Research Hypothesis: For a given population, X_1, X_2, X_3, X_4, and X_6 taken together are predictive of the criterion Y_1, over and above the overall mean as reflected by U.

Statistical Hypothesis: For a given population, X_1, X_2, X_3, X_4, and X_6 taken together are not predictive of the criterion Y_1, over and above the overall mean as reflected by U.

Full Model: $Y_1 = a_0U + a_1X_1 + a_2X_2 + a_3X_3 + a_4X_4 + a_6X_6 + E_1$

Restrictions: $a_1 = 0$; $a_2 = 0$; $a_3 = 0$; $a_4 = 0$; $a_6 = 0$

Restricted Model: $Y_1 = a_0U + E_2$

   where:

   Y_1 = the criterion;
   U = 1 for all subjects;
   X_1, X_2, X_3, X_4, and X_6 are either continuous or dichotomous pieces of information; and
   a_0, a_1, a_2, a_3, a_4, and a_6 are least squares weighting coefficients calculated so as to minimize the sum of the squared values in the error vectors, E_1 and E_2.

Degrees of freedom numerator = (m_1-m_2) = (6-1) = 5

Degrees of freedom denominator = $(N-m_1)$ = (N-6)

Note: If all predictors are continuous, $m_1 = 6$. If some predictors reflect categorical information, the linear dependencies must be ascertained.

tested. This hypothesis is presented as GRH 11.2 to illustrate the research hypothesis being tested. The authors would not recommend testing this particular hypothesis as it is not definitive.

The goal of control in the behavioral sciences

All behavioral science research is ultimately concerned with cause and effect--i.e., control. Many researchers initially react quite negatively to this statement, but when hard pressed will ultimately agree that cause and effect relationships are what they are looking for. Statisticians have traditionally indicated that cause and effect relationships cannot be ascertained from correlational studies but only from analysis of variance designs applied to experimental data. Familiarity with the multiple linear regression approach makes it clear that correlation and analysis of variance are both subsets of the general linear model, and that interpretations of cause and effect can only be made on logical grounds. If the logic is not defensible, then the cause and effect relationship is not tenable.

Unfortunately, many causal interpretations are based on significance levels rather than on the amount of variance accounted for (Byrne, 1974). Results that are highly significant (due to factors other than chance) may account for, say, only 1% of the variance in the criterion. A not too unusual bivariate outcome results in an R^2 less than .20. Of course, some phenomena are more highly accounted for than others, but few studies in the literature report results leading to an R^2 greater than .50. (It would be very interesting to force all researchers to report their R^2 values, and not let them hide behind statistical significance inflated by large sample sizes. All other factors being equal, as the sample size increases, the results will become more "significant," whereas the R^2 will simply more closely approximate the population R^2. Thus, increasing the number of subjects does not artificially inflate the R^2 value, although it does deflate the probability value.)

In the past, cause and effect has been interpreted on data which, although highly significant, yields small R^2 values. One of the positions of this text is to more fruitfully spend one's time by finding variables which will increase the R^2, rather than trying to understand how that small proportion of criterion variance has been predicted.

Of additional importance is the adequacy of the overlap of the criterion with the construct one is wishing to consider, as discussed at the end of Chapter Seven. Too often the criterion is equated with the construct, when in reality it must be realized that the criterion is only an approximation of the construct-- that is, there is only a partial overlap between the criterion and the construct.

If the unaccounted for criterion variance overlaps with the construct, then one would do well to continue trying to build up the R^2. Several consternating situations arise when the criterion only partly overlaps the criterion. It may be that the criterion variance accounted for is not overlapping the construct variance. Any interpretation regarding causality in this situation would be appropriate for the criterion, but not for the construct.

On the other hand, all of the common variance between the criterion and the construct may be accounted for. To attempt to build up the predictability of the criterion would in this case be fruitless. Interpretation of causality in this case may be of some value, but one must note that complete control (bringing about desired changes or "upsetting the prediction") of the construct would not be possible because of the lack of an isomorphic fit between criterion and construct. This situation calls for the selection of other criterion variables, either in lieu of the criterion, or in addition to the criterion.

Once the R^2 has approached 1.0 and the researcher has some faith in his criterion as a good measure of the construct, then one can begin considering how to control the construct. This implies that variables must be included in the model which can be manipulated. Non-manipulatable variables which must remain in the predictive model simply indicate differences in functional relationships for various populations. That is, if sex must remain in the model, then the functional relationship is different for boys than it is for girls. But if all of the predictor variables are non-manipulatable, then the model is simply describing a state of affairs, rather than indicating some possible ways of upsetting the state of affairs.

Given a choice between greatly overlapping predictor variables, some of which are more manipulatable than others, one would choose to include the more manipulatable variables in the model. Some variables are easier to manipulate than are others; but unfortunately, statistical procedures do not have ways of incorporating these differences. These decisions must be made on subjective bases rather than on objective bases.

Even though the R^2 is extremely high, the predictor variables may not be the causers of the criterion. The criterion and predictor variables may both be caused by some additional variable or set of variables. The technique of path analysis has been introduced as a possible means of assessing the causal sequence (Blalock, 1964). Those developments have not been entirely successful, but the intent is appropriate.

One of the real world advantages of multiple linear regression, that of allowing the use of correlated predictor variables, turns out to also be one of the problems in detecting control. When one of the predictor variables is changed, then one would expect that the values of the other (correlated) variables would also change. Indeed, the whole swarm could well change as the result of changing only one variable. If the criterion is increased, then this might well be a desired goal. But when control is being considered, one usually wants to know how much effect a certain change will have.

Multiple linear regression may give one the insight as to what variables to change but does not generally indicate what effects to expect. If the subjects have been randomly assigned to the independent variables, then these effects are implied by the multiple linear regression approach. But these kinds of designs do not adequately reflect the real world. When subjects are sampled from the real world, the predictor variables will be correlated, and hence the effects are not directly known. The effects

of changing predictor variables must wait for subsequent empirical manipulation. It has already been indicated that some variables are more easily manipulated than others. It might also be
the case that some kinds of subjects are more easily manipulated
than others. For this reason, the assessment of control must wait
for empirical verification.

Ascertaining causality is always a tentative proposition.
Empirical manipulation helps rule out other competing explainers,
but as Blalock (1964, p. 26) says:

> "no matter how elaborate the design, certain
> simplifying assumptions must always be made.
> In particular, we must at some point assume
> that the effects of confounding factors are
> negligible. Randomization helps to rule out
> some of such variables, but the plausibility
> of this particular kind of assumption is al-
> ways a question of degree. We wish to under-
> score this fact in order to stress the under-
> lying similarity between the logic of making
> causal inferences on the basis of experimental
> and nonexperimental designs."

As will be demonstrated in the next few sections, MLR can
assist in the various stages of one's research. Ultimately,
causers will be identified. But no statistical technique will
provide an understanding of why that variable is a cause. Theory,
common sense, and/or time and space, combined with statistical
results, nurture explanation and understanding.

A Proposed Research Strategy

Various research strategies were discussed in Chapter Nine,
although all of those were concerned with the sequencing of hypotheses. The present section puts the processes of data snooping,

	Stage	Emphasis on Low Probability	Emphasis on High R^2
Static Variables	1. Data Snooping (Hypothesis Generating)	No	Yes
	2. Hypothesis Testing (Static)	Yes	No
	3. Replication	No	Yes
Dynamic Variables	4. Manipulation (Dynamic Hypothesis Testing)	Yes	Yes
	5. Replication	Yes	Yes

Figure 11.1. Stages of a proposed research strategy and their relative emphasis upon probability and R^2.

hypothesis testing, replication, and manipulation into the perspective of an encompassing research strategy.

The strategy involves successively the use of data snooping to generate hypotheses (hopefully directional hypotheses), the testing of hypotheses on static variables, replication of that hypothesis testing, developing and testing hypotheses about the effects of manipulation of some of the previously static variables, and replication of the manipulated-variable hypotheses. Figure 11.1 displays the strategy schematically.

Data snooping

(Data snooping was discussed extensively in Chapter Nine.)

When one is data snooping, one has little idea about what he will find. When a large R^2 is observed, one may suspect that the set of variables used to obtain that R^2 is a good set of variables to investigate further. This can be a very exciting stage of research since most dramatic discoveries seem to have originated in a stage of this sort; but any discoveries at this stage must be tested on successive samples before one can reasonably say that a discovery was made. Serendipitous findings at this stage can be very valuable; one must simply realize that subsequent verification must be made.

Suppose for example, that a researcher has an interest in why ninth graders vary in their knowledge and achievement in the area of biological science. She has also chosen a criterion measure of this construct (realizing, of course, that it is not exactly the same as her construct); the criterion measure is a standardized biological science achievement test administered at the end of the ninth grade (after a year of biology instruction). Based on her own knowledge of the research area and on discussion with biology teachers and students, she decides to investigate the following variables to see if they will help to predict Biology achievement:

> Sex of student
>
> Socioeconomic status of student
>
> Time student has spent outdoors in earlier years
>
> Number of extra credit (outdoor) assignments
>
>> student completes

(Number of extra credits)^2

Size of biology class

Sex * size of biology class

The researcher does not have enough information at this point to state hypotheses about the ways that these variables relate to Biology achievement; she is data snooping.

She arranges to collect data on the criterion and predictor variables from the ninth grade students in a particular school. Just prior to administering the Biology achievement test, the school notices that- it does not have enough original copies of the test; so the school makes photocopies of the test for the final 1/4 of the students. Our researcher notices that the photocopies are not as legible as the originals. She therefore adds a new variable to her list to consider:

Legibility of the test

As a result of her data snooping, the researcher finds that the following model, containing only some of her suspected variables, does a good job of predicting Biology achievement, attaining a R^2 of .76:

Model 11.3: $B_1 = a_0 U + a_1 S_1 + a_2 A_1^2 + a_3 (X_1 * X_3) + a_4 (X_2 * X_3) +$

$a_5 L_1 + E_3$

where: B_1 = Biology achievement test score;
S_1 = Socioeconomic status (SES) of student on 5-point scale;
A_1^2 = (Number of extra credits)^2;
$X_1 * X_3$ = Size of biology class if student is male, 0 otherwise;
$X_2 * X_3$ = Size of biology class if student is female, 0 otherwise; and
L_1 = Legibility of the test.

The researcher is now ready to state and test hypotheses.

Hypothesis testing

An hypothesis cannot be made without the researcher relying on some combination of past research, theoretical orientation, and intuitive insight. Data snooping results can often aid in developing a supportable hypothesis. As discussed previously, one can in the hypothesis testing stage compare the difference in R^2 between the Full and Restricted Models to some preselected amount of difference, or one can rely on the probability values as proposed in this text. If only one restriction is being made on the Full Model, and if significance is obtained, one will know what is causing that significance. The goal of making an accurate causal statement will be enhanced if the hypothesis is tested in a model exhibiting a high R^2. But as can be seen in Figure 11.1, a high R^2 is not viewed as being necessary at this stage.

To return to the example of biology achievement, the researcher, based on her data snooping results, states several hypotheses, of which the following is one; and she collects new data on which to test them.

> Research Hypothesis: For ninth grade biology students, the score obtained on the Biology achievement test will be higher for boys if they are in small classes (under 20) rather than in large classes, and will be higher for girls if they are in large classes (over 20) rather than in small classes, over and above the effects of SES, A_1^2, and Legibility of the test.

The researcher then tests her hypothesis and finds that, at her chosen alpha level, her Statistical Hypothesis can be rejected; so her Research Hypothesis is accepted as tenable. Successively, she finds that all the variables in Model 11.3 are valuable.

Up to this point no variables have been manipulated, although some are potentially manipulatable. The model containing static (non-manipulated) variables (Model 11.3) was upheld by the hypothesis testing. The static hypothesis-testing phase now needs to be replicated.

Replication

As can be seen in Figure 11.1, the process of replication appears twice, once in the static phase and once in the manipulation phase. Since it is necessary to proceed from a static model to a dynamic one (containing manipulation) when hunting and testing for causal relationships, the researcher must make a decision concerning the extent and level of successful replication before manipulation. This decision is based largely on the potential cost of the manipulation intended--cost in money, time, and human resources. Rescheduling the large and small biology classes will cost money and may provoke parental/taxpayer protest. Thus, one may want to replicate the static hypothesis testing model several times before venturing to manipulate. The process of replication (before manipulation) should not be overly concerned with probability, but with obtaining a high R^2. (It would be ideal if approximately the same weighting coefficients could be obtained, but correlated predictors and small sample sizes usually operate against this goal.)

Manipulation

The first three stages above apply to static data. That is, within person, focal simuli, and context variables are measured as they exist, with no effort having been made to change them. Functional relationships are described so that the criterion variable may be predicted. One should be aware of the fact that a variable which is found to be predictive of a criterion in a static model may not be predictive in the same fashion as a dynamic variable. The extreme case would be a variable which simply cannot be manipulated. Another example would be a variable which, when manipulated, causes "side effects" which influences the criterion (either positively or negatively). On the other hand, variables which are found to be not predictive in a static situation may indeed be necessary causers in a dynamic situation. (Hickrod, 1971, discusses this and several other interesting aspects of regression with respect to school finance concerns.) As a consequence of the above realities some statisticians take a strong stand regarding the analysis of static variables. Werts (1970) states:

> "Obviously when partial regression weights are estimated from naturalistic data it will be far from certain that these coefficients will have any resemblance to what would be found under conditions of experimental control. By asking what would happen "if", we are in effect asking the question: What would the data be like if they weren't what they are?"

The authors of this text take a more moderate view. Researchers should be aware that what is discovered in a static situation may not be the case in a dynamic situation. But the researcher still investigates the static situation before the dynamic one because of the potential "cost" of creating the dynamic situation (manipulating variables under experimental control).

When an R^2 of 1.00 is obtained in a dynamic situation one can be confident that the causers have been isolated. One then needs to investigate the magnitude of the criterion scores. For the causers of increasing the criterion two units may have been discovered, but a two unit increase may not be sufficient. The discovered causers may not be sufficient to obtain a three unit increase. Hence other causers must be posited and investigated. The cycle of research has thus started again.

The manipulation step in scientific investigation can be thought of as changing subjects so that predictions can be upset. Johnny is predicted to be a bad reader by the use of the static data, but certain changes can be made so that the prediction becomes incorrect. A certain plot of land is predicted to be of low fertility, but certain changes may make that land desirable for planting of crops.

The stage of manipulation contains dynamic variables--those that have been willfully changed by the researcher. As has been emphasized previously, it is desirable to investigate manipulatable variables in the snooping and static hypothesis testing stages so that the move into the manipulation phase per se will be facilitated.

Our biology researcher wants to manipulate and test each predictor variable in Model 11.3. This is dynamic hypothesis testing and her hypotheses will hopefully be directional since, before manipulation, she has investigated her variables extensively.

Are all of the predictor variables in Model 11.3 manipulatable? Class size and the proportion of males to females in each class can be arranged by the school. And Legibility of the test is also directly manipulatable--one needs only to select certain levels of Legibility. The Number of extra credits may be more difficult to manipulate; some added inducement may be necessary to get students to complete more assignments (if the hypothesized relationship is positive). Or the researcher may feel that it is the content rather than the "credit" of those assignments that is of benefit and may remove them from extra credit and make them required.

Socioeconomic status is not readily manipulatable--SES is generally based on money, occupation, and education. Changing a subject's family circumstances may be impossible, so the researcher may wish to hypothesize critical elements or aspects of SES that might be varied. Nutrition, self-esteem, need achievement and teacher acceptance are surely related to Socioeconomic status and may also affect Biology achievement. Indeed, Biology achievement might be retarded by inadequate breakfast and/or a lack of desire to achieve. These may be the manipulatable variables which Socioeconomic status summarizes. If so, then a school breakfast program and training that instills need achievement may be manipulated to increase Biology achievement for students who would otherwise be predicted to perform poorly without such intervention.

Note that sex is not easily manipulated, but sex is placed in the model so as to interact with a manipulatable variable. It thus performs a useful predictive function without being manipulated. In this case, the researcher may not be concerned with sex as a manipulatable variable. She may, however, wish to consider what aspects of sex (underlying dynamic variables) that interact with class size to predict Biology achievement. Surely simple physiology has an explanatory role--but of unknown value; socialization variables, however, are more likely to be the underlying causitive factors which the variable of sex accounts for. For example, girls in general may have been taught, "Science is not really for girls." If so, a measure of Sex-role identity ought to be a better predictor of Biology achievement than the binary Sex variable. If it is a better predictor, a program designed to break down sex-role stereotypes ought to reduce the negative influence of being a girl on Biology achievement.

If one wishes to go beyond prediction to improvement, then it is important that static variables be examined carefully for their underlying dynamic causitive influences on criterion measures (e.g., Sex-role identity may underlie the effect on achievement of the binary Sex variable). An excellent presentation of this viewpoint is presented by McClelland (1973).

One may not always wish to do away with all non-manipulatable variables, however. They are valuable predictors in at least two situations: (1) They may be used until the underlying dynamic variable is found, and (2) They may be unchangeable but may interact with a dynamic variable (e.g., Sex interacting with Treatment). If one cannot manipulate, then one must rest on the tenuous grounds of theory. Theory always plays a large part, because the variables which are included in the analysis were purposeively put there--one may not call that theory, but it certainly guided one's actions.

Replication of manipulated findings

Once manipulations have been found to be successful (resulting in changes in the desired direction) then replications are called for to establish the magnitude of the effect of those manipulations. Researchers should put faith in causers only when manipulations have a consistent effect in the presence of a high degree of predictability.

Successful replication by the biology researcher should make her happy. Given that she has had success at each of the previous stages, she can be relatively assured that the results are valid. Additional replications on new samples of ninth grade students will give further credance to the findings, for chance may have been operating in all of the previously successful results.

What is more likely the case is that the biology researcher

did not perform research at some of the other stages. Because of lack of resources or time, research is often forced to proceed to the manipulation stage without prior success at the other stages. In these cases, replication is a more crucial stage. How many successful replications? There is no established answer to that question, although in part it depends upon how crucial it is that the criterion of interest be under control.

Figure 11.1 summarizes the stages of the proposed research strategy and shows their emphases upon probability and R^2. If success on the goal of either low probability or high R^2 is not obtained at any stage, activity must revert to the initial stage of data snooping.

Further reading

The multiple linear regression approach is an extremely flexible technique that opens up a host of secrets about statistics, once certain mechanical topics are mastered. New practical developments are constantly being made by those using multiple linear regression. More mathematical treatments of the technique than presented here can be found in Mendenhall (1968), Williams (1959), Draper and Smith (1966), and Ward and Jennings (1973). An elementary introduction for hypnotists, written by Starr (1971), contains minimal mathematical concepts.

A computer program written by Robert Mason is described in Appendix F. The valuable aspect of this program is that natural language research hypotheses can be key-punched, and fed into the program. The full and restricted models, and the F test will be generated automatically.

A Special Interest Group of the American Educational Research Association has been in existence for several years, and their communication "Multiple Linear Regression Viewpoints" is an unedited forum for new ideas. Many of the articles published in that journal have been referenced in this text. Since back issues of "Viewpoints" are not readily available in many libraries, the articles have been abstracted in Appendix E. "Viewpoints" affords a way for the reader to keep abreast of new regression developments. A nominal annual dues avails one of this journal. Further information on Viewpoints can be obtained from the editor, Isadore Newman, College of Education, University of Akron, Akron, Ohio, 44325.

APPENDIXES

REFERENCES

INDEXES

Appendix A

Program DPLINEAR

```
C     MULR05        DIVISION OF EDUCATIONAL RESEARCH SERVICES         00000
C                          UNIVERSITY OF ALBERTA                      00010
C     ...........................................................     00020
C                                                                     00030
C     PURPOSE*                                                        00040
C       CALCULATES CORRELATIONS AND CARRIES OUT REGRESSION ANALYSIS.  00050
C       FOR EACH REGRESSION MODEL, THE PROGRAM COMPUTES THE SQUARED   00060
C       MULTIPLE CORRELATION AND THE REGRESSION WEIGHTS BOTH          00070
C       STANDARDIZED AND RAW SCORE FORM FOR EACH VARIABLE SPECIFIED AS 00080
C       A PREDICTOR.                                                  00090
C       IN ADDITION TO THE FEATURES OF 'MULR04', THIS PROGRAM ALLOWS  00100
C       FOR ANALYSES OF VARIANCE.                                     00110
C                                                                     00120
C     CARD INPUT*                                                     00130
C       1. TITLE (20A4)                                              00140
C       2. PARAMETERS (4I5)                                          00150
C       3. FORMAT OF DATA (20A4)                                     00160
C       4. DATA (ACCORDING TO ABOVE FORMAT)                          00170
C       6. MODEL CARDS (2A4,A2,I3,F8.1,28I2,/,21X28I2)               00180
C       7. F-RATIO CARDS (2A4,A2,I3,8X2I2,2I4)                       00190
C       8. CARDS FOR SUMS OF SQUARES (2A4,A2,/,6A3,I2,F2.0,10I3)     00200
C       9. CARDS FOR ANOVA TABLE (2A4,A2,I3,E8.1,2I2,2I4)            00210
C      10. CARDS FOR MEANS (2A4,A2,I3,28I2,/,21X28I2)                00220
C      11. BLANK CARD (INDICATES END OF MODEL, F-RATIO, ETC. CARDS)  00230
C                                                                     00240
C     DESCRIPTION OF PARAMETERS*                                      00250
C       NOB    - NUMBER OF OBSERVATIONS                              00260
C       NVARIN - NUMBER OF VARIABLES INPUT                           00270
C       NVAPT  - NUMBER OF VARIABLES AFTER TRANSFORMATION (MAX-100)  00280
```

489

```
C   NFMT   - NUMBER OF FORMAT CARDS (USE ALL 80 COLUMNS)       00290
C            (MAX. 4, AT LEAST ONE ASSUMED)                    00300
C                                                              00310
C   DESCRIPTION OF MODEL, F-RATIO, ETC. CARDS*                 00320
C   MODEL CARDS - PROBID, NFLDS, STOPC, (MFLD(I),I=1,NFLDS)    00330
C      PROBID - IDENTIFICATION OF MODEL. MODELS SHOULD BE      00340
C               NUMBERED CONSECUTIVELY.                        00350
C      NFLDS  - NUMBER OF ELEMENTS IN MFLD.  THIS MUST BE AN ODD   00360
C               NUMBER (ADD MINUS SIGN TO HAVE RESIDUALS SHOWN)    00370
C      STOPC  - ITERATION STOP CRITERION.  ITERATION WILL      00380
C               CONTINUE UNTIL THE SQUARED MULTIPLE CORRELATION    00390
C               ON TWO SUCCESSIVE ITERATIONS DIFFERS BY LESS THAN  00400
C               STOPC. ( 0.00001 UNLESS OTHERWISE SPECIFIED.)  00410
C      MFLD   - VECTOR INDICATING VARIABLES TO BE INCLUDED IN  00420
C               MODEL.  ALL VARIABLES BETWEEN EACH PAIR OF     00430
C               ELEMENTS OF MFLD WILL BE INCLUDED IN THE MODEL.    00440
C               THE LAST ELEMENT IS THE CRITERION.             00450
C   F-RATIO CARDS - PROBID, NFLDS, STOPC, (MFLD(I),I=1,NFLDS)  00460
C      PROBID - IDENTIFICATION OF F-RATIO.                     00470
C      NFLDS  - MUST BE AN EVEN NUMBER *5                      00480
C      STOPC  - MAY BE LEFT BLANK.                             00490
C      MFLD   - MFLD(1) NUMBER CORRESPONDING TO FULL MODEL.    00500
C               MFLD(2) NUMBER CORRESPONDING TO RESTRICTED MODEL.  00510
C               MODEL 99 UNLESS OTHERWISE USED HAS A RSQ =0.   00520
C               MFLD(3) AND MFLD(4) TREATED AS ONE NUMBER      00530
C               REPRESENTING DF1                               00540
C               MFLD(5) AND MFLD(6) TREATED AS ONE NUMBER      00550
C               REPRESENTING DF2                               00560
C               IF NFLDS=8, DF1 AND DF2 ARE TAKEN AS THE NUMBER    00570
C               OF INDEPENDENT WEIGHTS, INCLUDING THE CONSTANT,    00580
C               IN THE FULL AND RESTRICTED MODELS RESPECTIVELY.    00590
C               OTHERWISE, THEY ARE RESPECTIVELY THE DEGREES OF    00600
C               FREEDOM OF THE NUMERATOR AND DENOMINATOR.      00610
C   SUMS OF SQUARES - PROBID, SOURCE, ICR, CON, (INRSQ(J),J=1,10)  00620
C      PROBID - THE WORD 'ANOVA' IN FIRST FIVE COLUMNS         00630
C      SOURCE - TITLE (DESCRIPTION OF SUMS OF SQUARES)         00640
C      ICR    - CRITERION VARIABLE NUMBER                      00650
C      CON    - CONSTANT FROM/TO WHICH R SQUARES ARE TO BE     00660
C               SUBTRACTED/ADDED                               00670
```

```
C     INRSQ   - NUMBERS OF R-SQUARES DEFINED BY MODEL CARDS TO        C0680
C               BE POOLED TO OBTAIN SUMS OF SQUARES.  PRECEDE BY      C0690
C               A MINUS SIGN IF R-SQUARE IS TO BE SUBTRACTED          C0700
C     ANOVA TABLE - PROBID, NFLDS, STOPC, (MFLD(I),I=1,NFLDS)         C0710
C     PROBID  - IDENTIFICATION FOR THIS ANOVA PROBLEM                 C0720
C     NFLDS   - SHOULD BE SET EQUAL TO '6'                            C0730
C     STOPC   - SHOULD BE A NEGATIVE NUMBER                           C0740
C     MFLD(1) - NUMBER OF SUMS OF SQUARES MODEL FOR NUMERATOR         C0750
C     MFLD(2) - NUMBER OF SUMS OF SQUARES MODEL FOR DENOMINATOR       C0760
C     MFLD(3) - AND MFLD(4) TREATED AS ONE NUMBER GIVING DEGREES      C0770
C               OF FREEDOM FOR NUMERATOR                              C0780
C     MFLD(5) - AND MFLD(6) TREATED AS ONE NUMBER GIVING DEGREES      C0790
C               OF FREEDOM FOR DENOMINATOR                            C0800
C     MEANS - PROBID, NFLDS,(MFLD(I),I=1,NFLDS)                       C0810
C     PROBID  - THE WORD 'MEAN' IN FIRST FOUR COLUMNS                 C0820
C     NFLDS   - NUMBER OF NUMBERS APPEARING IN MFLD                   C0830
C     MFLD    - TWO DIGIT NUMBERS APPEARING IN PAIRS REFERRING TO     C0840
C               THE CATEGORICAL VECTOR FOLLOWED BY THE PRODUCT        C0850
C               VECTOR (PRODUCT OF CATEGORICAL AND CRITERION)         C0860
C               CATEGORICAL VECTOR SPECIFIES THE MEMBERS OF THE       C0870
C               SUBGROUP FOR WHICH MEANS AND S.D.'S ARE DESIRED       C0880
C ...............................................                     C0890
C                                                                     C0900
C     SUBPROGRAMS REQUIRED*                                           C0910
C     DFTRAN, DFPRBF, DFPRNT, DFCRLB, DFREGR, CALMS, DFPRNT           C0920
C ...............................................                     C0930
C                                                                     C0940
C                                                                     C0950
      DIMENSION TITLE(20), FMT(80)                                    C0960     0001
      COMMON S(100),X(100),SS(100,100),SIDWT(100),WTS(101),RSQ(99)    C0970     0002
      NVMAX=100                                                       C0980     0003
1     READ(5,3,END=18)TITLE                                           C0990     0004
      DO2J=2,20                                                       C1000     0005
      IF(TITLE(1).NE.TITLE(J))GOTO4                                   C1010     0006
2     CONTINUE                                                        C1020     0007
      GOTO18                                                          C1030     0008
3     FORMAT(20A4)                                                    C1040     0009
4     WRITE(6,5)TITLE                                                 C1050     0010
5     FORMAT(1H1,30X,20A4)                                            C1060     0011
      READ(5,6)NOB,NVARIN,NVART,NFMT,NUNIT                                      0012
6     FORMAT(16I5)                                                             0013
```

```
0014        IF(NFMT.EQ.0) NFMT=1                                            01070
0015        IF(NUNIT.EQ.0) NUNIT=5                                          01075
0016        WRITE(6,7)NOB,NVARIN,NVART,NFMT,NUNIT                           01080
0017      7 FORMAT(1H0,//11X,22HNUMBER OF OBSERVATIONS I13/1H0,10X,25HNUMBER OF  01090
           * VARIABLES INPUT ,I10/1H0,10X,31HNUMBER OF VARIABLES TRANSFORMED,    01100
           *I4/1H0,10X,22HNUMBER OF FORMAT CARDS ,I13/                           01110
           *1H0,10X,'INPUT UNIT NUMBER',I18,/)                                   01115
0018        LFMT=20*NFMT                                                    01120
0019        READ(5,3)(FMT(I),I=1,LFMT)                                      01130
0020        WRITE(6,8)(FMT(I),I=1,LFMT)                                     01140
0021      8 FORMAT(  11X,6HFORMAT,3X,20A4,/(20X,20A4) )                     01150
     C        INITIALIZE                                                    01160
0022        RSQ(99)=0.                                                      01170
0023        WTS(101)=0                                                      01180
0024        REWIND 8                                                        01190
0025        NPER=0                                                          01200
0026        DO9J=1,NVART                                                    01210
0027        S(J)=0.                                                         01220
0028        DO9K=J,NVART                                                    01230
0029      9 SS(J,K)=0.                                                      01240
0030        N1=NVARIN+1                                                     01250
0031        JOBS=0                                                          01260
0032     10 JOBS=JOBS+1                                                     01270
0033        IF(NOB.EQ.0)GOTO11                                             01280
0034        IF(JOBS.GT.NOB)GOTO16                                          01290
0035     11 READ(NUNIT,FMT,END=16)(X(J),J=1,NVARIN)                        01300
0036        IF ( NOB .GT. 0 ) GO TO 21                                     01301
0037        DO 20 J = 1,NVARIN                                             01302
0038        IF ( X(J) .NE. 0 ) GO TO 21                                    01303
0039     20 CONTINUE                                                       01304
0040        GO TO 16                                                       01310
0041     21 IF(NVARIN.GE.NVART)GOTO13                                      01320
0042        DO12J=N1,NVART                                                 01330
0043     12 X(J)=0.                                                        01340
0044     13 CALLDFTRAN( X)                                                 01350
0045     14 NPER=NPER+1                                                    01360
0046        DO15J=1,NVART                                                  01370
            S(J)=S(J)+X(J)                                                 01380
            DO15K=J,NVART
```

```
15  SS(J,K)=SS(J,K)+X(J)*X(K)                                          01390
    WRITE(8) (X(K),K=1,NVART)                                          01400
    GOTO10                                                             01410
16  CALL DFCRL3(NPER,NVART,NVMAX,1,NVMAX+1,2*NVMAX+1)                  01420
17  NV2=(NVMAX+1)**2                                                   01430
    REWIND 8                                                           01440
    CALL DFREGR(NPER,1,NVMAX+1,2*NVMAX+1,NV2,NV2+NVMAX+1,NV2+2*NVMAX+1, 01450
   * NVMAX,NVART)                                                      01460
    GOTO1                                                              01470
18  STOP                                                               01480
    END                                                                01490

    SUBROUTINE DFTRAN(    X)                                           01500
C   DATA TRANSFORMATION                                               01510
    DIMENSION X(1)                                                    01520
C   INSERT DATA TRANSFORMATION CARDS BETWEEN HERE AND 'RETURN' CARD.  01530
    RETURN                                                            01540
    END                                                               01550

    SUBROUTINE DFREGR(NPER,LMEAN,LSIGMA,LCORR,LSTDWT,LWTS,LRSQ,NVMAX,  01560
   *NVAR)                                                              01570
C   REGRED FROM 8FEB/65 MODIFIED MAY/66 BY FLATHMAN, U OF M, DERS.    01580
C   ITERATIVE REGRESSION                                             01590
C
    IMPLICIT REAL * 8  ( A - H , O - Z )

C
C
    DIMENSION PROB10(3),INRSO(10),SOURCE(6),PRSS(50)                   01600
    REAL * 4  A(10500) , X(100)
```

```
      INTEGER*2 MFLD(50),MFLDL(27)                                              C1610
      DATA ANOVA,FMEAN,PLOT/'ANOV','MEAN','PLOT'/                               C1620
      EQUIVALENCE(A(10201),X(1))                                               C1630
      COMMON A                                                                  C1650
      DOIJ=1,50                                                                 C1660
    1 PKSS(J)=0.                                                                C1670
      MSI=0                                                                     C1680
      K6=0                                                                      C1690
    2 READ(5,2)PROBID,NFLDS,STOPC,(MFLD(I),I=1,26)                             C1700
    3 FORMAT(2A4,A2,I3,E8.1,2812)                                               C1710
      IF(IABS(NFLDS).GT.28)READ(5,4)(MFLD(I),I=29,56)                          C1720
    4 FORMAT(21X,29I2)                                                          C1730
      IF(PROBID(1).EQ.ANOVA)GOTO5                                               C1740
      IF(PROBID(1).EQ.FMEAN)GOTO57                                              C1750
      IF(PROBID(1).EQ.PLOT)GOTO58                                               C1760
      GOTO12                                                                    C1770
    5 READ(5,6)SOURCE,ICR,CON,(INRSQ(J),J=1,10)                                C1780
    6 FORMAT (6A3,I2,F2.0,10I3)                                                 C1790
      IICR=LSIGMA+ICR-1                                                         C1800
      SST=A(IICR)**2*FLOAT(NPER)                                                C1810
      NSI=MSI+1                                                                 C1820
      NCT=0                                                                     C1830
      DO7J=1,10                                                                 C1840
      IF(INRSQ(J).EQ.0)GOTO8                                                    C1850
      NCT=NCT+1                                                                 C1860
      IF(INRSQ(J).LT.0) W=(-1.)                                                 C1870
      IF(INRSQ(J).GT.0) W=1.                                                    C1880
      I=IABS(INRSQ(J))+LRSQ-1                                                   C1890
    7 PRSS(MSI)=PRSS(MSI)+(W*A(I))                                              C1900
    8 IF(CON.EQ.0.)GOTO9                                                        C1910
      PRSS(MSI)=CON+PRSS(MSI)                                                   C1920
    9 PRSS(MSI)=PRSS(MSI)*SST                                                   C1930
   10 WRITE(6,10)SOURCE,ICR,CON,(INRSQ(J),J=1,NCT)                             C1940
   10 FORMAT(/,1X,6A3,/,25X,9HCRITERION,1X,I5,/,25X,8HCONSTANT,2X,F5.1,        C1950
     */,20X,15HRSQ CUMBINATION,10(1X,I3))                                      C1960
   11 WRITE(6,11)SST,PKSS(MSI)                                                  C1970
   11 FORMAT(/,21X,3HSS TOTAL,F15.8,/,1X,26HSS ACCT BY RSQ COMBINATION,2       C1980
     *X,E15.8)                                                                 C1990
      GOTO2
```

```
12   IF(STOPC.LT.0.)GOTO14                                          C2000
     IRESID=0                                                       C2010
13   IF(NFLDS)13,18,19                                              C2020
     IRESID=1                                                       C2030
     NFLDS=IABS(NFLDS)                                              C2040
     GOTO19                                                         C2050
14   J=MFLD(1)                                                      C2060
     K=MFLD(2)                                                      C2070
     DF2=(MFLD(5)*100 +MFLD(6)                                      C2080
     DF1=MFLD(3)*100 +MFLD(4)                                       C2090
     AMS1=PRSS(J)/DF1                                               C2100
     AMS2=PRSS(K)/DF2                                               C2110
     FR=AMS1/AMS2                                                   C2120
     P=D+PKBF(DF1,DF2,FR)                                           C2130
     WRITE(6,15)PROBID                                              C2140
15   FORMAT(//11X244,A2,/12X61HMS NC.   SS SOURCE   -   MEAN SQUARE C2150
    *                    F        P)                                C2160
     WRITE(6,16)J,PRSS(J),AMS1,DF1,FR,P                             C2170
16   FORMAT(13X13,1XE15.3,3X,F15.3,2X,F/.0,2X,F7.2,1X,F8.3)         C2180
     WRITE(6,16)K,PRSS(K),AMS2,DF2                                  C2190
     WRITE(6,17)                                                    C2200
17   FORMAT(/,1X,80('*'))                                           C2210
     GOTO2                                                          C2220
18   RETURN                                                         C2230
19   IF(NFLDS-NFLDS/2*2)20,52,20                                    C2240
20   K5=NFLDS-1                                                     C2250
     IF(STOPC.EQ.0.0) STOPC=0.00001                                 C2260
     IDC=MFLD(NFLDS)                                                C2270
     WRITE(6,21)PROBID,STOPC,IDC,(MFLD(I),I=1,K5)                   C2280
21   FORMAT(1H0/10X,'....',2A4,A2,F20.8/12X' CRITERION',16/10X,' PREDI C2290
    *CTORS',16,2H -14,/(22X,16,2H -14))                             C2300
     NFLD1=NFLDS-1                                                  C2310
     DO22I=2,NFLD1,2                                                C2320
     M=I/2                                                          C2330
     NFLDL(M)=MFLD(I)                                               C2340
22   MFLD(M)=MFLD(I-1)                                              C2350
     DO23I=1,NVAR                                                   C2360
     J=I+LWTS-1                                                     C2370
     A(J)=0.                                                        C2380
```

```
C2390        J=I+LSTDWT-1
C2400   23   A(J)=0.
C2410        S=0.0
C2420        SIG2=0.0
C2430        RSQ=0.0
C2440        DEL=0.0
C2450        ITER=0
C2460        ID=1
C2470        NGRP=NFLDS/2
C2480   24   RSQL=0.0
C2490        DO29I=1,NGRP
C2500        KSTAR=MFLD(I)
C2510        KSTOP=MFLDL(I)
C2520        DO29J=KSTAR,KSTOP
C2530        IA=(LWTS-1)+J
C2540        IB=(LCORR-1)+((IC-1)*NVMAX)+J
C2550        IC=(LC(KR-1)+((IDC-1)*NVMAX+J)
C2560        A(IA)=A(IA)+(DEL*A(IC))
C2570        DEN=S-(A(IA)*A(IC))
C2580        IF(DEN)26,25,26
C2590   25   DELT= A(IC)
C2600        STEST=DELT*DELT
C2610        SIG2T=STEST
C2620        RSQT=STEST
C2630        GOTO27
C2640   26   DELT=((SIG2*A(IC))-(S*A(IA)))/DEN
C2650        STEST=S+(DELT*A(IC))
C2660        SIG2T=SIG2+(2.0*A(IA)*DELT)+(DELT*DELT)
C2670        RSQT=(STEST*STEST)/SIG2T
C2680   27   IF(RSQL-RSQT)28,29,29
C2690   28   SLAR=STEST
C2700        SIG2L=SIG2T
C2710        RSQL=RSQT
C2720        DELTL=DELT
C2730        IDLAR=J
C2740   29   CONTINUE
C2750        IF(RSQL-RSQ-STOPC)33,30,30
C2760   30   S=SLAR
C2770        SIG2=SIG2L
```

```
          RSQ=RSQL                                                         C2780
          DEL=DELTL                                                        C2790
          ITER=ITER+1                                                      C2800
          ID= IDLAR                                                        C2810
          IA = (LSTDWT-1) + ID                                             C2820
          A(IA)=A(IA)+DEL                                                  C2830
          IF(RSQ-1.)24,33,31                                              C2840
       31 WRITE(6,32)                                                      C2850
       32 FORMAT(///, RSQ IS GREATER THAN ONE, CHECK THIS MODEL CARD'/     C2860
        *'      AND AVOID LATER INTERPRETATIONS INVOLVING THIS MODEL')     C2870
          RSQL=9999.9                                                      C2880
          GOTO45                                                           C2890
       33 SDS2=S/SIG2                                                      C2900
          WRITE(6,34)RSQL,ITER                                            C2910
       34 FORMAT(/ 14X,5HRSQ =F11.8,30X,I5,1X,'ITERATIONS')                C2920
       DO35I=1,NGRP                                                        C2930
          KSTAR=MFLD(I)                                                    C2940
          KSTOP=MFLDL(I)                                                   C2950
       DO35J=KSTAR,KSTOP                                                   C2960
          IA=LSTDWT-1+J                                                    C2970
       35 A(IA)=A(IA)*SDS2                                                 C2980
          WRITE(6,36)                                                      C2990
       36 FORMAT(/10X' VAR. NUMBER        STD. WT.       ERROR'/)          C3000
       DO38I=1,NGRP                                                        C3010
          KSTAR=MFLD(I)                                                    C3020
          KSTOP=MFLDL(I)                                                   C3030
       DO38J=KSTAR,KSTOP                                                   C3040
          IA=LWTS-1+J                                                      C3050
          A(IA)=0.0                                                        C3060
       DO37L=1,NGRP                                                        C3070
          LSTAR=MFLD(IL)                                                   C3080
          LSTOP=MFLDL(IL)                                                  C3090
       DO37L=LSTAR,LSTOP                                                   C3100
          IB=LSTDWT-1+L                                                    C3110
          IC=LCORR-1+J+((L-1)*NVMAX)                                       C3120
       37 A(IA)=A(IA)+(A(IB)*A(IC))                                        C3130
          IC=LCORR-1+J+((IDC-1)*NVMAX)                                     C3140
          A(IA)=A(IA)-A(IC)                                                C3150
          IB=LSTDWT-1+J                                                    C3160
```

```
0118
0119
0120
0121
0122
0123
0124
0125
0126
0127
0128
0129
0130
0131
0132
0133
0134
0135
0136
0137
0138
0139
0140
0141
0142
0143
0144
0145
0146
0147
0148
0149
0150
0151
0152
0153
0154
0155
```

```
38  WRITE(6,39)J,A(IB),A(IA)
39  FORMAT(1X118,F13.8,F15.8)
    WRITE(6,40)
40  FORMAT(/ 9X'  VAR. NUMBER          WEIGHT'/)
    FK1=0.0
    DO43I=1,NGRP
    KSTAR=MFLD(I)
    DO43J=KSTAR,KSTOP
    IA=LSIGMA-1+J
    IB=LSTDWT-1+J
    IC=LSIGMA-1+IDC
    ID=LMEAN-1+J
    IE=LWTS-1+J
    IF(A(IA))42,41,42
41  A(IE)=0.0
    GOTO43
42  A(IE)=A(IB)*(A(IC)/A(IA))
    FK1=FK1+(A(ID)*(A(ID)/A(IA)))
43  WRITE(6,39)J,A(IE)
    ID = LMEAN-1+ IDC
    REGCD=A(ID)-(A(IC)*FK1)
    WRITE(6,44)REGCD
44  FORMAT ( 10X'CONSTANT=',F18.8//)
45  K5=LRSG+K5
    A(K5)=RSQL
    K6=K6+1
    IF(IRESID.NE.1)GOTO50
    WRITE(6,46)
46  FORMAT(/10X,'SUBJECT',5X,'OBSERVED',6X,'PREDICTED',7X,
   *'RESIDUAL',/,10X,'NUMBER',9X,'Y',I4X,'Y','Y'//)
    DO48IICB=1,NPER
    READ(8)(X(I),I=1,NVAR)
    PREY=0
    DO47I=1,NGRP
    KSTAR=MFLD(I)
    KSTOP=MFLD(I)
    DO47J=KSTAR,KSTOP
    IE=LWTS-1+J
```

```
47    PREY=PREY+A(IE)*X(J)                                    C3580
      PREY=PREY+REGCO                                         C3590
      DIF=X(IDC)-PREY                                         C3600
48    WRITE(6,49)IIDB,X(IDC),PREY,DIF                         03610
49    FORMAT(I15,3F15.4)                                      03620
      REWIND 8                                                03630
50    WRITE(6,51)                                             03640
51    FORMAT(/,1X,99('*'))
      GOTO2                                                   C3660
52    DF1=MFLD(5)*100+MFLD(4)                                 C3670
      DF2=MFLD(5)*100+MFLD(6)                                 C3680
      K8=MFLD(1)-1+LRSQ                                       03690
      K9=MFLD(2)-1+LRSQ                                       03700
      IF(NFLDS.NE.8)GOTO53                                    03710
      IF(A(K8).GE.1.0.OR.A(K9).GE.1.0)GOTO55
      FNPER=NPER                                              03730
      DF1=DF1-DF2                                             03740
      DF2=FNPER-DF1-DF2                                       03750
53    F=((A(K8)-A(K9))/DF1) / ((1.0-A(K8))/DF2)               03760
      P=DFPRBF(DF1,DF2,F)                                     03770
      I=DF1                                                   03780
      J=DF2                                                   03790
      DIR P = P / 2

      WRITE( 6 , 54 ) PROBID , A(K8) , A(K9) , I , F , A(K6) , DIR P
     1                      MFLD( 1 ) , P , MFLD( 2 ) , DIR P

C
C
54    FORMAT( 1H0 ,    2A4,A2 ,  '.....' ,  '.....' ,          27X
     1 '( RSQ F - RSQ R ) / DF1',7X,'(',F7.5,' - ',F7.5,' ) ',F7.5,' / ', I4 ,' /
     2 20X,     'F = ', 25(-),'  = ', 27(-),' = ', F10.6 ,'  ', I4 ,',27X,
     3 '(      1.0 - RSQ F ) / DF2      (   1.0  - ',F7.5,' ) / ', I4 ,'///
     4 20X,     'MODEL ',    I4 ,' = FULL MODEL'  /
     5 18X,     'NONDIRECTIONAL PROBABILITY =', F14.7 /
     6 20X,     'MODEL ',    I4 ,' = RESTRICTED MODEL' ,
     7 12X,     'DIRECTIONAL      PROBABILITY =', F14.7 ,  /
     8 60X,     '(IN HYPOTHESIZED DIRECTION)'        //
     9 1X,      100('*')  )                                   C3830
      GOTO2                                                   03840
55    WRITE(6,50)
```

```
56 FORMAT(1H0,10X,'THE CARD IN THIS LOCATION CANNOT BE INTERPRETED AS      C3850
   * ONE OF THE MODELS INVOLVED WAS IN ERROR.'//)                          C3860
      GOTO2                                                                C3870
57 CALL CALMS(MFLD,LMEAN,LSIGMA,NPER,NFLDS)                                C3880
      GOTO2                                                                C3890
58 CALL PLCTER(NPER,LMEAN,LSIGMA,NVAR,MFLD)                                C3900
      GOTO2                                                                C3910
      END                                                                  C3920

      SUBROUTINE CALMS(MFLD,LMEAN,LSIGMA,NPER,NFLDS)                       C3930
      DIMENSION GMEAN(56),GSDEV(56)                                        C3940
      INTEGER*2 MFLD(56)                                                   C3950
      COMMON A(10500)                                                      C3960
      FNPER=NPER                                                           C3970
      DO5N=1,NFLDS,2                                                       C3980
      J=MFLD(NM)                                                           C3990
      K=MFLD(NM+1)                                                         C4000
      JA=LMEAN+J-1                                                         C4010
      JB=LMEAN+K-1                                                         C4020
      GMEAN(NM)=A(JB)/A(JA)                                                C4030
      JC=LSIGMA+K-1                                                        C4040
      SS=FNPER*(A(JC)**2)                                                  C4050
      ANG=A(JA)*FNPER                                                      C4060
      FNG=FNPER-ANG                                                        C4070
      SCOR=A(JB)**2+FNG                                                    C4080
      SS=SS-SCOR                                                           C4090
      C=(GMEAN(NM)-A(JB))**2*ANG                                           C4100
      SS=SS-C                                                              C4110
      IF(SS.LT.0.)GOTO6                                                    C4120
      SS=SS/ANG                                                            C4130
      GSDEV(NM)=SQRT(SS)                                                   C4140
      WRITE(6,1)MFLD(NM),A(JA)                                             C4150
1 FORMAT(//,5X,6HVECTOR,4X15,5X,4HMEAN,4X,F7.2,1X,11HCATEGORICAL)          C4160
      WRITE(6,2)MFLD(NM+1),A(JB)                                           C4170
2 FORMAT(5X,6HVECTOR,4X15,5X,4HMEAN,4X,F7.2,1X,7HPRODUCT)                  C4180
```

```
0221

0222
0223
0224
0225
0226
0227

C001
C002
C003
C004
C005
C006
C007
C008
C009
C010
C011
C012
C013
C014
C015
C016
C017
C018
C019
C020
C021
C022
C023
C024
C025
C026
```

```
0027        WRITE(6,3)GMEAN(NM)                                          04190
0028      3 FORMAT(5X,29HCORRECTED MEAN PRODUCT VECTOR,F10.3)            04200
0029        WRITE(6,4)GSDEV(NM)                                          04210
0030      4 FORMAT(5X,29HCORRECTED SDEV PRODUCT VECTOR,F10.3)            04220
0031      5 CONTINUE                                                     04230
0032        RETURN                                                       04240
0033      6 WRITE(6,7)NM                                                 04250
0034      7 FORMAT(5X,25HNEG OR ZERO VARIANCE SET,I5)                    04260
0035        RETURN                                                       04270
0036        END                                                         04280
```

```
0001        FUNCTION DFPRBF (DN, DD, F)                                  04290
       C  RETURNS PROBABILITY OF F WITH DN AND DD DEGREES OF FREEDOM.    04300
       C  REQUIRES DFPGLF AND DFPBLF.                                    04310
0002        DIMENSION Y(6)                                              04320
0003        DFPRBF = 0.0                                                04330
0004        IF(F)1,1,2                                                   04340
0005      1 DFPRBF = 1.0                                                 04350
0006        GOTO8                                                       04360
0007      2 IF(DN*DD)8,3,3                                              04370
0008      3 Y(1) = .40546512                                           04380
0009        Y(2) = -1.2039728                                          04390
0010        Y(3) = .58778366                                           04400
0011        Y(4) = Y(2)                                                 04410
0012        Y(5) = Y(1)                                                 04420
0013        Y(6) = -.51082562                                          04430
0014        C = DFPGLF(DN) + DFPGLF(DD) - DFPGLF(DN+DD) - 69.693147/    04440
0015        H = SQRT (DD/(F*DN+DD))                                     04450
0016        H=ARSIN(H)/60.                                             04460
0017        X = 0.0                                                     04470
0018        DO7I=1,10                                                  04480
0019        IF(I-9)5,5,4                                                04490
0020      4 Y(6) = Y(2)                                                 04500
0021      5 DO7J=1,6                                                    04510
0022        X = X + H                                                   04520
```

```
0023        XS=SIN(X)                                                              C4530
0024        Z = Y(J) - C + (LD-1.) *ALOG (XS) + (DN/2. - .5) *ALOG (1.-XS*XS)       C4540
0025        IF(Z)7,6,6                                                              C4550
0026      6 DFPRBF = DFPRBF + EXP  (Z - 65.0)                                       C4560
0027      7 CONTINUE                                                               C4570
0028        DFPRBF = DFPRBF * H                                                     C4580
0029        RETURN                                                                 C4590
0030        END                                                                    C4600
```

```
0001        FUNCTION DFPGLF(X)                                                     C4610
0002        DFPGLF = .57236494                                                     C4620
0003        IF(X-1.0)3,3,1                                                         C4630
0004      1 L = X                                                                  C4640
0005        FL = L/2 * 2                                                           C4650
0006        IF(X-FL)4,4,2                                                          C4660
0007      2 DFPGLF = DFPGLF - (X-2.)*.69314718 + DFPGLF(X-2.)-DFPBLF(X/2.-1.5)     C4670
0008      3 RETURN                                                                 C4680
0009      4 DFPGLF = DFPBLF (X/2.0 - 1.0)                                          C4690
0010        RETURN                                                                 C4700
0011        END                                                                    C4710
```

```
0001        FUNCTION DFPBLF (X)                                                    C4720
0002        DFPBLF = 0.0                                                           C4730
0003        IF(X-2.0)2,1,1                                                         C4740
0004      1 Z = 1.0 / (X * X)                                                      C4750
0005        DFPBLF = (X+.5)*ALOG(X) - X+.91893853 + (.083333333-Z * (.0027778     C4760
          * - Z * (.00079365 - Z * (.00059524 - Z * .00084175)))) / X             C4770
0006      2 RETURN                                                                 C4780
0007        END                                                                    C4790
```

```
      SUBROUTINE DFPRNT(NR,NC,NRMAX,A,NUMHOL,TITLE)
C     PRINTS A TWO-DIMENSIONAL ARRAY @A@ WITH NR ROWS AND NC COLUMNS.
C     NRMAX IS THE MAXIMUM NUMBER OF ROWS DIMENSIONED FOR @A@.
C     FOR A LINEAR ARRAY, SET NR=NRMAX=1
      DIMENSION A(1),TITLE(20)
      N=(NUMHOL+3)/4
      WRITE(6,1)(TITLE(J),J=1,N)
    1 FORMAT(1H0/1H0,20A4)
      N=(NC-1)/10+1
      DO4K=1,N
      JA=K*10-9
      JB=K*10-K/N*(K*10-NC)
      WRITE(6,2)(J,J=JA,JB)
    2 FORMAT(1H0,3X,10I12)
      WRITE(6,3)
    3 FORMAT(1H )
      JA=(JA-1)*NRMAX
      JB=(JB-1)*NRMAX
      DO4I=1,NR
      JA=JA+1
      JB=JB+1
    4 WRITE(6,5)I,(A(J),J=JA,JB,NRMAX)
    5 FORMAT(I5,5X,10F12.4)
      RETURN
      END
```

```
      SUBROUTINE DFCRLB(NUM,NVAR,NVMAX,LMEAN,LSIGMA,LCORR)
C     MEANS,STANDARD DEVIATIONS,CORRELATIONS
C     PORTION OF CORRLB 8 FEB/65 MODIFIED MAY/66 BY FLATHMAN, U OF A.
      COMMON A(1)
C     COMPUTE R MATRIX
    1 FN=NUM
C     COMPUTE NON DIAGONAL ELEMENTS OF R MATRIX
      KM1=NVAR-1
    2 DO16I=1,KM1
```

```
  3   IPI=I+1                                                   05140
  4   DO16J=IPI,NVAR                                            05150
  5   ISI=LCORR-1+(I-1)*NVMAX+I                                 05160
  6   ISJ=LCORR-1+(J-1)*NVMAX+J                                 05170
  7   ISIJ=LCORR-1+(J-1)*NVMAX+I                                05180
  8   ISJI=LCORR-1+(I-1)*NVMAX+J                                05190
  9   IMI=LMEAN-1+I                                             05200
 10   IMJ=LMEAN-1+J                                             05210
 11   DEN= SQRT((FN*A(ISI)-A(IMI)*A(IMI))*(FN*A(ISJ)-A(IMJ)*A(IMJ)))   05220
 12   IF(DEN)15,13,15                                           05230
 13   A(ISIJ)=0.0                                               05240
 14   GOTO16                                                    05250
 15   A(ISIJ)=(FN*A(ISIJ)-A(IMI)*A(IMJ))/DEN                    05260
 16   A(ISJI)=A(ISIJ)                                           05270
C     COMPUTE MEAN AND SIGMA                                    05280
 17   DO23I=1,NVAR                                              05290
 18   IM=LMEAN-1+I                                              05300
 19   A(IM)=A(IM)/FN                                            05310
 20   IS=LSIGMA-1+I                                             05320
 21   II=LCORR-1+(I-1)*NVMAX+I                                  05330
 22   A(IS)= SQRT((A(II)/FN)-A(IM)*A(IM))                       05340
C     COMPUTE DIAGONAL ELEMENTS OF R MATRIX                     05350
 23   A(II)=1.                                                  05360
      CALLDFPRNT(1,NVAR,1,A(LMEAN),8,8HMEANS    )               05370
      CALLDFPRNT(1,NVAR,1,A(LSIGMA),20,20HSTANDARD DEVIATIONS ) 05380
      CALL DFPRNT(NVAR,NVAR,NVMAX,A(LCORR),12,12HCORRELATIONS)  05390
 24   RETURN                                                    05400
      END                                                       05410

      SUBROUTINE PLOTER(NPER,LMEAN,LSIGMA,NVAR,MFLD)            05420
      DIMENSION CLAB(11)                                        05430
      COMMON A(10200),VALUE(100)                                05440
      INTEGER*2 MFLD(56),ARRAY(51,101),X,Y,C,PLUS(2)/'.','.','+'/   05450
      INTEGER*2 FINVAL(11)/' ','1','2','3','4','5','6','7','8','9','*'/  05460
      X=MFLD(1)                                                 05470
```

```
0007            Y=MFLD(2)                                                              C5480
0008            C=MFLD(3)                                                              C5490
0009            WRITE(6,12)X,Y                                                         C5500
0010            IF(C.EQ.0)GOTO2                                                        C5510
0011            DO1I=4,28                                                              C5520
0012   1        IF(MFLD(I).NE.0) JEND=I                                                C5530
0013            WRITE(6,13)C,(MFLD(I),I=4,JEND)                                        C5540
0014   2        XMAX=A(X)+2.5*A(X+100)                                                 C5550
0015            XMIN=A(X)-2.5*A(X+100)                                                 C5560
0016            YMIN=A(Y)-2.5*A(Y+100)                                                 C5570
0017            YMAX=A(Y)+2.5*A(Y+100)                                                 C5580
0018            XRANGE=XMAX-XMIN                                                       C5590
0019            YRANGE=YMAX-YMIN                                                       C5600
0020            DO3K=1,101,10                                                          C5610
0021            Z=K-101                                                                C5620
0022   3        CLAB(K/10+1)=Z*XRANGE/100.0+XMAX                                       C5630
0023            WRITE(6,14)(CLAB(I),I=1,11,2),(CLAB(I),I=2,10,2)                       C5640
0024            DO4I=1,51                                                              C5650
0025            DO4J=1,101                                                             C5660
0026   4        ARRAY(I,J)=1                                                           C5670
0027            DO7J=1,APER                                                            C5680
0028            READ(8) (VALUE(I),I=1,NVAR)                                            C5690
0029            IF(C.EQ.0)GOTO6                                                        C5700
0030            DO5I=4,JEND                                                            C5710
0031            K=MFLD(I)                                                              C5720
0032            IF(VALUE(C).EQ.R)GOTO6                                                 C5730
0033   5        CONTINUE                                                               C5740
0034            GOTO7                                                                  C5750
0035   6        IF(VALUE(X).GT.XMAX.OR.VALUE(X).LT.XMIN.OR.VALUE(Y).GT.YMAX.OR.VAL     C5760
                *UE(Y).LT.YMIN)GOTO7                                                   C5770
0036            LX=(VALUE(X)-XMIN) /XRANGE*100.0+1.5                                   C5780
0037            LY=(YMAX-VALUE(Y))/YRANGE*50.0+1.5                                     C5790
0038            ARRAY(LY,LX)=ARRAY(LY,LX)+1                                            C5800
0039   7        CONTINUE                                                               C5810
0040            WRITE(6,8)                                                             C5820
0041   8        FORMAT(14X,20(5H.+...),'.+.')                                          C5830
0042            DO10I=1,51                                                            C5840
0043            Z=51-I                                                                 C5850
0044            RLAB=(Z-50.0)*YRANGE)/50.0+YMAX                                        C5860
```

```
0045      DO9J=1,101                                                              C5870
0046      IF(ARRAY(I,J).GT.11) ARRAY(I,J)=11                                      C5880
0047    9 ARRAY(I,J)=FINVAL(ARRAY(I,J))                                           C5890
0048      C=1                                                                     C5900
0049      IF(MOD(I,5).EQ.1) C=2                                                   C5910
0050   10 WRITE(6,11)RLAB,PLUS(C),(ARRAY(I,J),J=1,101),PLUS(C),RLAB              C5920
0051   11 FORMAT(1H F12.3,1X,103A1,F12.3)                                         C5930
0052      WRITE(6,3)                                                             C5940
0053   12 FORMAT(1H1,5X,16HPLOT OF VARIABLE,I4,25H (X AXIS) AGAINST VARIABLE    C5950
0054    * ,I4,10H (Y AXIS).)                                                     C5960
0055   13 FORMAT(6X,46HRESTRICTED TO THE OBSERVATIONS IN WHICH VARIABLE        C5970
0056    * I3,15H HAS VALUES OF 2013,/72X,3I5)                                    C5980
0057      WRITE(6,14)(CLAB(I),I=1,11,2),(CLAB(I),I=2,10,2)                       C5990
0058   14 FORMAT(1H0,6X,5(F12.3,8X),F12.3/17X,5(F12.3,8X))                       C6000
0059      REWIND 8                                                              C6010
          RETURN                                                                 C6020
          END                                                                    C6030
```

Appendix B

Miscellaneous Data
Transformation Notions

1. To generate nine mutually exclusive groups from a single continuous vector containing scores 1-9:

   X(1) has occupational levels punched as 1, 2, ..., 9.

   X(6) is the first available vector, 1 if occupational level
   is 1, 0 otherwise;

   X(7) = 1 if occupational level is 2, 0 otherwise;

   X(8) = 1 if occupational level is 3, 0 otherwise;

   .
   .
   .

   X(14) = 1 if occupational level is 9, 0 otherwise.

   A long way to accomplish the task:

   IF(X(1).EQ.1.0) X(6) = 1

   IF(X(1).EQ.2.0) X(7) = 1

   .
   .
   .

   IF(X(1).EQ.9.0) X(14) = 1

   A shorter way:

   L = X(1)

   X(L+5) = 1

   Shortest way:

   X(X(1)+5) = 1

2. To dichotomize a continuous variable:

   X(3) has IQ scores and you want to dichotomize above/below
      100, such that

   X(7) = 1 if IQ<100, 0 otherwise;

   X(8) = 1 if IQ≥100, 0 otherwise.

   The requisite transformation cards would be:

   IF(X(3).LT.100.0) X(7) = 1

   IF(X(3).GE.100.0) X(8) = 1

3. To trichotomize a continuous variable:

   X(3) has IQ scores and you want to generate:

   X(9) = 1 if IQ<90, 0 otherwise;

   X(10) = 1 if 90≤IQ≤110, 0 otherwise;

   X(11) = 1 if IQ>110, 0 otherwise.

   The requisite transformation cards would be:

   IF(X(3).LT.90.0) X(9) = 1

   IF(X(3).GE.90.0.AND.X(3).LE.110.0) X(10) = 1

   IF(X(3).GT.110.0) X(11) = 1

   Logical symbols which can be used in IF statements:

   LT   less than
   GT   greater than
   LE   less than or equal to
   GE   greater than or equal to
   OR   either true
   AND both true
   EQ   equal
   NE   not equal

4. To generate polynomials to investigate curvilinear relations:

   You suspect that there is an inverted U relationship between
   anxiety (X(6)) and performance (X(1)). You need to generate

the square of anxiety (X(7)) in order to allow for the hypothesized second degree curve.

The requisite transformation cards would be:

X(7) = X(6)*X(6)

or

X(7) = X(6)**2.0

When using whole number exponents, it is suggested to omit the decimal.

Other algebraic functions:

+   add
-    subtract
*   multiply
**   exponentiation
/   divide

5. To make a non-linear transformation:

You want to make a square root transformation of the criterion X(5). The requisite transformation card would be:

X(6) = DSQRT(X(5))

The above statement will place the square root of X(5) in X(6).

X(5) = DSQRT(X(5))

The above statement will replace the criterion score with the square root of the criterion.

Other non-linear functions:

DABS   absolute value   Note: when using a single
DLOG   natural antilogarithm   precision program,the "D"
DSIN   sine function   preceeding the function
DCOS   cosine function   symbol is omitted (e.g.
DATAN   arc tangent function   SQRT rather than DSQRT).

6. To generate a new set of data:

You have the following data on cards:

var   cols

```
X(1)   1-3     subject identification
X(2)   4-6     age in months
X(3)   7-9     mental age
X(4) 10-11     anxiety
```

You want

    a) the computer to generate the IQ (mental age/age in months)

    b) to print the subject identification of all subjects having a missing anxiety score

    c) to replace missing anxiety data with the mean of 15

    d) to generate a missing data vector

    e) to punch a new set of data containing:

```
cols
1-3     subject identification
4-6     IQ
7-8     anxiety
9       1 if anxiety score was missing; 0 otherwise
```

The requisite data transformation cards would be:

    K1 = X(1)

(a) K2 = X(3)/X(2)

    K3 = X(4)

    K4 = 0

    IF(K3.NE.0) GO TO 5

(b) PRINT 6, K1

   6 FORMAT(X, I3)

(c) K3 = 15

(d) K4 = 1

(e)5 PUNCH 7, K1, K2, K3, K4

   7 FORMAT(2I3, I2, I1)

7. To generate vectors for various ARH in the text:

ARH 7.1   X(16) = X(3)*X(3)

ARH 7.2   X(15) = X(5)*X(5)

ARH 8.2   X(15) = X(2)-X(3)

ARH 8.4   Two cards per subject are required to solve the
          repeated measure ARH. Therefore ARH 8.4 cannot be
          solved on the same run with other ARH. The data
          transformation statements that are required are
          somewhat complex. Below is one way of obtaining the
          results. Those statements generate vectors X(15)
          through X(80). The statement numbers on cards 3 and
          8 are to be punched in column 5. The number "71346"
          is just a random number. The last statement produces
          the 60 person vectors from the subject ID.

```
  IF(XI.EQ.0.71346) GO TO 3
  NOBS = 0
3 NOBS = NOBS+1
  XI = 0.71346
  IF(NOBS.LT.61) GO TO 6
  X(2) = X(3)
  X(15) = 1
6 X(16) = 1-X(15)
  X(17) = X(12)*X(15)
  X(18) = X(13)*X(15)
  X(19) = X(12)*X(16)
  X(20) = X(13)*X(16)
  X(X(1) + 20) = 1
```

ARH 10.1 X(15) = X(2) - (3*X(6))

ARH 10.2 X(16) = X(2) - (2*X(3))

ARH 10.3 X(17) = DSIN(X(1)*(6.283/12.0))

Appendix C

N = 60 Data Set Format

Columns	Variable	Format
1,2	Subject IDX_1	F2.0
3	Skip	
4,5	X_2	F2.0
6	Skip	
7,8	X_3	F2.0
9,10,11	Skip	
12,13	X_4	F2.0
14	Skip	
15,16	X_5	F2.0
17	Skip	
18	X_6	F1.0
19	X_7	F1.0
20	Skip	
21,22,23	X_8	F3.0
24,25,26	X_9	F3.0
27	Skip	
28	X_{10}	F1.0
29	X_{11}	F1.0
30	X_{12}	F1.0
31	X_{13}	F1.0
32	Skip	
33,34,35	X_{14}	F3.0
36	Skip	
37,38	X_{15}	F2.0
39	Skip	
40,41	X_{16}	F2.0

Note: Variables X_{15} and X_{16} used only with ARH 10.3.

```
01  4  4   26  8 10   82 97 0101   80 04 01
02  6  9   28  9 10   95101 0110   75 06 02
03  6  8   41 10 01  103104 0101   82 07 03
04  8 10   29 10 10  109106 0101   84 07 04
05 10 12   40 11 01   84 96 0101   85 06 05
06 11 13   30 11 10   93100 0110   82 06 06
07 12 14   39 12 01   98102 0110   84 05 07
08 14 15   32 13 10   99103 0101   90 05 08
09 16 15   37 14 01   87 98 0101   92 06 09
```

513

```
10 17 16    37 15 01 106105 0110    92 06 10
11 18 16    36 15 01 113112 0110    90 07 11
12 20 17    35 17 01 128125 0101    97 11 12
13 22 18    34 18 10  89 99 0101   100 17 13
14 24 18    33 58 01 115118 0110    96 18 14
15 26 20    32 21 01 125124 0110    98 18 15
16 27 20    32 22 01  91100 0110   100 18 16
17 28 20    31 23 01 116115 0110   101 18 17
18 28 21    36 55 10 117117 0101   105 18 18
19 29 20    30 24 01 123120 0101   106 17 19
20 30 21    29 53 01 126120 0101   108 17 20
21 31 21    28 26 01 120116 0110   103 17 21
22 32 22    26 52 01 129127 0110   105 18 22
23 32 22    37 27 10 120116 0101   110 19 23
24 33 22    24 51 01 114113 0101   112 25 24
25 34 22    37 52 10 100102 0110   107 28 25
26 36 23    38 28 10  98103 0110   110 29 26
27 38 24    38 49 10  96102 0101   115 30 27
28 40 24    38 48 10  85 95 0101   120 30 28
29 42 25    39 32 10 125128 0110   115 30 29
30 48 25    40 38 10  95102 0110   120 30 30
31  5  5    27  8 10  84 80 1010    75 04 01
32  7  8    27  9 10  85 82 1001    82 06 02
33  7  8    41 10 01 115123 1001    83 07 03
34  8 11    29 11 10  87 85 1010    80 07 04
35 10 12    41 11 01  95 95 1010    83 06 05
36 11 12    30 10 10  93 92 1010    84 06 06
37 12 13    39 11 01  99100 1010    85 05 07
38 14 15    32 14 10  95100 1001    90 05 08
39 16 16    37 13 01 118128 1001    92 06 09
40 17 16    37 14 01  88 86 1001    93 06 10
41 18 17    36 16 01  94 92 1010    90 07 11
42 20 18    35 16 01 102102 1010    92 11 12
43 21 17    34 19 10  89 97 1001    98 17 13
44 23 17    33 57 01 116122 1001   100 18 14
45 25 19    32 22 01  97 97 1001   103 18 15
46 27 20    32 21 01 103102 1001   105 18 16
47 28 21    31 24 01  90 89 1010   100 18 17
48 29 22    36 54 10  96 96 1010   102 18 18
49 30 22    30 24 01 108113 1010   103 17 19
50 30 21    39 54 10 100101 1010   104 17 20
51 31 22    28 27 01  91 89 1001   108 17 21
52 32 21    26 51 01 112116 1001   110 18 22
53 32 22    37 28 10 104104 1010   105 19 23
54 33 21    24 52 01  92 90 1001   112 25 24
55 33 22    37 51 10 120130 1001   110 28 25
56 35 23    38 29 10 122132 1001   113 29 26
57 37 24    38 50 10 128140 1010   110 30 27
58 39 24    38 49 10 124136 1010   113 30 28
59 41 25    39 33 10 126137 1001   114 30 29
60 46 25    40 39 10 130142 1001   124 30 30
```

Appendix D

Relationship between Analysis of Variance, Multiple Correlation, Analysis of Covariance, and Multiple Linear Regression

First will be discussed the defining characteristics of the four techniques: Analysis of Variance (ANOVA), Multiple Correlation (MC), Analysis of Covariance (ANCOVA), and Multiple Linear Regression (MLR). Then the weaknesses and strengths of each of these procedures will be pointed out.

ANOVA requires that all independent (predictor) variables be categorical. On the other hand, MC usage has usually been with only continuous predictor variables. The MC procedure does allow for categorical and powered terms, but few applications have done so. In some sense, ANCOVA is a mixture of the above two procedures in that both categorical and continuous variables are used. The variables tested are the categorical variables, with the (linear) continuous variables taking the role of confounding variables (covariates) whose effects need to be corrected for. The ANCOVA procedure does allow for categorical and powered terms as covariates (Winer, 1962), but few applications have done so. MLR encourages the use of whatever variables may account for criterion variance. The position of this text is that those variables will probably be continuous, but a pragmatic view is adopted. MLR subsumes all of the above procedures, and in that sense acquires all of their strengths and nullifies many of their weaknesses. Those

515

weaknesses are nullified because MLR goes beyond the other procedures. Those strengths and weaknesses will now be discussed.

Analysis of Variance (ANOVA)--strengths and weaknesses

ANOVA computational formula are well documented in most ad-
vanced statistical books. Canned computer programs exist at most
computer centers for their execution. Unfortunately, the various
designs limit the thinking of the researcher. One text may have 21
designs, another may have 32. Most researchers are unable to expand this limiting set of designs. Therefore, they begin to design
their studies within the established framework, rather than to ask
and get answers to the questions that are important to them. Often
the precise research hypothesis being tested is not evident. Furthermore, these research hypotheses are almost always non-directional. On the other hand, if a researcher has those kind of hypotheses in mind, then the canned ANOVA programs are usually much
easier to run than are the canned regression programs. Also, the
canned ANOVA programs test research hypotheses that many researchers have in mind who deal with experimental data. These research
hypotheses deal with variables which through the research design
have been made orthogonal by obtaining proportionality of cells;
therefore there is no problem in interpreting the results. Unfortunately, the process of making the cells proportional may have
distorted the real world--hence making the interpretations in-
applicable to the real world. Furthermore, missing data is a real
world problem which should not be ignored. Much work is being done

with the problem of nonorthogonality (Appelbaum and Cramer, 1973;
Gocka, 1973; Overall and Spiegel, 1969; Overall and Spiegel, 1973)
with the solutions being proposed in MLR models.

The treatment of the independent (predictor) variables is
also a weakness for three reasons. First, the independent variables must be categorized. Thus, one often finds an inherently
continuous variable such as I.Q. or Anxiety divided into artificial categories. Second, the available designs encourage the use
of a small number of variables. The goal of accounting for the
criterion variance is not one of the major goals. Third, little
attention is paid to the R^2 due to each variable or to the entire
set of variables--rather, attention is given to the probability of
chance operating. Lately there has been an emphasis on the total
predictability. Unfortunately, alternative indices have been developed (omega and eta squared) which are essentially equal to R^2.

Multiple Correlation (MC)--strengths and weaknesses

As indicated earlier, multiple correlation applications have
generally used the linear component of continuous variables. Unfortunately, few applications have considered the additional use
of categorical, interaction, or polynomial terms. So in this sense
there is no attempt to build prediction beyond the capability of
the linear terms.

The tests of significance are for either dropping out one
variable (GRH 11.1), or dropping out all variables (GRH 11.2). In
most presentations, two different tests of significance are pro-

vided, although the reader familiar with the general $\underline{F}$ test will recognize both of these cases as subsets of the general $\underline{F}$ test.

The MC technique is used frequently in prediction situations, and the weights of the variables often acquire more meaning than is statistically appropriate. Researchers will often order the variables in terms of importance, based upon the weighting coefficients. The fact that the weighting coefficients often change quite drastically when variables are added to or are deleted from the model (a phenomenon referred to as "bouncing betas") does not seem to alter this activity.

The R^2 is called the "coefficient of determination" and $(1-R^2)$ is referred to as the "coefficient of non-determination" or "coefficient of alienation." Unfortunately, though, most applications focus on the obtained R, rather than the R^2 value. (Since R is always greater than R^2 this may not be too surprising.)

It is interesting to note that users of MC are often admonished to use additional predictor variables which are uncorrelated with the other predictor variables already in the regression model. A good case for using correlated predictors was made in the inverted U section of Chapter Seven. McNeil and Spaner (1971) and Cramer (1974) make additional comments regarding correlated predictor variables. Their position is essentially that the variables one should use in a prediction model are the ones that yield a high R^2. The data itself will dictate which variables to include.

Stepwise regression was discussed in Chapter Nine, but a few comments are in order here since most MC applications would have

used a stepwise procedure. Most stepwise programs do not satisfactorily handle categorical variables if there are more than two
categories. The stepwise procedure developed by Williams (1974)
solves this problem. The AID-4 program developed by Koplyay (1972)
also applies stepwise notions to categorical data.

The forward stepwise procedure can ignore moderator variables, so the backward stepwise procedure is recommended. The backward stepwise procedure requires a lot of computer time, though.
The procedure does provide the researcher with a parsimonious set
of predictor variables--a very valuable gift. Most researchers,
though, get over zealous in interpreting the order in which variables were introduced into (or deleted from) the predictor set.
Weighting coefficients are either interpreted or applied to other
subjects from the same population. One must remember that the
stepwise procedures are a data snooping tool in that a multitude
of hypotheses are tested. The resulting model should be treated as
data upon which future hypotheses can be formulated, as discussed
in Chapter Eleven.

Analysis of Covariance (ANCOVA)--strengths and weaknesses

ANCOVA is an extension of ANOVA in the situation where groups
are thought to be initially different on variables (covariates),
and therefore an adjustment is called for. The covariates are almost always continuous linear variables, though they could be
interactions and polynomials. The canned computer programs that do
exist (Williams and Lindem, 1974) are usually for only one covari-

ate or for at most a two-way design. Many writers have shown how
ANCOVA can be accomplished within the MLR procedure (Williams,
1974; Huck, 1972; Ward and Jennings, 1973). Chapter Eleven of this
text makes the case that all least squares hypotheses are "analysis of covariance" hypotheses. The present authors choose to call
them "over and above" hypotheses because traditional analysis of
covariance has many connotations, such as assuming homogeneity of
regression lines, considering only linear terms as covariates, and
only testing categorical variables. One of the major values of MLR
is the big picture that it presents. On one hand, the relationships between the various traditional statistical techniques can
be easily grasped. On the other, the researcher realizes the infinity of restrictions that can be tested against any one model.
In a sense, the flexibility and comprehensiveness of MLR places
the burden on the researcher to ask the specific question that is
of interest. Once the hypothesis is stated, the remaining steps
are relatively simple.

Multiple Linear Regression (MLR)--strengths and weaknesses

An analysis using MLR must of necessity be accomplished
through the use of a computer. Unfortunately not all computer
systems have an hypothesis testing program, although most have a
stepwise version. Extensive reliance on the computer brings about
problems. First, some users will test models for which they have
not stated research hypotheses. Second, since it is so easy to
test hypotheses, some users will test more than they are really

interested in. Data snooping is appropriate as long as the results are replicated before causal interpretations are made. Third, reliance on the computer often leads to sloppy handling of data. We cannot emphasize enough the careful editing of one's data. One incorrect score can make a lot of difference in the outcome of the results.

Since we are biased we find few weaknesses with MLR. At times we have found it difficult to write models to reflect certain research hypotheses. Some users have had trouble in ascertaining the number of linearly independent vectors in models, but a little assistance has usually clarified the matter.

We find some researchers and some authors saying, "we did a multiple regression..." By now the reader should be aware that regression analysis must be specific to a particular hypothesis--thus the variables in the full and restricted models must be reflected in certain ways.

Appendix E

Abstracts of
Multiple Linear Regression Viewpoints

McNeil, K.A. The negative aspects of the eta coefficient as an

    index of curvilinearity. Viewpoints, 1970, 1, 7-17.

The objective of this paper is to discuss the eta coeffi-

cient and to point out some limitations and misconceptions about

the coefficient. Specifically, we will discuss the fact that:

1) the eta coefficient is a global measure of curvilinearity;

2) the eta coefficient has limited interpretability; 3) there are

a number of other curvilinear relationships that might be of more

significance and of more interpretability; 4) these other curvi-

linear relationships do not suggest nor encourage grouping of

data as does the eta coefficient; and 5) these other curvilinear

relationships may tend to be more amenable to replication than is

the eta coefficient.

Stone, L.A. and Skurdal, M.A. Estimation of product moment corre-

    lation coefficients through the use of the ratio of contin-

    gency coefficient to the maximal contingency coefficient.

    Viewpoints, 1970, 1, 19-25.

The authors attempt to demonstrate that the ratio, contin-

gency coefficient/maximal contingency coefficient (C/C_{max}), is

directly comparable to the product moment correlation coefficient.

Inspection of all of the computed C/C_{max} ratios, from 2 X 2 tables, showed that the ratios which correspond most closely to the product moment correlation coefficients were not always the ones which were associated with fourfold tables having dichotomies nearer to .50 - .50 proportions. However, the authors were lead to believe that the C/C_{max} ratios which best approximated the product moment correlation coefficients generally were from the fourfold tables where there was an approximate 50 - 50 split. The C/C_{max} ratio may be used as a "quick and dirty" estimate of the relationship measure provided by the product moment correlation model. No mathematical justification is offered for this contingency coefficient ratio.

Williams, J.D. Multiple comparisons in a regression framework.
    Viewpoints, 1970, 1, 26-39.

In using multiple regression as a problem solving technique, one problem that might arise is the overuse of a full model with several restricted models, without adjusting the probability level. This has long been a concern in statistics. Several multiple comparison procedures have been developed for different situations.

The intent of the present paper has been to extend some of the better known multiple comparison procedures (Duncan's Multiple Range Test, Dunn's "C" Test, and Scheffe's Test) to a multiple regression approach. The major change in the regression approach is to assess the result of multiple uses of a full model

to a correct distribution, rather than a straight-forward usage
of the $\underline{F}$ distribution.

Goff, A.F. and Houston, S.R. Concurrent validity of the Koppitz
    scoring system for the bender visual motor gestalt test.
    Viewpoints, 1971, 1, 45-52.

The study examined correlations between assessed visual-
motor perception, intelligence, and academic achievement. In addition, efficiency of prediction for criterion variables was investigated by employing two approaches of analysis: (a) regression model and (b) Bender $\underline{z}$ model. The following conclusions were
formulated on the basis of the obtained data and from the comparison of the two predictive models.

(1) The significant negative correlation found between age
and the Bender error score adds further substantiation to the
fact that the ability to correctly execute the Bender protocol
improves with increased age.

(2) The Bender $\underline{z}$ score correlated to a greater degree with
intelligence, reading, and arithmetic achievement than did the
Bender error score with the three specified variables.

(3) The obtained correlations of the Bender $\underline{z}$ score with the
three criterion variables agrees with the literature in directionality and in significance with assessed intelligence.

(4) However, efficiency is enhanced by using the Bender er-
ror score and age rather than the single variable of the Bender $\underline{z}$
score to predict achievement in reading and arithmetic and as-

sessed intelligence.

Jordan, T.E. Curvilinearity within early developmental variables.

 Viewpoints, 1971, 1, 53-77.

 Squared and cubed vectors were introduced into eighteen re-

gression models each applied to nine criteria. Data came from

study of several hundred children in the first three years of

life. Departure from linearity did not provide better accounts of

the relationship between five predictors and development at 12,

24, and 36 months of age. Illustrations of various patterns of

squared vectors and cubed vectors were presented from the data.

Reed, C.L., Feldhusen, J.F., and Van Mondfrans, A.P. Regression

 models in educational research. Viewpoints, 1971, 1, 78-88.

 Results of this study do not support the findings of Rock

(1965) that the interaction term regression was superior to the

quadratic form in predictive efficiency. The most efficient re-

gression model will depend upon: 1) how the variables and cri-

terion are related; 2) the reliability of the predictor variables;

and 3) the research question asked.

 The studies reviewed in this paper seem to indicate that

complex regression models are in some cases more efficient pre-

dictors of complex behavior than the most frequently assumed

first order model. When quadratic and interaction terms are sig-

nificant, however, interpretation is made more difficult. Still,

an attempt at interpretation seems somewhat better than ignoring the problem or assuming it does not exist.

McNeil, K.A. and Beggs, D.L. Directional hypotheses with the

    multiple linear regression approach. Viewpoints, 1971, 1,

       89-102.

    Two well known directional tests of significance are pre-

sented within the multiple linear regression framework. Adjust-

ments on the computed probability level are indicated. The case

for a directional interaction research hypothesis is defended.

Conservative adjustments on the computed probability level are

offered and a more precise computation is requested of statis-

ticians. Emphasis is placed more on the research question being

asked than on blind adherence to conventional fomulae.

McNeil, K.A. On the unit vector. Viewpoints, 1971, 2, 2.

    Short article presents several interpretations of the unit

vector. One of which conceptualizes the unit vector as any pre-

dictor to the zero power.

McNeil, K.A. On attenuating a multiple R. Viewpoints, 1971, 2, 3.

    A brief conceptual argument questioning the value of at-

tenuating a multiple R.

Duff, W.L.,Jr., Houston, S.R. and Bloom, S. A regression/princi-

    pal components analysis of school outputs. Viewpoints, 1971,

       2, 5-18.

The objective of this study is to identify the correlates of student performance and teacher retention in an inner-city elementary school district. The purpose is to provide urban school administrators with information necessary to cope with the special problems they face in organizing and administering their educational resources.

The study is divided into two parts: a descriptive section and an analytic section. In the descriptive section the writers are concerned with describing the inner urban school system. Here the data to be analyzed are presented and classical regression techniques are used to specify the three basic teacher retention and student performance models. In the second section the data are further analyzed in terms of the unique contribution of a priori specified subsets of predictor variables. This section ends with a comparison of a principal component regression approach to the a priori grouping of predictors used in the unique analysis.

Bolding, J.T. Empirical exercises for the study of multiple regression. Viewpoints, 1971, 2, 21-22.

Six short exercises are presented which utilize random data and illustrate several properties of MLR. Specifically, if ten random variables are added together to produce a criterion, then the 10 variables will yield an R^2 of 1.00 when predicting that criterion. Variations of the above situation are presented.

Connett, W. A note on multiple comparisons. Viewpoints, 1971, 2,

    23-24.

Two general methods are available for controlling the over-

all alpha level. One method is to adjust the F value required for

significance. This method in relation to multiple regression was

recently discussed in Viewpoints (Williams, 1970). The second way

to maintain an overall alpha level is by proper choice of the

alpha levels for the individual comparisons. The purpose of this

note is to review a method for determining the individual alpha

levels necessary to maintain some selected overall alpha level.

Williams, J.D. and Lindem, A.C. Setwise regression analysis-A

    new data-analytic tool. Viewpoints, 1971, 2, 25-27.

The setwise procedure drops one set at a time in a stepwise

fashion. There will be as many steps as there are sets. Statis-

tically, the steps are accomplished by an iterative procedure

that allows the R^2 term to be maximized at each stage in a back-

ward stepwise procedure. Once a set is discarded, the set is no

longer considered at later stages.

While the difficulty regarding the use of binary coded pre-

dictors has been at least partially solved, other difficulties in

regard to the stepwise procedure are also involved in the setwise

procedure; additionally, the setwise procedure has a new problem

unique to itself.

It has been pointed out several times that probability lev-

els in the stepwise procedure are usually violated. Further, when
k of the N variables have been dropped, the N - k remaining variables are not necessarily the set of N - k variables that would
yield the highest R^2 value. These criticisms would also be valid
in regard to the setwise procedure. Additionally, the differences
in the number of variables in a set will have some effect upon

when that set of variables would be dropped. Other things being
equal, a set with 6 variables will be retained longer than a set
with 3 variables. Notwithstanding these difficulties, if the setwise procedure is judiciously employed by researchers, then additional data analysis power can be obtained.

The program and sample printout are available on request.

Greenup, H. Watch that first step. Viewpoints, 1972, 2, 32-33.

Author discusses the interactive system at UCLA. "Biomed"
programs can be directed by (a) the 2250's typewriter-like keyboard, (b) an attached lightpen, or (c) some specially programmed
keys located near the keyboard.

McClaran, V.R. and Brookshire, W.K. A comment on multiple comparisons in a regression framework. Viewpoints, 1972, 2,
34-35.

Authors clarify an error in an earlier article in Viewpoints.
The error occurred in the Williams paper dealing with multiple
comparisons. One restricted model was inaccurate.

Newman, I. and Fry, J. A response to "a note on multiple compari-

    sons" and a comment on shrinkage. Viewpoints, 1972, 2, 36-39.

    Connett (1971) presented Kimball's (1951) formula for keep-

ing alpha levels constant when making a number of comparisons.

    A simplified formula, developed by the authors, for comput-

ing these alpha levels is presented along with a brief mathemati-

cal proof that Kimball's technique and that of the authors' are

approximately equivalent.

    The method simply takes the desired α level desired to keep

constant across a number of comparisons (α_0) and divides that

specific α_0 by the number of comparisons one wishes to make (N).

Authors suggest that when the ratio of subjects to variables is

10:1 or less, a shrinkage estimation should be used and reported.

If this is done, research based on multiple correlations will

tend to be more replicable and therefore more desirable and use-

ful.

Newman, I. A suggested format for the presentation of multiple

    linear regression. Viewpoints, 1972, 2, 42-45.

    Multiple regression, when presented in the literature, has

usually been formated in an idiosyncratic manner. Also, the

format rarely presents all the relevant information in a concise,

easy to inspect manner. Instead, one tends to find himself thumb-

ing through the articles to find the relevant information. Author

presents a format for the presentation of multiple regression models and the information required for their interpretation.

Connett, W.E., Houston, S.R. and Shaw, D.G. The use of factor

regression in data analysis. Viewpoints, 1972, 2, 46-49.

Suppose that it is desired to express the criterion variable as a function of a set of independent variables in which the
intercorrelations between the various independent variables is
near zero. The procedure involves restructuring the full regression model in such a way that the criterion variable is expressed
as a function of several mutually orthogonal factor variables.
This principal components-regression approach permits one then to
investigate the unique contribution of each of the factor variables to explaining the dependent variable.

The "factor regression" procedure begins with the complete
orthogonal factoring of the set of predictors. Factor scores are
then computed. If a regression model is cast, regressing some
criterion variable onto the set of factor score predictors, several interesting properties are noted. The beta weight for a predictor is equal to the validity for that predictor. The R^2 value
for any model is equal to the sum of the squares of the beta
weights for the model. The exclusion of a factor score variable
from the predictor set will result in a drop in R^2 equal to the
square of the beta weight for the variable dropped. And, perhaps
most important, the dropping of any predictor variable from the
predictor set will not affect the beta weights (predictive contribution in this case) of any of the other variables. These

properties are demonstrated with an example.

This procedure is not a substitute for the long established principals of statistical inference, those being hypothesis building and testing; rather, it provides data organization preliminary to hypothesis development and testing.

Williams, J.D., Maresh, R.T. and Peebles, J.D. A comparison of

raw gain scores, residual gain scores, and the analysis of

covariance with two modes of teaching reading. Viewpoints,

1972, 3, 2-16.

It should be abundantly clear from the 16 tables that the three approaches to psycho-educational change are different. While this set of data does not exhibit strong relationships between the dichotomous predictor and the various criteria, the use of the statistical significance approach would occasionally yield different interpretations. Perhaps the most objective comparison between the three measures would be the R^2 term (for the analysis of covariance, or adjusted means approach, $R^2_{FM} - R^2_{RM}$). Only one significant difference is found in the three measures. In Table 6, the raw gain is significant ($p < .05$), but, under exactly the conditions that would tend to make this occur, the vertical group was significantly smaller than the graded group on the pre-test, but this difference was almost erased on the post-test. In terms of the raw gains score, this produced a significant difference in favor of the vertical group.

Jennings, E. Linear models underlying the analysis of covariance,

    residual gain scores and raw gain scores. Viewpoints, 1972,

    3, 17-24.

    The problem of investigating "change" or "gain" that can be

attributed to "treatments" has been discussed extensively over a

number of years without a noticeable concensus emerging. Cronbach

(1970) has even suggested that many questions that appear to in-

volve "change" can be effectively resolved without reference to

the concept of change.

Koplyay, J.B. Automatic interaction detector AID-4. Viewpoints,

    1972, 3, 25-38.

    The primary value of AID-4 to the task scientist is its

ability to identify the maximum amount of variance in the criter-

ion which can be accounted for by the predictors available; it

relieves the task scientist of the trial-and-error task of at-

tempting to identify the various relevant combinations of lin-

ear and non-linear interaction terms presently required by the

multiple linear regression technique. The splitting process of

AID-4, being based upon maximizing the between sums-of-squares

and minimizing the within sums-of-squares, automatically takes

all present interactions into account, indicating the maximum

variance predictable in the criterion from the predictors. The

interactions and patterns or trends are identifiable from the

AID-4 output.

Newman, I. Some further considerations of using factor regression

    analysis. Viewpoints, 1972, 3, 39-41.

    Presents four major points concerning factoring the data be-

fore using MLR. When using factor regression procedures, it is

important to keep in mind that if one does not use all of the

factors (that is, accounting for 100% of the trace) he may be

overlooking a suppressor factor (suppressor variable).

    If one is interested in improving the multiple regression

equation by using factor techniques, there is only one way this

can be done--by using fewer factors than the number of original

variables. This will increase the df and also possibly decrease

shrinkage-estimates. However, when this is done one may be losing

information that can account for criterion variance by eliminat-

ing a factor that accounts for very little trace of the factored

matrix but is highly correlated with the criterion scores.

    Using only the factors that account for most of the trace

should be avoided when the predictor variables that are being

factored are likely to be highly reliable. Under these conditions

a variable that accounts for little of the trace variance may be

a good and highly reliable predictor of criterion variance.

Newman, I. and Fry, J. Proof that the degrees of freedom for the

    traditional method of calculating analysis of covariance

    and the multiple regression method are exactly the same.

    Viewpoints, 1972, 3, 42-45.

Authors provide a proof that the degrees of freedom for traditional analysis of covariance and for the "over and above" analysis with MLR are exactly the same.

Brebner, M.A. Conditions for no second order interaction in
    multiple linear regression models for three-factor analysis
    of variance. Viewpoints, 1972, 3, 46-57.

Author provides pictorial, algebraic, and regression models for investigating second order interaction in a three-way design.

Newman, I. and McNeil, K. A note on the independent variance of
    each criterion in a set. Viewpoints, 1972, 3, 58-60.

Frequently a researcher is interested in a number of criterion variables which may not be uncorrelated with each other. The chances are that these variables are likely to be significantly correlated with each other.

If one is interested in accounting for the independent piece of a criterion variable's variance, then the procedures outlined here should be used. By covarying other nonindependent criteria as explained in the paper, one can actually treat each test of each criterion as if the criterion variable was independent of all others.

Newman, I., Lewis, E.L. and McNeil, K.A. Multiple linear regres-
    sion models which more closely reflect bayesian concerns.
    Viewpoints, 1972, 3, 71-77.

The purpose of this paper is to discuss several relationships between the Bayesian approach and the multiple linear regression approach.

The testing of interaction is what distinguishes the regression procedure outlined in the present paper from that commonly
used in prediction studies. Too often, even plausible interactions are ignored and all subjects are lumped together and, hence,
treated as similar. Our conceptual theories have long ago turned

to distinct groupings, and it is about time that our statistical
procedures reflect this empirical possibility, whether the statistical procedures be Bayesian or multiple linear regression.
Until the Bayesian methodology has empirically been shown to be
more predictive than multiple regression analysis, the availability and relative mathematical simplicity of multiple regression analysis would seem to indicate preference for its utilization rather than the Bayesian approach.

Brookshire, W.K. and Bolding, J.T. Using coefficients of ortho-
        gonal polynomials as predictor variables in multiple regres-
        sion. Viewpoints, 1973, 4(1), 1-6.

Coefficients of Orthogonal Polynomials are presented by some
authors (Snedecor and Cochran) as a means of simplifying the computation required in trend analysis. Linear regression addicts
who are computer oriented can still make good use of such coding
in the analysis of complicated designs.

Consider a two factor design where the factors are assumed

to be quantitative with levels selected at equal intervals. Testing for main effects and trend analysis can both be simplified by the use of coefficients or orthogonal polynomials as predictor vectors.

McNeil, K.A. Testing an hypothesis about a single population mean with multiple linear regression. Viewpoints, 1973, 4(1), 7-14.

The recent emphasis on criterion referenced testing and on the explicit stating of objectives implies that more researchers will be testing hypotheses about a simple population mean. The generalized regression procedure is one way to test such an hypothesis. The appropriate regression models are presented in this paper.

Halasa, O. Identification of significant predictors of children's achievements and attendance. Viewpoints, 1973, 4(1), 15-22.

The identification of variables other than the treatment process, which are affecting the criterion measure variance has always been a problem. Most of the regression coefficients failed a statistical test of significance. Of the six predictors, the pre-test score evidenced consistent significant contributions to the criterion variance. The percent of predictable variance, however, indicates that a significant proportion of the variance remains unaccounted for.

Newman, I. and McNeil, K. Application of multiple regression
        analysis in investigating the relationship between the three
        components of attitude in Rosenberg and Hovland's theory for
        predicating a particular behavior. Viewpoints, 1973, 4(1),
        23-39.

        Multiple regression and factor analysis techniques were used
to investigate the relationship between the components of attitude and their differential predictive power. It was found that
the different components of attitude and their linear interaction
are more likely to be predictive for intimate rather than non-
intimate behaviors. The cognitive component was found to be significantly predictive of intimate behavior but not predictive for
non-intimate behavior. Out of the three measures used, the behavioral differential was the most predictive scale for both intimate and non-intimate behavior.

Ward, J.H.,Jr. Guidelines for reporting regression analyses.
        Viewpoints, 1973, 4(1), 40-41.

        Suggests including in a report of regression analyses:
1. General Comments
2. Regression Analysis Discussion
        a. natural language statements of the hypotheses
        b. identification of the assumed model
        c. hypotheses in terms of assumed model
        d. identification of the restricted model

e. results of the test

3. Vector definitions

4. Analyses

a. model specifications

b. model comparisons

5. Regression Computer Output

McNeil, K. Reaction to Ward's "Guidelines for reporting regres-

sion analysis" and some alternatives. Viewpoints, 1973,

4(1), 42-44.

Suggests guidelines other than Ward's:

1. Statement of research hypothesis

2. Statement of statistical hypothesis

3. Statement of alpha

4. Formulation of full model

5. Statement of restrictions

6. Formulation of restricted model

7. Definition of vectors

8. Reporting of the probability of the calculated F, and the

subsequent decision.

Newman, I. A revised "Suggested format for the presentation of

multiple regression analyses." Viewpoints, 1973, 4(1), 45-47.

Presents a format for displaying the hypotheses and results

for those hypotheses. This format could be used in conjunction

with either Ward's or McNeil's earlier stated formats.

Williams, J.D. Applications of setwise regression analysis.

    Viewpoints, 1973, 4(2), 1-7.

One of the earlier applications of the setwise technique was made by Grooters (1971). Grooters was interested in predicting costs per student credit hour in four state colleges. The input data were means by department for 16 variables, forming nine sets. Four of the sets were single variables, four sets were logical sets and one set was formed among mutually exclusive binary sets similar to the religious set described earlier. Such a situation typically involves a linear dependency within the total set. To remove the dependency, any one of the variables within the set can be excluded, and the analysis can be performed. Using the setwise technique, Grooters was able to isolate a rather intriguing result; student costs are in some measure higher in departments that have a higher incidence of outside of school professional activity (consulting, speaking, local community work, artistic endeavors outside the college setting, etc.). Yet another interesting result was that the average salary paid per member in the department was the first set to drop out.

Cummins, M.E. Utilization of multiple regression analysis in

    changing the verbal behavior patterns of elementary class-

    room teachers through self-evaluation. Viewpoints, 1973,

    4(2), 8-17.

Multiple regression analysis was used to investigate the effectiveness of a self-evaluation technique in changing the verbal behavior patterns of elementary classroom teachers. Three ratios were employed: the I/D ratio, or ratio of indirect to direct statements by the teacher; the TRR, or teacher response ratio which eliminates questioning and lecturing from the total I/D ratio; and the PIR, or the ratio of pupil talk-response to pupil talk-initiated. The self-evaluation technique for changing the verbal behavior patterns of elementary classroom teachers was not found to be significant for any of the three measures.

Pohlmann, J.T. Incorporating cost information into the selection
    of variables in multiple regression analysis. Viewpoints,
    1973, 4(2), 18-26.

    The problem of finding the best regression equation is con-
sidered from the standpoint of predictor costs. Typically, variables are selected for inclusion in prediction equations on the
basis of their unique contribution to the prediction of a criterion. A method is presented whereby losses due to lack of predictability and predictor costs are combined in a loss function.
The best predictor set is then chosen that simultaneously mini-
mizes losses incurred in measuring the predictors and losses incurred from lack of predictability of a criterion variable.

Lewis, E.L. and Mouw, J.T. The use of contrast coding to simplify
    anova and ancova procedures in multiple linear regression.
    Viewpoints, 1973, 4(2), 27-44.

    MLR is a powerful and flexible technique for handling data

APPENDIX E

543

analysis. The present paper presents a discussion of the use of

"contrast coding" in performing analysis of variance and analysis

of covariance procedures in MLR. Contrast coding provides a meth-

od for coding a nominal variable in the set of predictor vectors

in MLR so that such vectors reflect a set of orthogonal compari-

sons. As a result, one is able to test hypotheses concerning more

specific research questions than those usually tested in more

traditional MLR coding procedures. By adding more components to

the general linear model, contrast coding provides a relatively

simple and logical basis for extending analysis of variance to

its various subclassifications.

Newman, I. Variations between shrinkage estimation formulas and

the appropriateness of their interpretation. Viewpoints,

1973, 4(2), 45-48.

This is a discussion paper dealing with the use of shrink-

age, different methods for estimating shrinkage, and the accu-

racy of shrinkage estimates when variables are preselected as in

stepwise regression and when variables are not preselected.

Pyle, T.W. The analysis of split-plot and simple hierarchical

designs using multiple linear regression. Viewpoints, 1973,

4(3), 1-9.

For those researchers who are interested in using MLR as a

general data analytic technique, it is of some importance to

recognize what types of models and restrictions on models gener-

ate $\underline{F}$ ratios that correspond to classical hypothesis testing procedures. One of these is the splitplot factorial design and another is the completely randomized hierarchical design. What will
be demonstrated here is that MLR models can be constructed for
these designs which generate $\underline{F}$ ratios that are equivalent to
those obtained by traditional computing formulas.

Buzahora, R.C. and Williams, J.D. An empirical comparison of
     residual gain analysis and the analysis of covariance. View-
     points, 1973 4(3), 10-17.

An extensive comparison of the analysis of covariance and
the residual gain analysis was made. Using subtests of the Iowa
Tests of Basic Skills, the School Attitude Inventory and School
Sentiment Index independently at each grade level, grades 3-8,
258 analyses were compared. While the analysis of covariance
gave some indication of being more powerful, this result was not
uniform over the various analyses.

Houston, J.A. and Houston, S.R. Judgement analysis and porno-
     graphy. Viewpoints, 1973, 4(3), 18-31.

Judgement Analysis (JAN) was used as a methodology for determining what is pornographic by testing this technique with
three groups concerned with this issue. These groups included
doctoral students majoring is Psychology, Counseling and Guidance
(PCG) at the University of Northern Colorado, lawyers, and police officers from the city of Greeley, Colorado. JAN proved to

be an effective technique in the identification of policies. The problem of what is pornographic is indeed a complex one as evidenced by the many specific categorical and complex policies present in the PCG judges, lawyers, and police officers.

Houston, S.R. Identifying faculty policies of teaching effectiveness. Viewpoints, 1973, 4(3), 32-38.

Judgement Analysis (JAN) was employed to capture the teacher effectiveness policy (ies) of College of Education faculty at the University of Northern Colorado. Fifty-seven judges evaluated 60 hypothetical faculty members, each on four characteristics. Results indicated that possibly three different judgemental systems or policies existed.

Brebner, M.A. Multiple linear regression models for analysis of covariance including test for homogeneity of regression slope: Part 1--oneway designs with one covariate. Viewpoints, 1973, 4(3), 39-50.

The objective of the paper is to describe the multiple linear regression models which correspond exactly in all respects to the classical analysis of covariance and tests for homogeneity of regression slope. It is the author's opinion that the approach pioneered by Bottenberg and Ward gives a much clearer and more meaningful understanding of analysis of covariance than the classical method.

Pohlmann, J.T. and Newman, I. Regression effects when the assumption of rectilinearity is not tenable. Viewpoints, 1973, 4(3), 51-59.

When analyzing data which deals with repeated testing, one may find that extreme scorers on a pre-test regress away from the mean upon post-test, contrary to what one would expect from the regression effect. This paper discusses regression effects and presents the argument that when contrary results occur, they are indicative of violation of an underlying assumption of rectilinearity for the Pearson r. Therefore it is recommended that one should look for non-linear relationships when interpreting such data. In addition, three methods for determining if non-linear relationships exist in data are suggested and briefly discussed.

Olson, G.H. A general least squares approach to the analysis of repeated measures designs. Viewpoints, 1973, 4(3), 60-69.

Two procedures for computing general least squares analyses of repeated measures designs are discussed. The first procedure, appropriate for small N, is a straight-forward application of the usual regression approach to the analysis of variance. The second approach, appropriate for large N, also utilizes the regression approach but requires some minor calculation in addition to that typically performed by most computer programs.

Williams, J.D. A note on contrast coding vs. dummy coding. Viewpoints, 1974, 4(4), 1-5.

A comparison is made between the contrast coding system for solution to the analysis of variance design presented by Lewis and Mouw (1973), and the use of dummy coding for solution to the analysis of variance designs. Some of the limitations and advantages of each approach are given.

Klein, M. and Newman, I. Estimated parameters of three shrinkage estimate formuli. Viewpoints, 1974, 4(4), 6-11.

This paper examines the shrinkage formuli of Wherry, McNemar and Lord in relation to overcorrection. A table is given which shows the number of times that each formula resulted in a negative value of R^2 for different numbers of variables and sample sizes.

McNeil, K. and McShane, M. Complexity in behavioral research as viewed within the multiple linear regression approach. Viewpoints, 1974, 4(4), 12-15.

The paper attempts to clarify the notion of complexity in research. Perhaps, in the past, researchers have over simplified the variables and their interrelationships. Thinking of interaction variance as bad variance or as producing an undesired result, or lamenting the complexity of the phenomena under consideration does little or nothing to advance research in the behavioral sciences. The authors examine two views of complexity which seem to exist: (1) complexity as indicated by the number of predictor variables needed to account for a criterion behavior,

and (2) complexity as indicated by the nature of the predictor
variables. The authors argue that the second view of complexity
is not valid, and that the multiple linear regression technique
provides an easy way to index the first view.

Wyshak, G. Modification of multiple regression when an independent variable is subtracted from the dependent variable.
Viewpoints, 1974, 4(4), 16-20.

Behavioral scientists are often concerned with regressing
some dependent or outcome variable, Y, on a number of independent
or explanatory variables, X_i, i = 1,2,...,k. (Model 1). If Y is
a final score or measurement and X_1 an initial score, the investigator may be interested in some measure of change, say Y -
X_1, and its relation to the several explanatory variables including X_1. (Model 2). Analyses would be based on two multiple regression equations, one relating to the regression of Y on X_1,
X_2,...,X_k; and the other to the regression of (Y - X_1) on the
same X's.

The note calls attention to the fact that one analysis would
suffice for the two models because the regression coefficients,
the total sums of squares, deviations sums of squares and regression sums of squares are readily obtained for Model 2 once the
calculations have been made under Model 1.

St. Pierre, R.G. Vargen: A multiple regression teaching program.
Viewpoints, 1974, 4(4), 21-29.

VARGEN (Variable Generator) creates sets of data with known statistical properties by generating user-specified variables which are functions of uniformly distributed random numbers for each of a group of subjects. The user specifies the relative size and location of from one to nine predictor variables and one criterion variable within a ten by ten matrix (hereafter called the "universe") and, therefore, the amount of variance accounted for by each variable. A visual display is produced showing the size and location within the universe of any five variables.

Williams, J.D. and Lindem, A.C. Regression computer programs for
    setwise regression and three related analysis of variance
    techniques. Viewpoints, 1974, 4(4), 30-46.

Four computer programs using the general purpose multiple linear regression program have been developed. Setwise regression analysis is a stepwise procedure for sets of variables; there will be as many steps as there are sets. COVARMLT allows a solution to the analysis of covariance design with multiple covariates. A third program has three solutions to the two-way disproportionate analysis of variance: (a) the method of fitting constants, (b) the hierarchical model and (c) the unadjusted main effects solution. The fourth program yields three solutions to the two-way analysis of covariance, with or without proportionality, and with multiple covariates. The three solutions are similar to those described for a two-way analysis of variance with disproportionate cell frequencies.

Mason, R.L. and McNeil, K.A. MASHIT--for ease in regression pro-

gram communication. Viewpoints, 1974, 4(4), 47-64.

This regression system is an intermediate result of a pro-

ject to develop a comprehensive regression computer system as a

foundation for a complete statistical man-machine interface. The

outstanding features of the system can be condensed into two

principle concepts. First, the program dynamically allocates

core--resulting in no limit on title cards, question cards, etc.

Secondly, "English type" user commands are used in a free format

mode to save computer instruction time. The resulting system is

two phase constructed in such a manner that additional capabili-

ties can be added efficiently.

Fanning, F.W. and Newman, I. The development and demonstration

of multiple regression models for operant conditioning ques-

tions. Viewpoints, 1974, 4(4), 65-87.

Based on the assumption that inferential statistics can make

the operant conditioner more sensitive to possible significant

relationships, regression models were developed to test the sta-

tistical significance between slopes and Y-intercepts of the ex-

perimental and control group subjects. These results were then

compared to the traditional operant conditioning eyeball tech-

nique analysis.

One major advantage of using the regression procedure, rath-

er than the traditional eyeball technique is that probability

estimates can be attributed to the accuracy of the statements.

Another advantage of the regression procedure used is the ability to test the curvilinear relationships above and beyond linear ones. which is not feasible with the eyeball technique on multiple baseline analysis. Similarly, one cannot test to see if the slopes of the control group are significantly different statistically.

In addition, as demonstrated in this paper, we can also test to see if the functional relationship of one treatment is significantly different from the functional relationship of some other treatment (across some area of interest).

Haynes, J.R. and Swanson, R.G. Method for comparison of non-
      independent multiple correlations. Viewpoints, 1974, 5(1), 1-6.

The technique for the comparison of related Rs is based on the residual criterion scores rather than directly on the R or R^2 values. The procedural steps are: (a) generate predicted criterion scores for each set of predictor variables; (b) for each set of predicted scores, obtain the absolute difference between each predicted and corresponding actual criterion scores; (c) analyze these difference scores by a single classification analysis of variance for repeated measures. If a significant F is obtained from the ANOVA, some or all of the multiple correlations would be considered different. Multiple comparisons to determine which Rs differ from each other are then conducted using techniques such as the Newman-Keuls studentized range statistical test.

Maola, J.F. The multiple regression approach for analyzing dif-

    ferential treatment effects--the reversed gestalt model.

    Viewpoints, 1974, 5(1), 7-10.

This paper was written in order to demonstrate a method of
using multiple regression for determining the effect of independent differences of providing treatment. The procedures involve
a model of comparing differential treatment effects to the total
treatment and was, therefore, named the "reversed gestalt model"
because of its theoretical base.

Byrne, J. The use of regression equations to demonstrate caus-

    ality. Viewpoints, 1974, 5(1), 11-22.

A universal objective of scientists involved with explanations of behavior or phenomenon is to demonstrate their knowledge of what is causing it. Explanations of causality usually
entail knowledge of a number of elements or underlying variates
that interrelate to produce the phenomenon.

An R^2 of 1.00 means that 100% of the change that takes
place in the phenomenon has been numerically connected to changes
that take place among the variables and experimental manipulations controlled by the scientist. As the R^2 value approaches 0.0,
it means that more and more things outside of the comprehension
of the scientist are appearing which cause the phenomenon.

Pyle, T.W. Classical analysis of variance of completely within-

    subject factorial designs using regression techniques. Viewpoints, 1974, 5(1), 23-32.

The use of regression analysis in analyzing data obtained

from between-subject experimental designs is well-documented and

easily obtainable. In addition, some texts cover the classical

analysis of variance (ANOVA) of single-factor repeated measure-

ments designs, and an identification has also been made of re-

gression models which yield F ratios (for main and interaction

effects) equivalent to those obtained by standard computing

formulas for split-plot factorial and simple hierarchical de-

signs (Pyle, 1973). However, documentation for the standard ANOVA

of completely within-subject factorial designs using regression

analysis is not available as far as this author is aware. The

purpose of this article is to provide such documentation by means

of an example.

Gloeckler, T.L. Use of multiple linear regression on analysis

   of intelligence test score changes for visually handicapped

   adults. Viewpoints, 1974, 5(1), 33-39.

Multiple linear regression was used to analyze complex data

assessing longitudinal changes in IQ test performance of visually

handicapped adults. Results indicated: (1) patterns of perfor-

mance similar to those found in sighted populations, and (2) no

influence on IQ changes by a variety of ontological factors.

Maola, J.F. Causality and prediction: similar but not synonymous.

   Viewpoints, 1974, 5(2), 1-2.

This short note was meant to provide the reader with: (1) an understanding of the semantic difference of prediction and causality, (2) the value of Byrne (1974) for presenting a rationale for reporting R^2, and (3) a brief encounter with the limitations of R^2.

Williams, J.D. Regression solutions to the A*B*S design. View-points, 1974, 5(2), 3-9.

An alternative regression solution to the repeated measures design (A*B*S) is given, contrasting to a solution given earlier by Pyle (1974). The solution given here can be completed without altering the criterion measures to find the A and B effects. Also, the present solution can be translated quite easily into experimental design terminology.

St. Pierre, R.G. Possible relationships between predictor and criterion variables. Viewpoints, 1974, 5(2), 10-27.

One way to investigate the possible relationships between predictor and criterion variables is to use VARGEN, which, as described in the paper, allows the student to create his own data sets. He is free to vary the amount of overlap between variables and thus acquire a feel for what is meant by common and unique variance and by overlap between variables, as well as to see the effect of varying relationships between predictors and criterion on a regression analysis.

Houston, S.R. and Bolding, J.T. Regression Chi-square: Testing

   for a linear trend in proportions in a 2*2 contingency table.

   Viewpoints, 1974, 5(2), 28-31.

   The usual chi-square test for a 2 X C contingency table can

fail to produce statistical significance when, in fact, a signi-

ficant linear trend in proportions is present in the data. The

ordinary chi-square test lacks power in the case when the vari-

ables can be considered ordered classifications. A regression

chi-square test is described and illustrated with hypothetical

data in which the usual chi-square test produced non-significant

results even though a significant linear trend was present in

the data.

Williams, J.D. Four-way disproportionate hierarchical models.

   Viewpoints, 1974, 5(2), 32-40.

   A solution for the four-way disproportionate fixed effects

hierarchical model is given. Because there are 15 effects to be

ordered, and the ordering will have a major impact on each dif-

ferent effect, the n-way hierarchical model should be used only

when there is a strong a priori ordering suggesting itself to

the researcher.

Houston, S.R. and Ohlson, E.L. Issues in teaching multiple linear

   regression for behavioral research. Viewpoints, 1974, 5(2),

   41-44.

Several questions, issues and approaches to teaching a basic course in Multiple Linear Regression (MLR) are raised in this article. These include the following: 1) MLR as a generalized procedure; 2) Use of matrix algebra in MLR; 3) Redundant models in MLR; 4) Orthogonal coding; and 5) Data analysis and MLR. Suggestions and recommendations are made for the issues raised.

Deitchman, R., Newman, I., Burkholder, J. and Sanders, R.E. An application of the higher order factorial analysis designs with disproportionality of cells. Viewpoints, 1974, 5(2), 45-57.

The results of the application of several techniques available for analyzing data when there is disproportionality of cells is reported. How accurately they answer the research questions asked is also discussed.

The authors imply that since the techniques make different assumptions, that they are actually testing slightly different questions.

Appendix F

Mini Manual for Program MASHIT

MASHIT (Mason's Automatic Statistical Hypothesis Interpreter and Tester) is a computer program written in high level programming languages to facilitate the interaction between the computer and the researcher. The primary unique aspect of MASHIT is that the program performs regression analysis based on conversation language research hypotheses. Ultimately, other statistical techniques will be incorporated into the system; however, the current version handles that wide range of research hypotheses that can be tested using Multiple Linear Regression. Indeed, any least squares hypothesis concerned with a single criterion can be tested with MASHIT. The program readily handles "analysis of covariance" questions.

After studying several available regression programs, a composite of shortcomings was compiled. There was no one system that offered all the features that the researchers and students desired to test their hypotheses. In addition, the only systems that allowed free form input were the interactive terminal programs, whereas the majority of the researchers must cope with "batch mode" computer systems. MASHIT is a system directed toward researchers desiring ease and flexibility in accessing a "batch mode" computer system. The following is a list of the outstanding program features:

557

(a) Analysis of natural language regression questions.
(b) Free format (no column restrictions with the one exception of any optional FORTRAN transformation statements desired).
(c) Virtually unlimited number of variables.
(d) Virtually unlimited number of models.
(e) Virtually unlimited number of research questions.
(f) Virtually unlimited number of title cards.
(g) Any size variable labels.
(h) The program will dichotomize all "A" or "I" field (discrete) variables. The user does not have to keep track of variables newly created by the computer.
(i) Any FORTRAN transformation statements allowed.
(j) No parameter card necessary.
(k) All double precision calculations.
(l) Multiple returns from transformation subroutine.

MASHIT was written for the IBM 360/370 series computers. Constructed in two parts, the program first reads and analyzes the researcher's instructions, and passes this information to the second stage which performs the regression and test calculations.

The first stage actually creates the second stage program resulting in a "tailored" regression program for each individual user. Storage sizes can be expanded or contracted to fit the researcher's program requirement and the desired machine region (core) request. This flexibility of the program does not affect the number of variables which can be processed, however, the machine CPU time is decreased as the core storage size is increased.

In order for a computer program to interpret natural language, established criteria must be met. They are: (1) variables must be prelabeled if referenced by labels in a hypothesis, (2) only specific phrases from a list of keywords can be used, and (3) certain syntactical rules must be followed. These rules are discussed in a later section.

The flexibility of the program allows the user to input his "control deck" as though it were written in a manner similar to a paragraph. There are no column restrictions with the one exception of FORTRAN transformations. MASHIT searches for keywords and labels; therefore, blanks are placed between coded words.

The program reads the entire "control deck" as if it were one long card. Therefore, coding can skip from card to card, even with the option of inserting blank cards in the control deck. Slashes, periods, or question marks are delimiters indicating the end of one type of control information within the control deck. There are presently eight types of control cards as follows:

    (a) Title Card(s)
    (b) Label Card(s)
    (c) Transformation Card(s)
    (d) Special Command Card(s)
    (e) Question Card(s)
    (f) Model Card(s)
    (g) Test Card(s)
    (h) Format Card(s)

All the control cards are optional with the exception of the format card. If the format is the only card included, MASHIT prints the means, standard deviations and the correlations. Rules for each type control card is included in Mason (1973b). To facil-itate the remainder of the discussion, example deck setups are shown to illustrate the features of MASHIT.

Example 1

This is a rather limited application of MASHIT. One question is asked without the use of labels. The program recognizes the word "PREDICT" and uses the variable that follows as the criterion. The first variable and the unit vector are employed as predictor variables. Since there are no covariates, the restricted model is inferred to have an R^2 value of zero. The number of subjects and numbers of linearly independent vectors are calculated and the F̲ test is evaluated. Following the six lines below is the printout as generated by the program.

```
// (Job Card)
// EXEC MASHIT
DOES X1 PREDICT VAR 2?
(2F3.0)
     Data Cards
/*
```

NUMBER OF OBSERVATIONS-------------------- 12

NUMBER OF VARIABLES READ------------------ 2

NUMBER OF VARIABLES AFTER TRANSFORMATION-- 2

NUMBER OF VARIABLES - CONTINUOUS---------- 2

NUMBER OF VARIABLES - DISCRETE------------ 0

INPUT UNIT NUMBER------------------------- 5

FORMAT (2F3.0)

VARIABLE NUMBER	TYPE OF VARIABLE	NUMBER OF DIFFERENT VALUES	MEAN ****	STANDARD DEVIATION	VARIABLE NAME
1	CONTINUOUS		31.50000	19.80951	
2	CONTINUOUS		56.63333	26.73273	

```
                            CORRELATION MATRIX
                            ******************
                      1          2
==================================================================
VARIABLE 1
             1.00000
        ----------------------------------------------------------
VARIABLE 2
            -0.48373    1.00000
        ----------------------------------------------------------



*****************************************************************

      DOES X1 PREDICT VAR 2 ?

*****************************************************************


CRITERION NUMBER =     2          INDEPENDENT VECTORS  =  2
MODEL R-SQUARE...= 0.23399746     NUMBER OF ITERATIONS =  1
                 **********

VARIABLE          RAW SCORE              VARIABLE
 NUMBER            WEIGHTS                 NAME
------------------------------------------------------------------
   1              -0.65279252
REGRESSION
 CONSTANT   =    77.39629787

*****************************************************************


      FULL MODEL........= MODEL FROM ABOVE

      RESTRICTED MODEL...= ZERO RSQ MODEL

      (RSQ F - RSQ R) / DF1    (0.23400 - 0.0   ) / 1
F =   -------------------- = ------------------------ = 3.054787
      ( 1.0  - RSQ F) / DF2    ( 1.0    - 0.23400) / 10   ********

      NONDIRECTIONAL  PROBABILITY  =    0.1084337

      DIRECTIONAL    PROBABILITY  =    0.0542168
      (IN HYPOTHESIZED DIRECTION)

*****************************************************************
```

Example 2

   This example illustrates the use of the "title," "model,"
"label," and "test" cards. Note that the title is two cards in
length and each variable label is placed on a separate card. This
is not necessary as a title can be any number of cards in length
and labels only have to be separated by a delimiter.

```
// (Job Card)
// EXEC MASHIT
THIS STUDY USES LABEL CARDS, TWO MODEL CARDS, AND
A TEST CARD /
LABELS:
  SPEED FOR 100 YARDS:
  AGE:
  MOTIVATION /
MODEL A: AGE AND MOTIVATION PREDICTING SPEED FOR
          100 YARDS /
MODEL B: AGE PREDICTING SPEED FOR 100 YARDS /
TEST MODEL A AGAINST MODEL B /
(3F3.0)
      Data Cards
/*
```

   The structure in this example is similar to that used in
other hypothesis testing regression programs with the exception
that the variables have been labeled and referenced in natural
language in the delineation of the models. This technique is used
when testing many restricted models against the same full model.
A simpler structure is available especially if only one F test
were being computed. By use of the "question" card as in example
1 with the attachment of a covariate phrase, one question can replace two models and a test as in the following:

```
DOES MOTIVATION PREDICT SPEED
FOR 100 YARDS OVER AND ABOVE AGE?
```

   Since any least squares hypothesis can be phrased in "covari-

ance" terminology as above, the "question" card has much potential
for researchers. The printed output is shown on the next pages.

****

    THIS STUDY USES LABEL CARDS, TWO MODEL CARDS, AND

    A TEST CARD /

****

    NUMBER OF OBSERVATIONS-------------------- 12

    NUMBER OF VARIABLES READ----------------- 3

    NUMBER OF VARIABLES AFTER TRANSFORMATION-- 3

    NUMBER OF VARIABLES - CONTINUOUS---------- 3

    NUMBER OF VARIABLES - DISCRETE------------ 0

    INPUT UNIT NUMBER------------------------ 5

    FORMAT (3F3.0)

	NUMBER OF			
VARIABLE TYPE OF	DIFFERENT	MEAN	STANDARD	VARIABLE
NUMBER VARIABLE	VALUES	****	DEVIATION	NAME
1 CONTINUOUS		490.16667	285.29570	SPEED FOR 100 YARDS
2 CONTINUOUS		484.41667	265.58597	AGE
3 CONTINUOUS		642.08333	274.91831	MOTIVATION

                      CORRELATION MATRIX
                      ****
                    1   2   3
==
VARIABLE 1
          1.00000

VARIABLE 2
            0.26106   1.00000

VARIABLE 3
            0.07457   0.21081   1.00000

****

    MODEL A : AGE AND MOTIVATION PREDICTING SPEED FOR 100 YARDS /

****

CRITERION VARIABLE NAME = ...   SPEED FOR 100 YARDS

CRITERION NUMBER =   1   INDEPENDENT VECTORS   =   3
MODEL R-SQUARE...= 0.06855198   NUMBER OF ITERATIONS =   2
                   ****

VARIABLE   RAW SCORE   VARIABLE
 NUMBER   WEIGHTS   NAME
----

    2   0.27580396   AGE
    3   0.02121845   MOTIVATION
REGRESSION
  CONSTANT =   342.93861958

****

****

      MODEL B : AGE PREDICTING SPEED FOR 100 YARDS /

****

CRITERION VARIABLE NAME = ...   SPEED FOR 100 YARDS

CRITERION NUMBER =   1   INDEPENDENT VECTORS   =   2
MODEL R-SQUARE...= 0.06815249   NUMBER OF ITERATIONS =   1
                   ****

VARIABLE NUMBER	RAW SCORE WEIGHTS	VARIABLE NAME

| 2 | 0.28043419 | AGE |

REGRESSION
CONSTANT   =   354.31967218

****

****

TEST MODEL A AGAINST MODEL B /

****

FULL MODEL.........=   MODEL A

RESTRICTED MODEL...=   MODEL B

$$F = \frac{(RSQ\ F - RSQ\ R)\ /\ DF1}{(1.0\ -\ RSQ\ F)\ /\ DF2} = \frac{(0.06855 - 0.06815)\ /\ 1}{(1.0\ -\ 0.06855)\ /\ 9} = \frac{0.003860}{*******}$$

NONDIRECTIONAL PROBABILITY   =   0.9505624

DIRECTIONAL   PROBABILITY   =   0.4752812
(IN HYPOTHESIZED DIRECTION)

****

Example 3

Here is featured the input of nominal data, FORTRAN transformations, and a "covariate" question. Three variables are read
and a fourth is created as the square of the third variable.
Also, the A-format for variable two implies that it is discrete.
MASHIT then automatically constructs and maintains the mutually

exclusive group membership vectors. When variable two is referenced in the research question, the group membership vectors are
substituted.

```
// (Job Card)
// EXEC MASHIT
     X(4) = X(3) ** 2
IS VAR 1 PREDICTED BY X2
GIVEN KNOWLEDGE OF X(3), X4?
(F5.0, A1, 4X, F5.0)
     Data Cards
/*
```

Notice in the printout which follows that the program reports the number of observations, how many variables were read,
how many were created by transformations and the number of mutually exclusive vectors that resulted.

NUMBER OF OBSERVATIONS-------------------- 15

NUMBER OF VARIABLES READ------------------ 3

NUMBER OF VARIABLES AFTER TRANSFORMATION-- 4

NUMBER OF VARIABLES - CONTINUOUS---------- 3

NUMBER OF VARIABLES - DISCRETE------------ 1

INPUT UNIT NUMBER------------------------- 5

NUMBER OF DICHOTOMIZED VARIABLES---------- 2

FORMAT (F5.0, A1, 4X, F5.0)

VARIABLE NUMBER	TYPE OF VARIABLE	NUMBER OF DIFFERENT VALUES	MEAN ****	STANDARD DEVIATION	VARIABLE NAME
1	CONTINUOUS		41.66667	29.12883	
2	DISCRETE	2			

| 3 | CONTINUOUS | 53.20000 | 21.22326 |
| 4 | CONTINUOUS | 3280.66667 | 2415.94944 |

		CREATED BY			
VARIABLE	TYPE OF	VARIABLE	MEAN	STANDARD	VARIABLE
NUMBER	VARIABLE	NUMBER	****	DEVIATION	VALUE

| 5 | DICHOTOMOUS | 2 | 0.66667 | 0.47140 | A |
| 6 | DICHOTOMOUS | 2 | 0.33333 | 0.47140 | B |

CORRELATION MATRIX
****

	1	2	3	4	5	6

VARIABLE 1

    1.00000

VARIABLE 2

    ******** ********

VARIABLE 3

    -0.38035 ******** 1.00000

VARIABLE 4

    -0.32130 ******** 0.96756 1.00000

VARIABLE 5

    -0.03560 ******** -0.10662 -0.17225 1.00000

VARIABLE 6

    -0.03560 ******** 0.10662 0.17225 -1.00000 1.00000

****

    IS VAR 1 PREDICTED BY X2

    GIVEN KNOWLEDGE OF X(3), X4?

****

CRITERION NUMBER =   1   INDEPENDENT VECTORS = 4
MODEL R-SQUARE...= 0.18111067   NUMBER OF ITERATIONS = 5
                 ****

VARIABLE NUMBER	RAW SCORE WEIGHTS	VARIABLE NAME
2 - 5	3.08248870	VALUE = A
2 - 6	0.0	VALUE = B
3	-1.55778456	
4	0.00946970	
REGRESSION CONSTANT =	91.41887092	

****

CRITERION NUMBER =   1   INDEPENDENT VECTORS = 3
MODEL R-SQUARE...= 0.17883439   NUMBER OF ITERATIONS = 2
                  ****

VARIABLE NUMBER	RAW SCORE WEIGHTS	VARIABLE NAME
3	-1.49366987	
4	0.00882175	
REGRESSION CONSTANT =	92.18867231	

****

FULL MODEL.........= MODEL FROM ABOVE

RESTRICTED MODEL...= MODEL FROM ABOVE

$$F = \frac{(RSQ\ F - RSQ\ R)\ /\ DF1}{(1.0\ -\ RSQ\ F)\ /\ DF2} = \frac{(0.18111 - 0.17883)\ /\ 1}{(1.0\ -\ 0.18111)\ /\ 11} = 0.030577$$

****

NONDIRECTIONAL PROBABILITY =   0.8583112

DIRECTIONAL   PROBABILITY =   0.4291556
(IN HYPOTHESIZED DIRECTION)

****

Further Documentation

MASHIT was developed by Robert L. Mason as part of a doctoral dissertation under the direction of Dr. Keith McNeil. The
dissertation (Mason, 1973a) has complete documentation. Also, a
65 page "MASHIT" user's guide (Mason, 1973b) is available from
Robert Mason, Computer Center, Iowa Western Community College,
2700 College Road, Council Bluffs, Iowa 51501.

References

Mason, R.L. The development of an automated statistical hypothe-
    sis interpretor and tester. Unpublished doctoral disserta-
    tion, Southern Illinois University, 1973a.

Mason, R.L. "MASHIT" user's manual, unpublished manuscript,
    1973b.

References

Andrews, F., Morgan, J. and Sonquist, J. Multiple Classification
Analysis. Ann Arbor, Michigan: Institute for Social Research,
The University of Michigan, 1967.

Appelbaum, M.I. and Cramer, E.M. Some problems in the non-orthogonal analysis of variance. No. 120, The L. L. Thurstone
Psychometric Laboratory, University of North Carolina, 1973.

Bell, R.Q. Stimulus control of parent or caretaker behavior by
offspring. Developmental Psychology, 1971, 4, 63-72.

Bender, J.A., Kelly, F.J., Pierson, J.K. and Kaplan, H. Analysis
of the comparative advantages of unlike exercises in relation to prior individual strength levels. The Research
Quarterly, 1968, 39, 443-448.

Blalock, H.M. Causal Inference in Nonexperimental Research. Chapel
Hill: University of North Carolina Press, 1964.

Bottenberg, R.A. and Ward, J.H. Applied Multiple Linear Regression. Lackland Air Force Base, Texas: Aerospace Medical
Division, AD 413128, 1963.

Brazelton, T.B. Infants and Mothers. New York: Delacorte Press,
1969.

Byrne, J. The use of regression equations to demonstrate causality. Multiple Linear Regression Viewpoints, 1974, 5(1),
11-22.

Cahen, L.S. and Linn, R.L. Regions of significant criterion differences in aptitude--treatment--interaction research,
American Educational Research Journal, 1971, 8, 521-530.

Campbell, D.T. and Stanley, J.C. Experimental and quasi-experi-
mental designs for research on teaching. In N. L. Gage (Ed.),
Handbook of Research on Teaching. Chicago: Rand McNally,
1963.

571

Castaneda, A., Palermo, D.S. and McCandless, B. Complex learning and performance as a function of anxiety in children and task difficulty. Child Development, 1956, 27, 328-332.

Cattell, R.B. (Ed.) Handbook of Multivariate Experimental Psychology. Chicago: Rand McNally and Company, 1966.

Christal, R.E. Selecting a harem--and other applications of the policy capturing model. The Journal of Experimental Education, 1968, 36(1), 24-27.

Cochran, W.G. and Cox, G. Experimental Designs. New York: John Wiley, 1957.

Cohen, J. Approximate power and sample size determination for common one-sample and two-sample hypothesis tests. Educational and Psychological Measurement, 1970, 30, 811-832.

Connett, W.E., Houston, S.R. and Shaw, D.G. The use of factor regression in data analysis. Multiple Linear Regression Viewpoints, 1972, 2, 46-49.

Cooley, W.W. and Lohnes, P.R. Multivariate Procedures for the Behavioral Sciences. New York: Wiley, 1962.

Cramer, E.M. Brief report: The use of highly correlated predictors in regression analysis. Multivariate Behavioral Research, 1974, 9, 241-244.

Denny, J.P. Effects of anxiety and intelligence on concept formation. Journal of Experimental Psychology, 1966, 72, 596-602.

Draper, N. and Smith, H. Applied Regression Analysis. New York: Wiley, 1966.

DuCette, J. and Wolk, S. Ability and achievement as moderating variables of student satisfaction and teacher perception. The Journal of Experimental Education, 1972, 41, 12-17.

Edwards, A.L. Statistical Methods for the Behavioral Sciences. New York: Holt, Rinehart, and Winston, 1964.

Ennis, R.H. On causality. Educational Researcher, 1973, 2(6), 4-11.

Foster, G. Some multiple linear regression models for the evaluation of school programs. Paper presented at the meeting of the American Educational Research Association, Chicago, February 1968.

Friedman, H. Magnitude of experimental effect and table for its rapid estimation. Psychological Bulletin, 1968, 70, 245-251.

Gagne, R.M. The Conditions of Learning. New York: Holt, Rinehart, and Winston, 1965.

Glass, G.V., Peckham, P.D. and Sanders, J.R. Consequences of failure to meet assumptions underlying the analysis of variance and covariance. Review of Educational Research, 1972, 42, 237-288.

Glass, G.V., Willson, V. and Gottman, J. Design and Analysis of Time-Series Experiments. Boulder: Colorado Associated University Press, 1974.

Gocka, E.F. Regression analysis of proportional cell data. Psychological Bulletin, 1973, 80, 25-27.

Goldberg, L.R. Parameters of personality inventory construction and utilization: a comparison of prediction strategies and tactics. Multivariate Behavioral Research Monographs, 1972, 72, 2.

Guilford, J.P. Creativity. American Psychologist, 1950, 5, 444-454.

Guilford, J.P. Three faces of the intellect. American Psychologist, 1959, 14, 469-479.

Guilford, J.P. Intelligence: 1965 model. American Psychologist, 1966, 21, 20-26.

Hickrod, G.A. Local demand for education: A critique of school finance and economic research circa 1959-1969. Review of Educational Research, 1971, 41, 35-50.

Houston, S.R. and Bolding, J.R. Regression chi square: Testing for a linear trend in proportions in a 2*2 contingency table. Multiple Linear Regression Viewpoints, 1974, 5(2), 28-31.

Huberty, C.J. Regression analysis and 2-group discriminant analysis. The Journal of Experimental Education, 1972, 41(1), 39-41.

Huck, S.W. The analysis of covariance: increased power through reduced variability. The Journal of Experimental Education, 1972, 41(1), 42-46.

Jennings, E.E. Fixed effects analysis of variance by regression analysis. Multivariate Behavioral Research, 1967, 2, 95-108.

Katahn, M. Interaction of anxiety and ability in complex learning situations. Journal of Personality and Social Psychology, 1966, 3, 475-478.

Kelly, F.J., Beggs, D.L., McNeil, K.A., Eichelberger, T. and Lyon, J. Research Design in the Behavioral Sciences: Multiple Regression Approach. Carbondale: Southern Illinois University Press, 1969.

Kelly, F.J., Newman, I. and McNeil, K.A. Suggested inferential statistical models for research in behavioral modification. The Journal of Experimental Education, 1973, 41(4), 54-63.

Kerlinger, F.N. and Pedhazur, E.J. Multiple Regression in Behavioral Research. New York: Holt, Rinehart, and Winston, 1973.

Klein, M. and Newman, I. Estimated parameters of three shrinkage estimate formuli. Multiple Linear Regression Viewpoints, 1974, 4(4), 6-11.

Koplyay, J.B. Automatic interaction detector AID-4. Multiple Linear Regression Viewpoints, 1972, 3, 25-38.

Korner, A.F. Some hypotheses regarding the significance of individual differences at birth for later development. In J. Hellmuth (Ed.), Exceptional Infant, Volume 1, The Normal Infant. Seattle: Special Child Publications, 1967.

Kunce, J.T. Prediction and statistical overkill. Measurement and Evaluation in Guidance, 1971, 4, 38-42.

McClelland, D.C. Testing for competence rather than for "Intelligence." American Psychologist, 1973, 28, 1-14.

McGuire, C. The textown study of adolescence. Texas Journal of Science, 1956, 8, 264-274.

McNeil, J.T. The refinement and prediction of a measure of infant development. Unpublished doctoral dissertation, Southern Illinois University, 1974.

McNeil, J.T. and McNeil, K.A. Manual of Program DPLINEAR for Use with McNeil, Kelly, McNeil: Testing Research Hypotheses Using Multiple Linear Regression. Carbondale: Southern Illinois University Press, 1975.

McNeil, K.A. The negative aspects of the eta coefficient as an index of curvilinearity. Multiple Linear Regression Viewpoints, 1970a, 1, 7-17.

McNeil, K.A. Meeting the goals of research with multiple linear regression. Multivariate Behavioral Research, 1970b, 5, 375-386.

McNeil, K.A. Some notions about the use of statistical hypothesis testing. Paper presented at the meeting of the Southwest Psychological Association, San Antonio, May 1971.

McNeil, K.A. and Beggs, D.L. Directional hypotheses with the multiple linear regression approach. Paper presented at the meeting of the American Educational Research Association, New York, February 1971.

McNeil, K.A. and Kelly, F.J. Express functional relationships among data rather than assume "intervalness." The Journal of Experimental Education, 1970, 39(2), 43-48.

McNeil, K.A. and Lewis, E.L. A rejoinder to prediction and statistical overkill. Measurement and Evaluation in Guidance, 1972, 5, 360-365.

McNeil, K.A. and McNeil, J.T. Statistical interaction as viewed within the multiple regression approach. Paper presented at the meeting of the American Educational Research Association, New Orleans, February 1973.

McNeil, K. and McShane, M. Complexity in behavioral research as viewed within the multiple linear regression approach. Multiple Linear Regression Viewpoints, 1974, 4(4), 12-15.

McNeil, K.A. and Spaner, S.D. Brief Report: Highly correlated predictor variables in multiple regression models. Multivariate Behavioral Research, 1971, 6, 117-125.

McNemar, Q. Psychological Statistics. (3rd ed.) New York: John Wiley, 1962.

Mendenhall, W. Introduction to Linear Models and the Design and Analysis of Experiments. Belmont, California: Wadsworth, 1968.

Morgan, J. Can a computer beat the horses? Look, 1971, 35, 33.

Mosteller, F. and Tukey, J.W. Data analysis, including statistics. In G. Lindzey and E. Aronson (Ed.) The Handbook of Social Psychology, Volume Two. Reading, Massachusetts: Addison-Wesley, 1968.

Namboodiri, N.K. Experimental designs in which each subject is used repeatedly. Psychological Bulletin, 1972, 77, 54-64.

Newman, I. Some further considerations of using factor regression analysis. Multiple Linear Regression Viewpoints, 1972, 3, 39-41.

Newman, I. Variations between shrinkage estimation formulas and the appropriateness of their interpretation. Multiple Linear Regression Viewpoints, 1973, 4(2), 45-48.

Newman, I., Lewis, E.L. and McNeil, K.A. Multiple linear regression models which more closely reflect bayesian concerns. Multiple Linear Regression Viewpoints, 1972, 3, 71-77.

Newman, I. and Fry, J. A response to "a note on multiple comparisons" and a comment on shrinkage. Multiple Linear Regression Viewpoints, 1972, 2, 36-39.

Novick, M.R. and Jackson, P.H. Bayesian guidance technology. Review of Educational Research, 1970, 40, 459-494.

Overall, J.E. and Spiegel, D.K. Concerning least squares analysis of experimental data. Psychological Bulletin, 1969, 72, 311-322.

Overall, J.E. and Spiegel, D.K. Comment on "regression analysis of proportional cell data." Psychological Bulletin, 1973, 80, 28-30.

Pohlmann, J. Incorporating cost information into the selection of variables in multiple regression analysis. Multiple Linear Regression Viewpoints, 1973, 4(2), 18-26.

Pyle, T.W. The analysis of split-pot and simple hierarchical designs using multiple linear regression. Multiple Linear Regression Viewpoints, 1973, 4(3), 1-9.

Pyle, T.W. Classical analysis of variance of completely within-subject factorial designs using regression techniques. Multiple Linear Regression Viewpoints, 1974, 5(1), 23-32.

Sarason, I.G. and Palola, E.G. The relationship of test and general anxiety, difficulty of task, and experimental instructions to performance. Journal of Experimental Psychology, 1960, 59, 185-191.

Saunders, D.R. Moderator variables in prediction. Educational and Psychological Measurement, 1956, 16, 209-222.

Shine, L.C. A design combining the single-subject and multi-subject approaches to research. Educational and Psychological Measurement, 1973, 33, 763-766.

Spaner, S.D. Application of multivariate techniques to developmental data: the derivation, verification, and validation of a predictive model of twenty-four month cognitive data. Unpublished doctoral dissertation, Southern Illinois University, 1970.

Starr, F.H. The remarriage of multiple regression and statistical inference: a promising approach for hypnosis researchers. The American Journal of Clinical Hypnosis, 1971, 13, 175-197.

Suppes, P., Hyman, L. and Jerman, M. Linear structural models for response and latency performance in arithmetic. Technical Report 100. Stanford, California: Institute for Mathematical Studies in the Social Sciences, 1966.

Taylor, J. A personality scale of manifest anxiety. Journal of Abnormal and Social Psychology, 1953, 48, 285-290.

Veldman, D.J. FORTRAN Programming for the Behavioral Sciences. New York: Holt, Rinehart, and Winston, 1967.

Walberg, H.J. Generalized regression models in educational research. American Educational Research Journal, 1971, 8, 71-91.

Ward, J.H. and Jennings, E.E. Introductions to Linear Models. Englewood Cliffs, N.J.: Prentice-Hall, 1973.

Webster, W.J. and Eichelberger, R.T. A use of the regression model in educational evaluation. The Journal of Experimental Education, 1972, 40(3), 91-96.

Werts, C.E. The partitioning of variance in school effects studies: A reconsideration. American Educational Research Journal, 1970, 7, 127-132.

Werts, C.E. and Linn, R.L. Lord's paradox: A generic problem. Psychological Bulletin, 1969, 72, 423-425.

Whiting, J.W.M. and Child, I.L. Child Training and Personality. New Haven: Yale University Press, 1953.

Williams, E.J. Regression Analysis. New York: Wiley, 1959.

Williams, J.D. A regression approach to experimental design. The Journal of Experimental Education, 1970, 39(1), 83-90.

Williams, J.D. Two-way fixed effects analysis of variance with disproportionate cell frequencies. Multivariate Behavioral Research, 1972, 7, 57-83.

Williams, J.D. Regression Analysis in Educational Research. New

York: MSS Information Corporation, 1974.

Williams, J.D. and Lindem, A.C. Regression computer programs for
    setwise regression and three related analysis of variance
    techniques. Multiple Linear Regression Viewpoints, 1974,
    4(4), 30-46.

Winer, B.J. Statistical Principles in Experimental Design. New
    York: McGraw-Hill, 1962.

Wood, D.A. and Langevin, M.J. Moderating the prediction of grades
    in freshman engineering. Journal of Educational Measurement,
    1972, 9, 311-320.

Name Index

Subject Index